Welding and Joining of Metallic Materials: Microstructure and Mechanical Properties

Welding and Joining of Metallic Materials: Microstructure and Mechanical Properties

Editors

Reza Beygi
Mahmoud Moradi
Ali Khalfallah

Basel • Beijing • Wuhan • Barcelona • Belgrade • Novi Sad • Cluj • Manchester

Editors

Reza Beygi
Arak University
Arak
Iran

Mahmoud Moradi
University of Northampton
Northampton
UK

Ali Khalfallah
University of Coimbra
Coimbra
Portugal

Editorial Office
MDPI AG
Grosspeteranlage 5
4052 Basel, Switzerland

This is a reprint of articles from the Special Issue published online in the open access journal *Crystals* (ISSN 2073-4352) (available at: https://www.mdpi.com/journal/crystals/special_issues/D36F0198NS).

For citation purposes, cite each article independently as indicated on the article page online and as indicated below:

Lastname, A.A.; Lastname, B.B. Article Title. *Journal Name* **Year**, *Volume Number*, Page Range.

ISBN 978-3-7258-2495-3 (Hbk)
ISBN 978-3-7258-2496-0 (PDF)
doi.org/10.3390/books978-3-7258-2496-0

© 2024 by the authors. Articles in this book are Open Access and distributed under the Creative Commons Attribution (CC BY) license. The book as a whole is distributed by MDPI under the terms and conditions of the Creative Commons Attribution-NonCommercial-NoDerivs (CC BY-NC-ND) license.

Contents

About the Editors vii

Preface ix

Ali Khalfallah, Mahmoud Moradi and Reza Beygi
Welding and Joining of Metallic Materials: Microstructure and Mechanical Properties
Reprinted from: *Crystals* 2024, *14*, 839, doi:10.3390/cryst14100839 1

Qunbing Zhang, Lina Ren, Xiaowei Lei, Jiadian Yang, Kuo Zhang and Jianxun Zhang
Effect of Laser Heat Input on the Microstructures and Low-Cycle Fatigue Properties of Ti60 Laser Welded Joints
Reprinted from: *Crystals* 2024, *14*, 677, doi:10.3390/cryst14080677 6

Vasyl Lozynskyi, Bohdan Trembach, Egidijus Katinas, Kostiantyn Sadovyi, Michal Krbata, Oleksii Balenko, et al.
Effect of Exothermic Additions in Core Filler on Arc Stability and Microstructure during Self-Shielded, Flux-Cored Arc Welding
Reprinted from: *Crystals* 2024, *14*, 335, doi:10.3390/cryst14040335 20

Kamel Touileb, Rachid Djoudjou, Abousoufiane Ouis, Abdeljlil Chihaoui Hedhibi, Sahbi Boubaker and Mohamed M. Z. Ahmed
Particle Swarm Method for Optimization of ATIG Welding Process to Joint Mild Steel to 316L Stainless Steel
Reprinted from: *Crystals* 2023, *13*, 1377, doi:10.3390/cryst13091377 51

Hyeri Go, Taejoon Noh, Seung-Boo Jung and Yoonchul Sohn
Microstructural Optimization of Sn-58Bi Low-Temperature Solder Fabricated by Intense Pulsed Light (IPL) Irradiation
Reprinted from: *Crystals* 2024, *14*, 465, doi:10.3390/cryst14050465 72

Stavroula Maritsa, Stavros Deligiannis, Petros E. Tsakiridis and Anna D. Zervaki
Experimental and Computational Study of Microhardness Evolution in the HAZ for Al–Cu–Li Alloys
Reprinted from: *Crystals* 2024, *14*, 246, doi:10.3390/cryst14030246 82

Leonardo Oliveira Passos da Silva, Tiago Nunes Lima, Francisco Magalhães dos Santos Júnior, Bruna Callegari, Luís Fernando Folle and Rodrigo Santiago Coelho
Heat-Affected Zone Microstructural Study via Coupled Numerical/Physical Simulation in Welded Superduplex Stainless Steels
Reprinted from: *Crystals* 2024, *14*, 204, doi:10.3390/cryst14030204 96

Aditya M. Mahajan, K. Vamsi Krishna, M. J. Quamar, Ateekh Ur Rehman, Bharath Bandi and N. Kishore Babu
Structure–Property Correlation between Friction-Welded Work Hardenable Al-4.9Mg Alloy Joints
Reprinted from: *Crystals* 2023, *13*, 1119, doi:10.3390/cryst13071119 113

Mustafa Elmas, Oğuz Koçar and Nergizhan Anaç
Study of the Microstructure and Mechanical Property Relationships of Gas Metal Arc Welded Dissimilar Protection 600T, DP450 and S275JR Steel Joints
Reprinted from: *Crystals* 2024, *14*, 477, doi:10.3390/cryst14050477 127

Kamel Touileb, Elawady Attia, Rachid Djoudjou, Abdejlil Chihaoui Hedhibi, Abdallah Benselama, Albaijan Ibrahim and Mohamed M. Z. Ahmed
Effect of Microchemistry Elements in Relation of Laser Welding Parameters on the Morphology 304 Stainless Steel Welds Using Response Surface Methodology
Reprinted from: *Crystals* **2023**, *13*, 1138, doi:10.3390/cryst13071138 . . . . **147**

Sipokazi Mabuwa and Velaphi Msomi
Investigation of the Microstructure and Mechanical Properties in Friction Stir Welded Dissimilar Aluminium Alloy Joints via Sampling Direction
Reprinted from: *Crystals* **2023**, *13*, 1108, doi:10.3390/cryst13071108 . . . . **167**

About the Editors

Reza Beygi

Reza Beygi is a Researcher at "Instituto de Engenharia Mecanica e Gestao Industrial" (INEGI) in Portugal. He is a member of the Advanced Joining Process Unit (AJPU) of INEGI. His research areas pertain to joining, welding and metallurgy. Dr. Beygi has conducted several academic research studies as well as industrial projects in the field of welding, the failure analysis of materials and materials engineering. He has been a Professor at the University of Arak since 2015, and he served as a Principal Investigator at INEGI during the period of 2020-2023. Dr Beygi has supervised several MSc theses and has taught main courses on welding, physical metallurgy and materials science. He is the author and co-author of more than 60 papers in the field of welding and metallurgy.

Mahmoud Moradi

Mahmoud Moradi is currently a Senior Lecturer and Programme Leader of MSc Advanced Design & Manufacturing at the University of Northampton in the UK. He has actively been conducting research on laser material processing for 18 years. His main scientific contributions have been made in the following areas: laser materials processing, non-traditional machining, welding technology, additive manufacturing, 3D/4D printing, and statistical modeling and optimization via the design of experiments (DOE). He has published more than 180 papers in international and national journals and conferences and 9 book chapters in the areas of his interests. He has been a scientific committee member, keynote speaker, workshop presenter, chairman, and chair of panels for various international conferences. He is an Editorial Board Member and Guest Editor of several reputable journals. He was featured in the list of the "World Ranking of Top 2% Scientists in the World" in the 2024, 2023, 2022 & 2021 & 2020 databases based on a Stanford University report.

Ali Khalfallah

Ali Khalfallah is a Full Professor in the Department of Mechanical Engineering of the Higher Institute of Applied Sciences and Technology at the University of Sousse in Tunisia (DME-UST). Previously, he was the Head of the Department of Mechanical Engineering. He received his PhD degree from the University of Tunis El-Manar (Tunisia) in 2004. In 2015, he obtained a diploma of the "Habilitation a Diriger des Recherches - HDR" in Mechanical Engineering from the University of Sousse in Tunisia. In 2019, he joined the Centre of Mechanical Engineering Materials and Processes (CEMMPRE) in the Department of Mechanical Engineering, the Faculty of Sciences and Technology, the University of Coimbra (FCTUC) in Portugal. Prof. Khalfallah's research activities focus on various topics, including the multidisciplinary modeling of the mechanical behavior of material simulation of the friction stir welding process and the identification of constitutive material parameters using a combination of computational physics, artificial intelligence, and multi-scale simulations, recently exploring the tribology of coatings. He is the author and co-author of over 65 publications (i.e., book, book chapters, papers in international journals), a keynote speaker at international conferences, and the owner of an issue patent. In the Google Scholar indexing database, the publications of Prof. Khalfallah have received more than 470 citations and an h-index of 12. Prof. Khalfallah has been invited to join the editorial board of several international journals and serve as a reviewer for many international journals and conferences. He has also received the outstanding reviewers' award (2023-IOP).

Preface

Welding and joining technologies for metallic materials have played a fundamental role in the development of numerous industries, from aerospace and automotive to energy and beyond. These processes enable the creation of complex components and are essential to ensuring the mechanical integrity and reliability of structures. The relationship between the microstructure of welded joints and their mechanical properties holds the key to optimizing joint strength, durability, and overall performance, particularly in the demanding conditions encountered in modern applications.

In recent years, rapid advancements in welding techniques and the emergence of new materials have expanded the scope of research needed to optimize these processes. However, these innovations also present new challenges, particularly in understanding the interactions between welding parameters, microstructural evolution, and mechanical performance. Bridging the gap between fundamental scientific research and practical industrial applications requires an ongoing, in-depth investigation.

The reprint Welding and Joining of Metallic Materials: Microstructure and Mechanical Properties addresses this need by bringing together a collection of ten research articles. Through a combination of experimental studies, numerical simulations, and process optimizations, the authors provide valuable insights into recent advances in welding technologies.

This reprint is divided into three key areas of research. The first set of papers focuses on process optimization and the development of advanced welding techniques, offering novel approaches to enhancing weld quality, energy efficiency, and mechanical properties. The second set deals with the microstructural evolution of welded joints and its impact on material performance, mainly in relation to strength and corrosion resistance. Finally, the third set of studies addresses the complexities of welding dissimilar materials, highlighting strategies to overcome the challenges associated with joining dissimilar alloys.

The Guest Editors express their gratitude to the authors for their significant contributions and to the Peer Reviewers for ensuring the quality of each paper. This reprint would not have been possible without their commitment and expertise.

Reza Beygi, Mahmoud Moradi, and Ali Khalfallah
Editors

Editorial

Welding and Joining of Metallic Materials: Microstructure and Mechanical Properties

Ali Khalfallah [1,*], Mahmoud Moradi [2] and Reza Beygi [3,4]

1. CEMMPRE, Department of Mechanical Engineering, University of Coimbra, Rua Luís Reis Santos, 3030-788 Coimbra, Portugal
2. Faculty of Arts, Science and Technology, University of Northampton, Northampton NN1 5PH, UK; mahmoud.moradi@northampton.ac.uk
3. Institute of Science and Innovation in Mechanical and Industrial Engineering (INEGI), Rua Dr. Roberto Frias, 4200-465 Porto, Portugal; rbeygi@inegi.up.pt
4. Department of Materials Engineering and Metallurgy, Faculty of Engineering, Arak University, Arak 38156-8-8349, Iran
* Correspondence: ali.khalfallah@dem.uc.pt; Tel.: +351-933-411-080

Citation: Khalfallah, A.; Moradi, M.; Beygi, R. Welding and Joining of Metallic Materials: Microstructure and Mechanical Properties. *Crystals* 2024, 14, 839. https://doi.org/10.3390/cryst14100839

Received: 20 September 2024
Accepted: 25 September 2024
Published: 27 September 2024

Copyright: © 2024 by the authors. Licensee MDPI, Basel, Switzerland. This article is an open access article distributed under the terms and conditions of the Creative Commons Attribution (CC BY) license (https://creativecommons.org/licenses/by/4.0/).

The study of welding and joining technologies for metallic materials has long been fundamental to advancing numerous industries, including aerospace, automotive, and energy [1,2]. These techniques not only enable the fabrication of complex components but also serve as a critical factor in the mechanical performance and reliability of structures [3,4]. A thorough understanding of the microstructure and mechanical properties of welded joints is essential for improving joint strength, durability [5], and overall material performance, particularly in demanding environments [6,7].

In recent years, technological advancements in welding processes [8] and the development of new materials [9] have prompted an increasing need for in-depth research to optimize these techniques [10,11]. In this context, the Special Issue "*Welding and Joining of Metallic Materials: Microstructure and Mechanical Properties*" brings together a collection of ten innovative research articles. These contributions focus on elucidating the relationships between welding parameters, the resulting microstructures, and the mechanical properties of both similar and dissimilar material joints.

This Special Issue aims to bridge the gap between fundamental scientific research and practical industrial applications, addressing the challenges posed by modern welding processes. By exploring advanced welding techniques, investigating the microstructure of various alloys, and evaluating the performance of welded joints under different conditions, the published articles provide valuable insights into how these processes can be refined and adapted for enhanced performance.

The ten contributions can be broadly categorized into three key research themes. This Editorial discusses their scientific significance, offering readers a cohesive overview of the advancements in welding and joining technologies presented in this Special Issue.

1. Process optimization and advanced welding techniques

Three papers explore the optimization of welding processes and the development of advanced techniques to enhance weld quality, energy efficiency, and mechanical properties.

- Contribution (1): Effect of Laser Heat Input on the Microstructures and Low-Cycle Fatigue Properties of Ti60 Laser Welded Joints [12]. This study investigates the effects of varying laser heat input on the microstructure and mechanical properties of Ti60 titanium alloy welded joints. The results reveal that different heat inputs significantly affect the morphology of the weld zone (WZ), reducing porosity in Y-type WZs, which enhances low-cycle fatigue resistance. The research contributes valuable insights into optimizing laser welding parameters for improved weld quality.

- Contribution (2): Effect of Exothermic Additions in Core Filler on Arc Stability and Microstructure during Self-Shielded Flux-Cored Arc Welding (FCAW) [13]. This paper explores the use of exothermic additives in flux-cored arc welding to enhance arc stability and deposition efficiency. By optimizing the CuO/Al and CuO/C ratios in the core filler, this study demonstrates how filler composition influences arc stability and welding current, providing a route to more energy-efficient welding processes. The mathematical models developed in this work further enhance the prediction of welding parameters.
- Contribution (3): Particle Swarm Method for Optimization of ATIG Welding Process to Join Mild Steel to 316L Stainless Steel [14]. This article presents the optimization of activating tungsten inert gas (ATIG) welding using particle swarm optimization (PSO) to join mild steel to 316L stainless steel. The findings show that the optimized flux composition significantly improves weld penetration and hardness, without compromising the mechanical properties of the joint. This study offers a promising solution for improving the quality of dissimilar material welds.

2. Microstructure and mechanical properties of welded joints

Four contributions focus on the intricate relationship between welding-induced microstructural changes and their effects on the mechanical properties of welded joints, shedding light on material behavior under different conditions.

- Contribution (4): Microstructural Optimization of Sn-58Bi Low-Temperature Solder Fabricated by Intense Pulsed Light (IPL) Irradiation [15]. This paper introduces IPL soldering as a novel method for optimizing the microstructure of Sn-58Bi solder. By regulating the irradiation time, this study demonstrates how the solder microstructure evolves from immature to refined, impacting mechanical properties such as hardness. This technique offers potential for improving low-temperature soldering processes in electronics manufacturing.
- Contribution (5): Experimental and Computational Study of Microhardness Evolution in the HAZ for Al–Cu–Li Alloys [16]. This paper focuses on simulating the microhardness evolution in the heat-affected zone (HAZ) of Al-Cu-Li alloy welds. By replicating the thermal cycles experienced during welding, the research explores the dissolution and coarsening of strengthening precipitates in the HAZ, contributing to a deeper understanding of how welding affects the mechanical properties of advanced aluminum alloys.
- Contribution (6): Heat-Affected Zone Microstructural Study via Coupled Numerical/Physical Simulation in Welded Superduplex Stainless Steels [17]. This work introduces a hybrid approach, combining physical and numerical simulations to investigate the HAZ of super-duplex stainless steel (SDSS). The authors successfully simulate thermal cycles using a Gleeble® machine and compare the results with actual welds, providing valuable insights into how thermal management during welding can preserve the microstructural balance between ferrite and austenite, essential for maintaining corrosion resistance and mechanical integrity.
- Contribution (7): Structure–Property Correlation between Friction-Welded Work-Hardenable Al-4.9Mg Alloy Joints [18]. The research evaluates the microstructure and mechanical properties of AA5083 H116 joints produced by rotary friction welding. This study reveals that grain refinement in the dynamically recrystallized zone (DRZ) enhances joint strength, although slight reductions in ductility are observed. These findings are particularly relevant for applications in aerospace and automotive sectors where high-strength aluminum alloys are frequently used.

3. Welding of dissimilar materials

Three papers address the complexities and challenges involved in joining dissimilar materials, focusing on optimizing welding parameters and techniques to produce strong and high-performance joints.

- Contribution (8): Study of the Microstructure and Mechanical Property Relationships of Gas Metal Arc Welded Dissimilar Protection 600T, DP450, and S275JR Steel Joints [19]. This paper addresses the challenges of welding dissimilar steels through gas metal arc welding (GMAW). The findings reveal that the mechanical properties of the joints, such as tensile strength and hardness, can be optimized by selecting appropriate welding parameters. In particular, DP450/S275JR dissimilar joints exhibit the highest joint efficiency, making this technique highly applicable to industrial processes.
- Contribution (9): Effect of Microchemistry Elements in Relation to Laser Welding Parameters on the Morphology of 304 Stainless Steel Welds Using Response Surface Methodology [20]. This study investigates the effect of sulfur content on the weld bead morphology of AISI 304 stainless steel by using laser welding. The response surface methodology (RSM) analysis reveals how small variations in sulfur content and welding parameters, such as power and focus point, can significantly influence weld pool formation and mechanical performance.
- Contribution (10): Investigation of the Microstructure and Mechanical Properties in Friction Stir Welded Dissimilar Aluminium Alloy Joints via Sampling Direction [21]. This paper investigates the microstructure and mechanical properties of dissimilar aluminum alloys joined by friction stir welding (FSW), with an emphasis on the effect of sampling direction. This work shows that the longitudinal joints exhibit higher tensile strength and elongation compared to the transverse joints, providing critical insights into the optimization of FSW for dissimilar aluminum alloys.

In essence, this Special Issue on *Welding and Joining of Metallic Materials: Microstructure and Mechanical Properties* brings together cutting-edge research that spans advanced welding techniques, microstructural analysis, and the challenges of welding dissimilar materials.

This Special Issue reflects the incredible effort and dedication of the authors who shared their valuable findings. We are deeply grateful to each contributor for their insights, and to the peer reviewers for their hard work in upholding the quality of these papers. A big thank you also goes to everyone else involved in making this possible.

The contributions to this Special Issue are listed as follows:

Contribution (1)—Zhang, Q.; Ren, L.; Lei, X.; Yang, J.; Zhang, K.; Zhang, J. Effect of Laser Heat Input on the Microstructures and Low-Cycle Fatigue Properties of Ti60 Laser Welded Joints. *Crystals* **2024**, *14*, 677.

Contribution (2)—Lozynskyi, V.; Trembach, B.; Katinas, E.; Sadovyi, K.; Krbata, M.; Balenko, O.; Krasnoshapka, I.; Rebrova, O.; Knyazev, S.; Kabatskyi, O.; et al. Effect of Exothermic Additions in Core Filler on Arc Stability and Microstructure during Self-Shielded, Flux-Cored Arc Welding. *Crystals* **2024**, *14*, 335.

Contribution (3)—Touileb, K.; Djoudjou, R.; Ouis, A.; Hedhibi, A.C.; Boubaker, S.; Ahmed, M.M.Z. Particle Swarm Method for Optimization of ATIG Welding Process to Joint Mild Steel to 316L Stainless Steel. *Crystals* **2023**, *13*, 1377.

Contribution (4)—Go, H.; Noh, T.; Jung, S.-B.; Sohn, Y. Microstructural Optimization of Sn-58Bi Low-Temperature Solder Fabricated by Intense Pulsed Light (IPL) Irradiation. *Crystals* **2024**, *14*, 465. https://doi.org/10.3390/cryst14050465.

Contribution (5)—Maritsa, S.; Deligiannis, S.; Tsakiridis, P.E.; Zervaki, A.D. Experimental and Computational Study of Microhardness Evolution in the HAZ for Al–Cu–Li Alloys. *Crystals* **2024**, *14*, 246.

Contribution (6)—da Silva, L.O.P.; Lima, T.N.; Júnior, F.M.d.S.; Callegari, B.; Folle, L.F.; Coelho, R.S. Heat-Affected Zone Microstructural Study via Coupled Numerical/Physical Simulation in Welded Superduplex Stainless Steels. *Crystals* **2024**, *14*, 204.

Contribution (7)—Mahajan, A.M.; Krishna, K.V.; Quamar, M.J.; Rehman, A.U.; Bandi, B.; Babu, N.K. Structure–Property Correlation between Friction-Welded Work and Hardened Al-4.9Mg Alloy Joints. *Crystals* **2023**, *13*, 1119.

Contribution (8)—Elmas, M.; Koçar, O.; Anaç, N. Study of the Microstructure and Mechanical Property Relationships of Gas Metal Arc Welded Dissimilar Protection 600T, DP450 and S275JR Steel Joints. *Crystals* **2024**, *14*, 477.

Contribution (9)—Touileb, K.; Attia, E.; Djoudjou, R.; Hedhibi, A.C.; Benselama, A.; Ibrahim, A.; Ahmed, M.M.Z. Effect of Microchemistry Elements in Relation of Laser Welding Parameters on the Morphology 304 Stainless Steel Welds Using Response Surface Methodology. *Crystals* **2023**, *13*, 1138.

Contribution (10)—Mabuwa, S.; Msomi, V. Investigation of the Microstructure and Mechanical Properties in Friction Stir Welded Dissimilar Aluminium Alloy Joints via Sampling Direction. *Crystals* **2023**, *13*, 1108.

As welding technologies evolve and new materials are introduced, the interplay between microstructure and mechanical properties will continue to be a focal point of research. We believe this collection of articles offers a clear snapshot of the latest breakthroughs and makes a significant contribution to advancing our knowledge in this vital area. We hope the findings presented in this reprint will inspire further innovations in the field of welding and joining.

Funding: This study received no external funding.

Conflicts of Interest: The authors declare no conflicts of interest.

References

1. Lawal, S.L.; Afolalu, S.A. Innovations in TIG and MIG Welding Technologies: Recent Developments and Future Trends. In Proceedings of the 2024 International Conference on Science, Engineering and Business for Driving Sustainable Development Goals (SEB4SDG), Omu-Aran, Nigeria, 2 April 2024; pp. 1–6.
2. Shim, J.Y.; Park, M.W.; Kim, I.S. An Overview of Resistance Element Welding with Focus on Mechanical and Microstructure Joint and Optimization in Automotive Metal Joints. *J. Weld. Join.* **2023**, *41*, 37–48. [CrossRef]
3. Bamberg, P.; Gintrowski, G.; Reisgen, U.; Schiebahn, A. Robustness and Reliability Assessment of Single-Sided Spot Welding as a Process for Sheet to Closed Profile Joining for Body in White Vehicle Structures. *Weld. Int.* **2022**, *36*, 331–343. [CrossRef]
4. Zhou, S.; Hao, S.; Liang, L.; Chen, B.; Li, Y. A Fatigue Reliability Assessment Model for Welded Structures Based on the Structural Stress Method. *Adv. Mech. Eng.* **2023**, *15*, 168781322211474. [CrossRef]
5. Rominiyi, A.L.; Mashinini, P.M. A Critical Review of Microstructure and Mechanical Properties of Laser Welded Similar and Dissimilar Titanium Alloy Joints. *J. Adv. Join. Process.* **2024**, *9*, 100191. [CrossRef]
6. John Raj Kumar, P.R.; Baskaran, A.; Thirumalaikumarasamy, D.; Sonar, T.; Ivanov, M.; Pavendhan, R. Microstructure and Mechanical Properties of Double Pulse TIG Welded Super Austenitic Stainless Steel Butt Joints. *Mater. Test.* **2024**, *66*, 1379–1387. [CrossRef]
7. Li, G.; Wang, Y.; Liang, Y.; Gao, P.; Liu, X.; Xu, W.; Yang, D. Microstructure and Mechanical Properties of Laser Welded Ti-6Al-4V (TC4) Titanium Alloy Joints. *Opt. Laser Technol.* **2024**, *170*, 110320. [CrossRef]
8. Chinakhov, D.A. Additive Technologies, Advanced Joining Technology and Study of Weld Joints. *Metals* **2023**, *13*, 1873. [CrossRef]
9. Dias, F.; Cipriano, G.; Correia, A.N.; Braga, D.F.O.; Moreira, P.; Infante, V. Joining of Aluminum Alloy AA7075 and Titanium Alloy Ti-6Al-4V through a Friction Stir Welding-Based Process. *Metals* **2023**, *13*, 249. [CrossRef]
10. Kumar, R.; Ghosh, B.; Nigam, R.; Mukherjee, A.; Das, S. Optimization of Process Parameters of Gas Metal Arc Welding Using Taguchi Method. *J. Mines Met. Fuels* **2024**, *71*, 6–14. [CrossRef]
11. Wang, Y.; Li, T.; Liu, Y.; Li, X. Research on Welding Parameter Optimization and Automatic Control Based on Machine Learning Algorithm. In Proceedings of the 2024 Asia-Pacific Conference on Software Engineering, Social Network Analysis and Intelligent Computing (SSAIC), New Delhi, India, 10 January 2024; pp. 226–231.
12. Zhang, Q.; Ren, L.; Lei, X.; Yang, J.; Zhang, K.; Zhang, J. Effect of Laser Heat Input on the Microstructures and Low-Cycle Fatigue Properties of Ti60 Laser Welded Joints. *Crystals* **2024**, *14*, 677. [CrossRef]
13. Lozynskyi, V.; Trembach, B.; Katinas, E.; Sadovyi, K.; Krbata, M.; Balenko, O.; Krasnoshapka, I.; Rebrova, O.; Knyazev, S.; Kabatskyi, O.; et al. Effect of Exothermic Additions in Core Filler on Arc Stability and Microstructure during Self-Shielded, Flux-Cored Arc Welding. *Crystals* **2024**, *14*, 335. [CrossRef]
14. Touileb, K.; Djoudjou, R.; Ouis, A.; Hedhibi, A.C.; Boubaker, S.; Ahmed, M.M.Z. Particle Swarm Method for Optimization of ATIG Welding Process to Joint Mild Steel to 316L Stainless Steel. *Crystals* **2023**, *13*, 1377. [CrossRef]
15. Go, H.; Noh, T.; Jung, S.-B.; Sohn, Y. Microstructural Optimization of Sn-58Bi Low-Temperature Solder Fabricated by Intense Pulsed Light (IPL) Irradiation. *Crystals* **2024**, *14*, 465. [CrossRef]
16. Maritsa, S.; Deligiannis, S.; Tsakiridis, P.E.; Zervaki, A.D. Experimental and Computational Study of Microhardness Evolution in the HAZ for Al–Cu–Li Alloys. *Crystals* **2024**, *14*, 246. [CrossRef]

17. da Silva, L.O.P.; Lima, T.N.; Júnior, F.M.d.S.; Callegari, B.; Folle, L.F.; Coelho, R.S. Heat-Affected Zone Microstructural Study via Coupled Numerical/Physical Simulation in Welded Superduplex Stainless Steels. *Crystals* **2024**, *14*, 204. [CrossRef]
18. Mahajan, A.M.; Krishna, K.V.; Quamar, M.J.; Rehman, A.U.; Bandi, B.; Babu, N.K. Structure–Property Correlation between Friction-Welded Work and Hardened Al-4.9Mg Alloy Joints. *Crystals* **2023**, *13*, 1119. [CrossRef]
19. Elmas, M.; Koçar, O.; Anaç, N. Study of the Microstructure and Mechanical Property Relationships of Gas Metal Arc Welded Dissimilar Protection 600T, DP450 and S275JR Steel Joints. *Crystals* **2024**, *14*, 477. [CrossRef]
20. Touileb, K.; Attia, E.; Djoudjou, R.; Hedhibi, A.C.; Benselama, A.; Ibrahim, A.; Ahmed, M.M.Z. Effect of Microchemistry Elements in Relation of Laser Welding Parameters on the Morphology 304 Stainless Steel Welds Using Response Surface Methodology. *Crystals* **2023**, *13*, 1138. [CrossRef]
21. Mabuwa, S.; Msomi, V. Investigation of the Microstructure and Mechanical Properties in Friction Stir Welded Dissimilar Aluminium Alloy Joints via Sampling Direction. *Crystals* **2023**, *13*, 1108. [CrossRef]

Disclaimer/Publisher's Note: The statements, opinions and data contained in all publications are solely those of the individual author(s) and contributor(s) and not of MDPI and/or the editor(s). MDPI and/or the editor(s) disclaim responsibility for any injury to people or property resulting from any ideas, methods, instructions or products referred to in the content.

Article

Effect of Laser Heat Input on the Microstructures and Low-Cycle Fatigue Properties of Ti60 Laser Welded Joints

Qunbing Zhang [1,*], Lina Ren [2,3], Xiaowei Lei [4], Jiadian Yang [5], Kuo Zhang [1] and Jianxun Zhang [2]

1. School of Materials Engineering, Xi'an Aeronautical University, Xi'an 710077, China; qqhrzhang@163.com
2. State Key Laboratory for Mechanical Behavior of Materials, Xi'an Jiaotong University, Xi'an 710049, China; ren.lina@hotmail.com (L.R.); jxzhang@mail.xjtu.edu.cn (J.Z.)
3. Western Titanium Technologies, Co., Ltd., Xi'an 710201, China
4. School of Physical Science and Technology, Northwestern Polytechnical University, Xi'an 710072, China; xiaowei_lei@nwpu.edu.cn
5. Gui Zhou Aviation Technical Development, Co., Ltd., Guiyang 550081, China; yangjiadian08@163.com
* Correspondence: qunbing_zhang@126.com; Tel.: +86-18700969645

Abstract: In this paper, the effects of laser heat input on the microstructures, tensile strength, and fatigue properties of Ti60 laser welded joints were investigated. The results show that with the increase in laser heat input, the macro morphology of the weld zone (WZ) changes from the Y-type to X-type. In the Y-type WZ, the porosity defects are almost eliminated. In contrast, there are a lot of porosity defects in the lower part of the X-type WZ. The microstructure of the base metal (BM) comprises equiaxed α phases, and β phases are mainly distributed at the boundaries of α phases. The heat-affected zone (HAZ) is comprised of α phases and acicular α′ phases, while the WZ mainly contains acicular α′ phases. With the increase in laser heat input, the quantity of the α phase gradually decreases and the acicular α′ phase gradually increases in the HAZ, and the size of the acicular α′ phase in the WZ gradually decreases. Due to the different microstructures, the hardness of BM is lower than the HAZ and WZ under different laser heat input conditions. In the tensile tests and low-cycle fatigue tests, the welded joints are fractured in BM. The porosity defects do not have decisive effects on the tensile and low-cycle fatigue properties of Ti60 laser welded joints.

Keywords: Ti60 titanium alloy; laser welding; porosity defect; microstructure; low-cycle fatigue property

1. Introduction

Titanium alloys are important structural materials in the aerospace field due to their low density, high specific strength, and superior corrosion resistance [1–4]. In order to achieve the light weight of titanium alloy components, welding has become a necessary processing method [5,6].

Compared with other titanium alloys, Ti60 alloy has excellent comprehensive mechanical properties and fatigue strength [7,8]. At present, the research on the welding of Ti60 mainly focuses on electron beam welding, friction welding and brazing. Li et al. [9] carried out electron beam welding on TA15/Ti60 alloy and found that there were no welding defects such as undercut and unwelded areas in the WZ. The microstructure of the WZ was mainly columnar grains. Song et al. [10,11] studied the tensile property of an electron beam welding joint of a Ti60/GH3128 alloy. It was revealed that the tensile fracture mode of the joint was brittle fracture. Guo et al. [12,13] analyzed the linear friction welding behavior of Ti60. They found that an element diffusion layer with a width of about 1 μm was formed at the interface of the welded joint, and the grains on both sides of the interface connected in a eutectic way. The tensile strength of the welded joint was higher than that of the BM. Liu et al. [14] studied the inertial radial friction welding behavior of Ti60/TC18. They found that the joint was tensile fractured at BM, and the fracture analysis consistently revealed a quasi-cleavage fracture mode. Zhao et al. [15] analyzed the microstructure and mechanical properties of a C/C-SiC/Ti60 brazed joint. It was revealed with an increase in the brazing temperature, the joint strength increased and then decreased. Both the block Ti2(Ni,Cu) at low brazing temperature and the overreaction of SiC at a high brazing temperature deteriorated the joints. Wang et al. [16] studied the microstructure and mechanical properties of a Ti2AlNb/Ti60 brazed joint and found that the fracture mode of the original brazed joints predominantly exhibited cleavage fracture. However, after homogenization treatment at 600 °C for 1 h, the fracture mode shifted primarily to intergranular brittle fracture.

Compared with electron beam welding, friction welding and brazing, laser welding has the advantages of a flexible processing capacity, fast welding speed, small HAZ, and easy automation, thereby, it is widely used in aerospace, petrochemical, nuclear energy, biomedicine, and other fields [17–19]. Li et al. [20] studied the influence of laser welding parameters on a TC4 alloy. The results indicated that the defocused position had the largest effect, followed by the laser power and welding speed. The optimal welding parameters were a laser power of 2.3 kW, welding speed of 0.04 m/s and defocused position of 0 mm. Zhu et al. [21] analyzed the influence of laser welding parameters on a Ti-4Al-2V alloy. They found that negative defocusing helped to increase the depth of penetration, but it was more prone to undercut defects. Porosity defects were prone to forming in the middle and bottom parts of the fusion zone due to rapid cooling. The mechanical properties of the joints were significantly affected by the laser power [22]. Wang et al. [23] studied the influences of laser power and welding speed on low-alloy high-strength steel. The results indicated that the laser power was a decisive factor in the weld formation: excessive laser power lead to an unstable behavior of the molten pool and key hole, while insufficient laser power resulted in a lack of penetration. Cai et al. [24] studied the influence of parameters on the porosity defects, weld formation, and properties of the joints. Results showed that increasing the laser power and decreasing the welding speed were conducive to improve the formation of welds and reduce porosity.

Clearly, the previous studies mainly focused on the effects of the laser power and scanning rate on the microstructure and properties of welding joints, but the essence is the interaction of these two parameters, that is, the influence of heat input. Particularly, the design of heat input is crucial once it lies in the range that the shape of weld zone changes significantly. However, the relevant work on Ti60 alloys is still insufficient, and further clarifications on the relationships between heat input, microstructure, and fatigue performance are indispensable. Therefore, in this paper, laser welding is performed on Ti60, and the effects of laser heat input on the microstructures and low-cycle fatigue properties of the welded joint are studied. The research results can provide a meaningful reference for the regulation and optimization of the microstructure and mechanical properties of Ti60 welded joints.

2. Materials and Methods

The experimental BM is forged Ti60 titanium alloy comprising α + β phases, and the chemical composition and mechanical properties are shown in Tables 1 and 2, respectively. First, the forged Ti60 was cut into plates with a thickness of 5 mm. Then, the plates were etched for 2 min using 90 mL of pure water + 6 mL of HNO_3 + 4 mL of HF to remove the surface stains. Afterward, the surfaces of the plates were wiped with dehydrated ethanol and dried naturally. These treated specimens were welded using a YLS-4000 fiber laser with a YASKAWA welding robot. The welding process and parameters are shown in Figure 1a and Table 3.

Table 1. Chemical composition of the Ti60 base metal (wt%).

Al	Sn	Zr	Mo	Si	Nd	C	Ta	Ti
5.3	4.0	2.0	0.8	0.3	1.0	0.09	0.30	Bal.

Table 2. Mechanical properties of the Ti60 base metal at room temperature.

Temperature	Density (kg/m³)	E (MPa)	Poisson's Ratio	σ_b (MPa)	$\sigma_{0.2}$ (MPa)
25 °C	4.53	114	0.31	1050	960

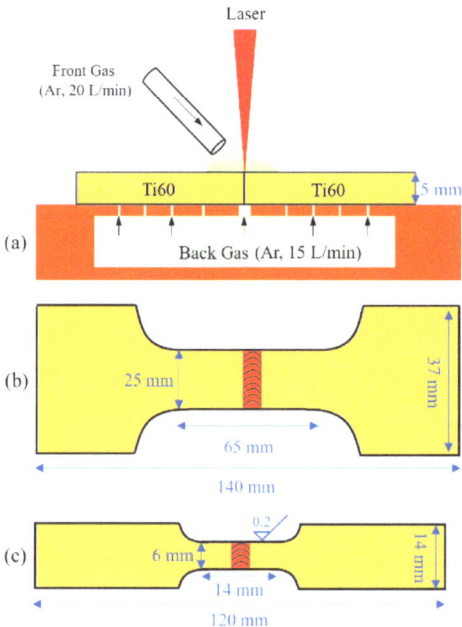

Figure 1. Schematic diagrams: (**a**) welding process; (**b**) tensile test specimen; and (**c**) fatigue test specimen.

Table 3. Laser welding parameters.

Specimen Number	Laser Power (kw)	Welding Speed (mm/min)	Laser Heat Input (J/mm)	Laser Defocusing (mm)	Shielding Gas
1#	3	90	2000	0	Ar (99.99%)
2#	3.5	90	2333		
3#	3.5	80	2625		
4#	4	80	3000		

After welding, the welded plates were machined into metallographic, tensile, and fatigue specimens using electrical discharge wire cutting equipment and a grinding machine. The metallographic specimen of the welded joint was polished with sandpaper and a silica suspension, etched with 80 mL of pure water + 15 mL of HNO_3 + 5 mL of HF etching solution for 3 min, and then the surface was wiped with anhydrous ethanol and finally dried with a hair dryer. The tensile specimens were prepared according to ISO 4136:2022, and the low-cycle fatigue specimens were prepared in accordance with ISO 12106:2017. After processing, the final thicknesses of the tensile and fatigue specimens were about 3 mm, and their dimensions are illustrated in Figure 1b,c. The low-cycle fatigue test equipment was a PLS-100 electro-hydraulic servo static and dynamic testing machine, the control mode was stress control, the waveform was a triangular wave, the stress ratio was R = −1, the frequency was 0.125 Hz, and the stress amplitude was 900 MPa. The tensile test equipment was a UTM5305 electronic universal testing machine, and the tensile rate was 5 mm/min. The microhardness test equipment was a TMVS-1 Vickers hardness tester, the load was 200 g, and the load holding time was 15 s. In order to eliminate test errors, all the mechanical property tests were repeated three times, and their average values were taken as the experimental results.

Finally, a HIROX-200 optical microscope (OM), a JSM-6510A scanning electron microscope (SEM), and a JEM-2100 transmission electron microscope (TEM) were used to characterize the macro morphology, microstructure, and fracture morphology of the welded joints.

3. Results and Discussion

The microstructure of Ti60 BM is shown in Figure 2. As can be seen in Figure 2a, its microstructure is an α + β dual-phase structure; β phases are mainly distributed at the boundaries of α phases, and the average size of α phases is about 13 µm. In the magnified image, the β phases present lamellar structure features, as shown in Figure 2b. According to the TEM images of BM in Figure 2c,d, it can be seen that the width of the β phase is about 200 nm, and α phases are distributed between β phases.

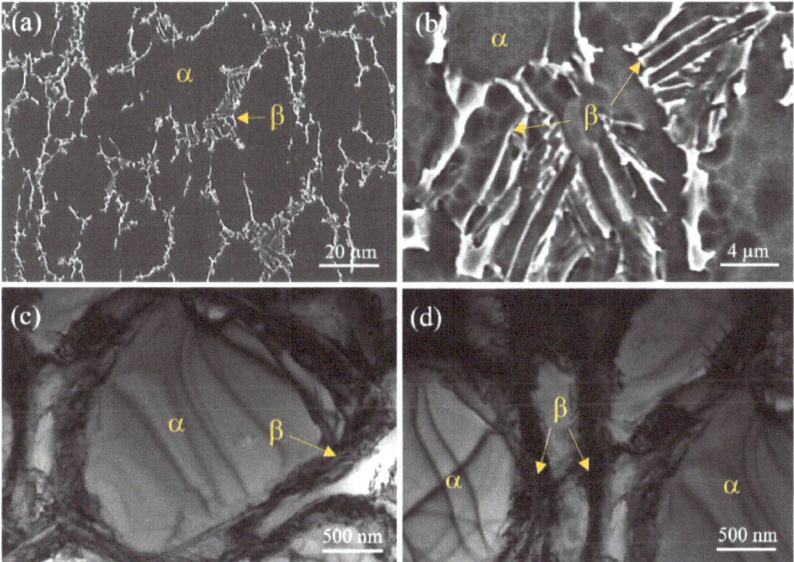

Figure 2. Microstructure of the Ti60 base metal: (**a**,**b**) SEM images and (**c**,**d**) TEM images.

The macroscopic morphologies of Ti60 laser welded joints are shown in Figure 3. It can be seen that with the increase in the laser heat input, the macro morphology of the WZ changes from the Y-type to X-type. There are basically no porosity defects in the Y-type WZ, while lots of welding pores emerge in the middle and lower parts of the X-type WZ.

The macro morphology types of the WZs are mainly caused by the keyhole effect of laser deep penetration welding. Due to the high laser energy density, the liquid metal in the molten pool is vaporized, and a keyhole is formed under the impact of metal vapor. When the depth of the keyhole is smaller than the thickness of the specimen, the metal vapor can only eject outwards from the upper surface of the specimen. Under this circumstance, the metal vapor drives the liquid metal to move upward along the inner wall of the keyhole, forming Marangoni vortices, as shown in Figure 4a. Under the action of Marangoni vortices, the width of the upper part of the weld is larger than those of the middle and lower parts, forming the Y-type WZ. In contrast, when the depth of the keyhole exceeds the thickness of the specimen, the bottom of the molten pool is penetrated by the laser, and the metal vapor in the keyhole simultaneously ejects outward from the upper and lower surfaces of the specimen, forming Marangoni vortices in the upper and lower parts of the molten pool, respectively, as shown in Figure 4b. As a consequence, the widths of the upper and lower parts of the weld are larger than that the middle part, forming the X-type WZ.

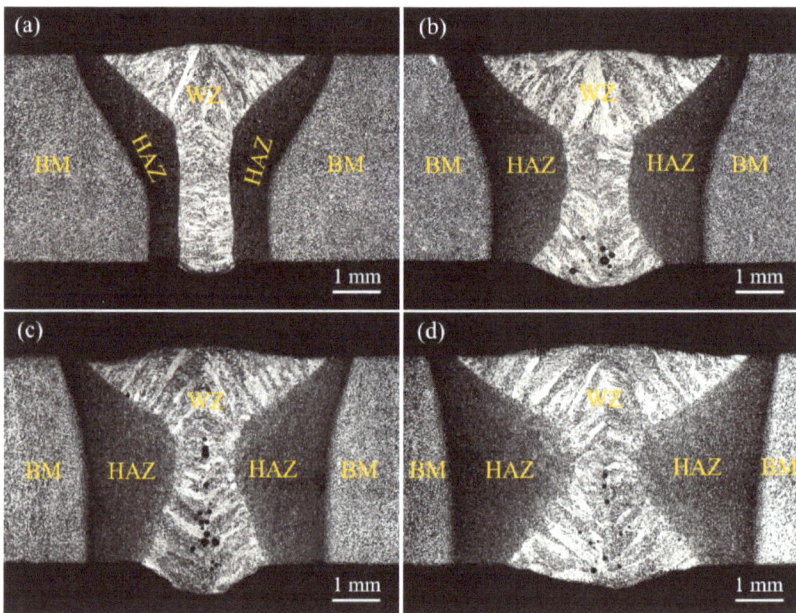

Figure 3. Effect of the laser heat input on the macroscopic morphologies of the welded joints: (**a**) 2000 J/mm; (**b**) 2333 J/mm; (**c**) 2625 J/mm; and (**d**) 3000 J/mm.

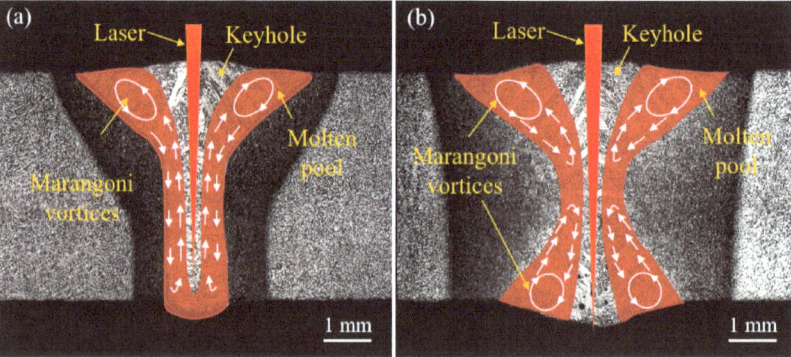

Figure 4. Schematic diagram of the influence of the laser heat input on the macroscopic morphology of the joint: (**a**) Y-type and (**b**) X-type.

The difference in the number of welding pores is due to the different ejection directions of metal vapor between the Y-type and X-type WZs. During the cooling process of the molten pool, the liquid metal of the Y-type WZ flows along the keyhole wall to the bottom of the keyhole, and the metal vapor escapes along the center of the keyhole, rendering the formation of a pore-free weld after solidification.

For the X-type WZ, the escape mode of the metal vapor in the keyhole becomes complicated. As shown in Figure 5, the liquid metal at the upper part of the molten pool flows down along the keyhole wall under the action of gravity during the post-welding cooling process, blocking the narrow zone in the middle of the X-type WZ. In this context, the keyhole is divided into two independent zones, as shown in Figure 5b. In the upper part of the keyhole, the metal vapor ejects upward along the keyhole, so that after cooling, there is almost no porosity defect in the upper part of the WZ, as shown in Figure 3b–d. However, in the lower part of the keyhole, the upward escape channel of metal vapor is blocked, and so the metal vapor has to eject through the opening on the lower surface of the specimen. Meanwhile, the liquid metal in the lower part of the molten pool flows downward under the action of gravity, resulting in the narrowing of the porosity escape channel on the lower surface, as shown in Figure 5c. Due to the rapid cooling rate of the weld pool in laser welding, the metal vapor that has not escaped in time is trapped in the liquid molten pool and dispersed into several bubbles. After the solidification of the molten pool, pore defects form at the lower part of the X-type WZ, as shown in Figure 5d.

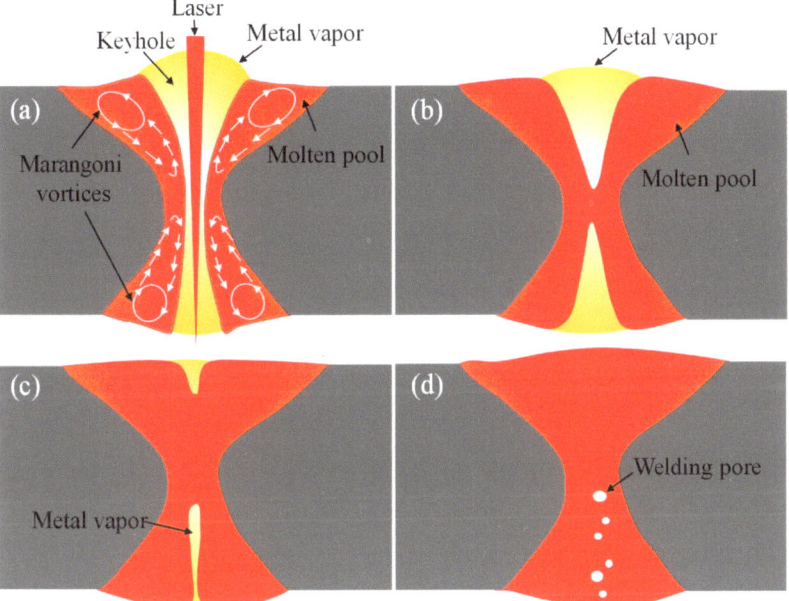

Figure 5. Schematic diagram of the pore formation process of the X-type laser welded joint: (**a**) keyhole of X-type WZ; (**b**) the keyhole divided into two independent zones; (**c**) the lower part of the molten pool flows downward under the action of gravity; (**d**) pore defects form at the lower part of the X-type WZ.

The difference in the macroscopic color contrast between the BM, HAZ, and WZ is caused by their distinct microstructures. In Figure 6, we compare the microstructures of the welded joint (2000 J/mm) with that of the BM. It should be noted that, considering the similarity of the microstructures in the grain and phase features aspect, here, we only present the microstructures of the welded joint with a laser heat input of 2000 J/mm.

Figure 6a shows the microstructure of the BM. As described in Figure 2, it comprises equiaxed α-phase grains, β phases are mainly distributed in the grain boundaries, and the average grain size is about 13 μm. Figure 6b shows the boundary line between the BM and HAZ, where the left side is the BM and the right side is the HAZ. The closer to the WZ, the more heat input and higher temperature during the welding process. The temperature on the left side of the line is lower than the α→β phase transition temperature, and the microstructure is basically unchanged. The temperature on the right side of the line is higher than the phase transition temperature of α→β, but the laser welding speed is so fast that only a small amount of α phase transformed into β phase, and then the β phase transformed into a needle-like α′ phase due to the rapid cooling process.

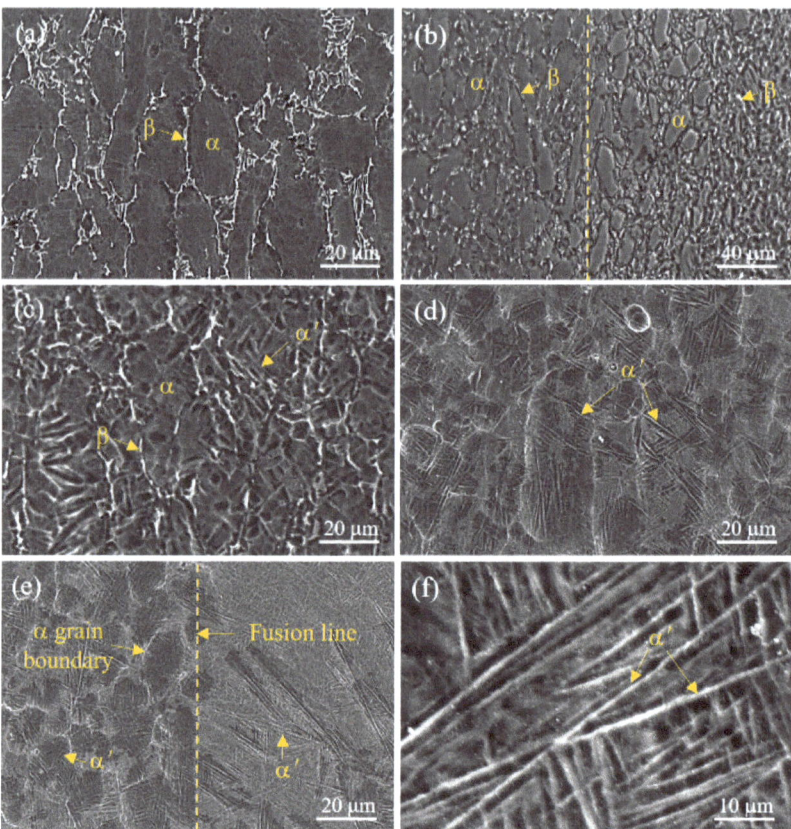

Figure 6. SEM microstructure of the Ti60 laser welded joint: (**a**) BM; (**b**) boundary between BM and HAZ; (**c**) HAZ far from the WZ; (**d**) HAZ near to the WZ; (**e**) boundary between HAZ and WZ; and (**f**) WZ.

Figure 6c,d shows the microstructures of the far-HAZ (far away from the WZ) and near-HAZ (near to the WZ) regions. With the decrease in the distance to WZ, the temperature gradually surpasses the α→β phase transition temperature, and more and more α phases are transformed into β phases. During the rapid cooling process after welding, more and more acicular α′ phases are formed. Additionally, the average grain size of the HAZ is the same as that of the BM, due to the pinning effect of β phases at the boundary of α grains. Figure 6e shows the boundary between the HAZ and WZ. The microstructures on both sides of the fusion line are comprised of acicular α′ phases. However, the microstructure of the HAZ (on the left side of the fusion line) still clearly retains the grain boundary outline of the original α phase. In contrast, the grain boundaries of WZ (on the right side of the fusion line) are virtually invisible. At the same time, the numbers and sizes of the acicular α′ phases of the HAZ and WZ are nearly identical. As shown in Figure 6f, the microstructure in the center of the WZ comprises acicular α′ phases, but the size and quantity of α′ phases are larger than those of the HAZ.

The microhardness curve of the Ti60 laser welded joint (2000 J/mm) is shown in Figure 7. It can be seen that the hardness of the BM is about 350 HV. The microhardness of the HAZ is higher than the BM, and the average value is 379 HV. The microhardness of the WZ is the highest, with an average value of 412 HV. The different hardnesses of the three zones are attributed to their distinct microstructures, as displayed in Figures 2, 3 and 6, where the BM is an equiaxed α grain + grain boundary β phase, the HAZ comprises an equiaxed α grain + grain boundary β phase + acicular α′ phase, and the closer to the WZ, the more acicular α′ phase in the HAZ. The microstructure of the WZ is acicular α′ phase. As is well known, the crystal structure of α phase is close-packed hexagonal (HCP) and the β phase is body-centered cubic (BCC). Since that BCC has 12 slip systems and HCP only has 3 slip systems, the β phase is more prone to deform, thereby the hardness of β phase is lower than the α phase. In addition, the acicular α′ phase contains more dislocations than does the equiaxed α phase, and thus the hardness of acicular α′ is higher than the equiaxed α phase owing to dislocation strengthening. In summary, from the BM to WZ, the content of the β phase gradually decreases and the content of acicular α′ phase gradually increases; so, the microhardness is gradually enhanced.

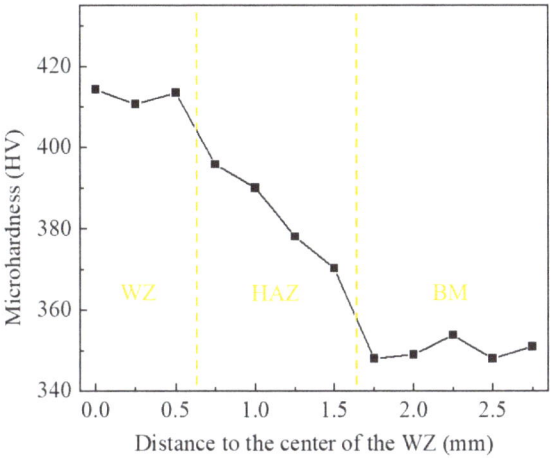

Figure 7. Microhardness curve of the Ti60 laser welded joint.

In order to explore the influence of the laser heat input on the microstructure and mechanical properties of Ti60 welded joint, the microstructures of the HAZ and WZ and the tensile properties and fatigue properties of the welded joints under different laser heat inputs are compared, and the results are shown in Figures 8–12.

Figure 8 shows the microstructure of the HAZ under different laser heat inputs. It can be seen that with the increase in the laser heat input, the β phases distributed at the boundaries of α grains gradually disappear, and the number and size of the acicular α′ phases in the α grains gradually increase. This is because with the increase in the laser heat input, the holding time of the α→β phase transition temperature of the HAZ is increased, and more α phases transform into β phases. In the following post-welding process, due to the fast cooling, the β phases transform into acicular α′ phases. It is understandable that the greater the laser heat input, the slower the cooling rate, and so the size of the needle-shaped α′ phase increases.

Figure 9 shows the microstructure of the WZ under different laser heat inputs. It can be seen that the WZs are all comprised of α phases + acicular α′ phases. With the increase in the laser heat input, the α phase content gradually decreases, and the content and the size of acicular α′ phases gradually increase. This is because with the increasing laser heat input, the cooling rate after welding decreases, and the acicular α′ phase has more sufficient time to grow; so, the size increases while the quantity decreases.

Figure 10 shows the effect of the laser heat input on the microhardness of WZ and HAZ. It can be seen that with the increase in the laser heat input, the hardness of the HAZ increases, but the hardness of the WZ increases first and then decreases. The main reason for this phenomenon is that, as is depicted in Figure 8, with the increase in the laser heat input, the fraction of α phases in the HAZ gradually decreases, and the content of acicular α′ phases gradually increases. The acicular α′ phases can increase the strength of the material, and so the microhardness of the HAZ increases continuously. In contrast, for the WZ shown in Figure 9, with the increase in the laser heat input, the content and size of the acicular α′ phases are gradually elevated. This will have two opposite effects: the higher content of the acicular α′ phases enhances the microhardness, but the coarser acicular α′ phases reduce the microhardness. Under the mutual interactions of these two effects, the increasing laser heat input causes the microhardness of the WZ to peak at 2600 J/mm.

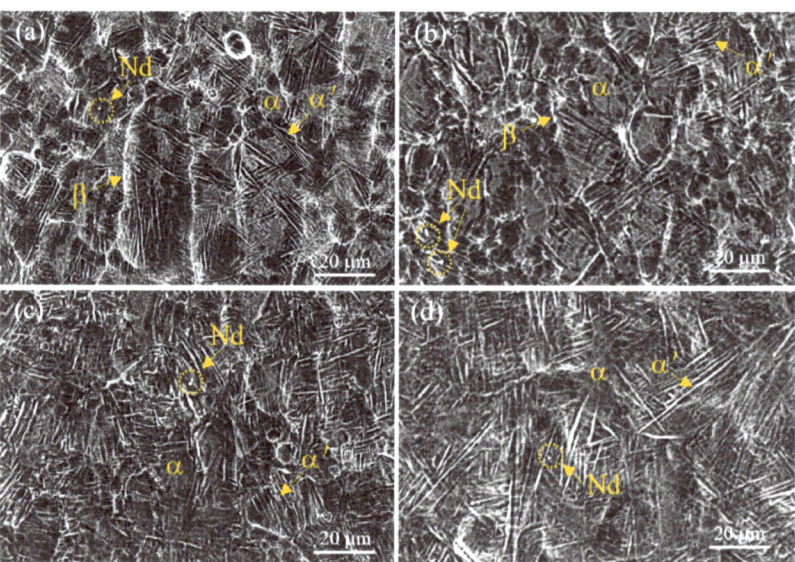

Figure 8. Effect of the laser heat input on the microstructures of the HAZ: (**a**) 2000 J/mm; (**b**) 2333 J/mm; (**c**) 2625 J/mm; and (**d**) 3000 J/mm.

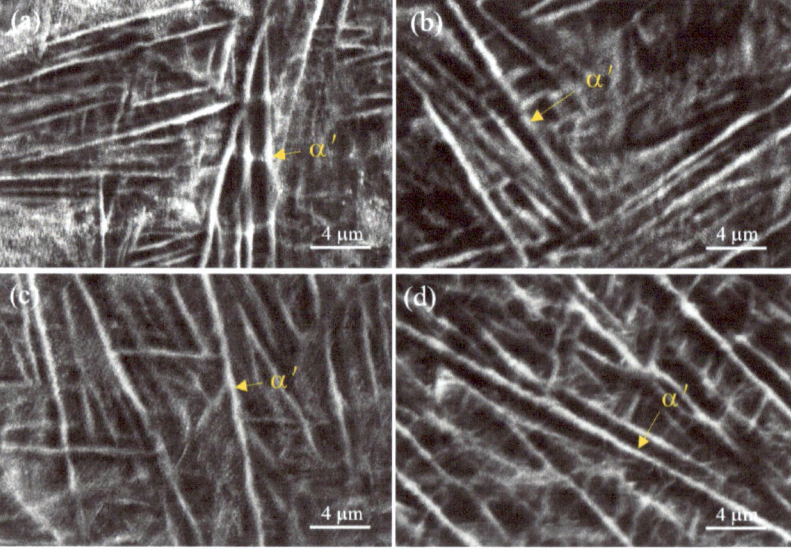

Figure 9. Effect of the laser heat input on the microstructures of the WZ: (**a**) 2000 J/mm; (**b**) 2333 J/mm; (**c**) 2625 J/mm; and (**d**) 3000 J/mm.

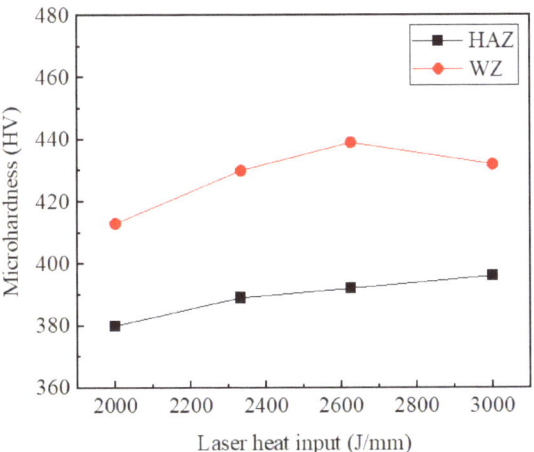

Figure 10. Effect of the laser heat input on the microhardness of the HAZ and WZ.

As shown in Table 4, the tensile and low-cycle fatigue fracture locations of the Ti60 laser welded joint are located at the BM under different laser heat inputs. Additionally, with the increase in the laser heat input, the tensile strength and low-cycle fatigue life both increase first and then decrease. According to the microhardness test results in Figure 7, the hardness values of the WZ with different heat inputs are higher than those of the BM. Clearly, the harder WZ can withstand a higher strength, and the softer BM can withstand a lower strength during tensile and low-cycle fatigue tests, leading to the prior fracture of the BM.

Table 4. Fracture location and strength of the welded joints after tensile and fatigue tests.

Specimen Number	Laser Heat Input (J/mm)	Tensile Strength (MPa)	Tensile Fracture Location	Low-Cycle Fatigue Life (cycle)	Low-Cycle Fatigue Fracture Location
1#	2000	987	BM	5702	BM
2#	2333	1045	BM	6614	BM
3#	2625	964	BM	6380	BM
4#	3000	917	BM	4438	BM

The tensile fracture morphology of the Ti60 laser welded joint (2000 J/mm) is shown in Figure 11. As previously known, the WZ has the highest hardness, followed by the HAZ, and the hardness of BM is the lowest. In the process of the tensile test, due to the low hardness, the BM is easier to deform and fracture. It can be seen from Figure 11b that the fracture surface is smooth, which is caused by tangential stress that is perpendicular to the tensile direction. Inside the tensile fracture, the surface exhibits mainly the dimple feature, but there are also some characteristic morphologies similar to the grain boundaries of equiaxed grains and some lamellar features similar to shutters, as shown in Figure 11c,e. Compared with the microstructure of the BM (Figure 2), it is confirmed that these features are consistent with the equiaxed α grains and the lamellar β phases in the BM, as shown in Figure 11d,e. As mentioned above, the β phase is BCC crystal with more slip systems than the HCP α phase. Therefore, in the process of the tensile test, the plastic deformation is mainly concentrated at the β phases. As a consequence, the β phases distributed around the equiaxial α grains form a topography similar to the grain boundaries on the fracture surface, and the lamellar β phases form a topography similar to shutters.

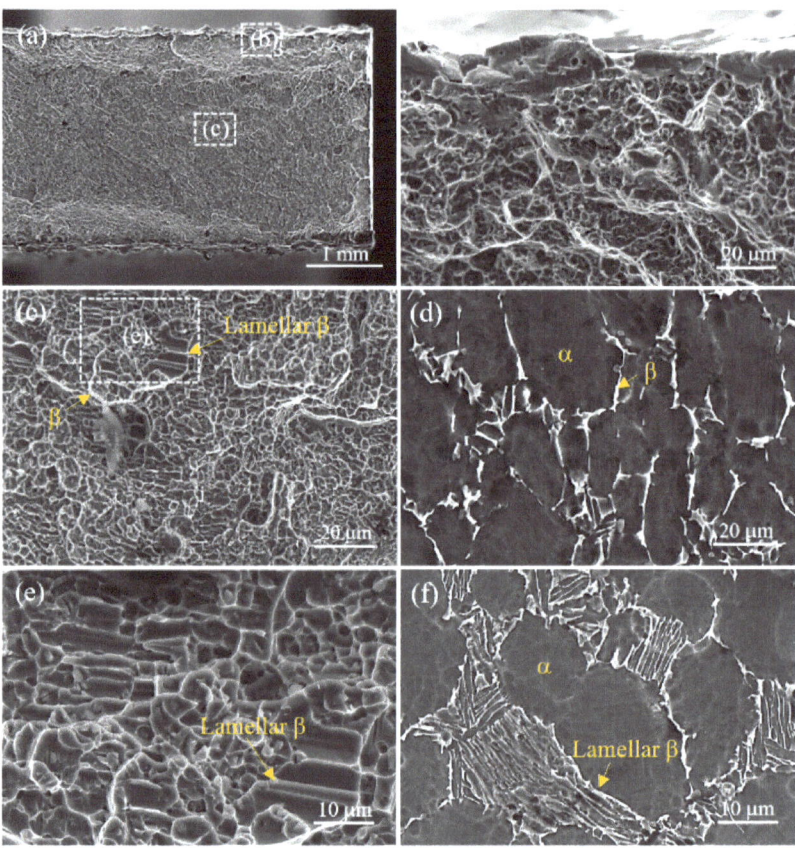

Figure 11. Fracture morphology of the tensile specimen of the Ti60 laser welded joint: (**a**) macrofracture surface; (**b**) the edge of the fracture surface; (**c**) the inside of the fracture surface; (**d**) the equiaxed α grains in the BM; (**e**) the enlarged view of the inside of the fracture surface; and (**f**) the lamellar β phases in the BM.

The low-cycle fatigue fracture morphology of the Ti60 laser welded joint (3000 J/mm) is shown in Figure 12. As can be seen in Figure 12a, the fracture surface is relatively flat, indicating that the low-cycle fatigue property of the material is relatively uniform. As can be seen in the morphology of the fatigue crack initiation zone in Figure 12b, the surface of the crack initiation zone is relatively flat, and no metallurgical defects such as welding pores, inclusions and processing defects can be found on the fracture surface. As is mentioned in Figure 3, many porosity defects exist in the WZ. The fact that the specimen is fractured at the BM suggests that the pore defects in the WZ do not induce low-cycle fatigue fracture in the present case. Under the actions of tensile and compressive stresses caused by the low-cycle fatigue test, slip bands are produced and extrude the surface of the fatigue specimen; then, micro-cracks form on the surface. Figure 12c shows the fracture morphology of the fatigue crack expansion zone. It can be seen that a large number of fatigue striations appear in the expansion zone, indicating that the plasticity of the material is good. In the process of fatigue crack propagation, tensile stress induces the expansion of the crack tip, while the compressive stress compresses the crack tip in the manner of work hardening, and then fatigue striations form on the fracture surface. Every fatigue striation represents one fatigue cycle. At the same time, there are also some tearing edges around the fatigue striations. Compared with the microstructure of the BM (Figure 6a), it can be found that the outlines of tearing edges are similar to the boundaries of equiaxial α phase grains, and the outlines of fatigue striations are similar to the equiaxial α phase grains. Figure 12d shows the morphology of the final fracture zone of the low-cycle fatigue specimen. With the continuous expansion of the fatigue crack, the bearing capacity of the fatigue specimen decreases continuously. When the crack length reaches a critical value, the tensile stress loaded on the fatigue specimen (900 MPa) exceeds its bearing limit, then the specimen undergoes tensile fracture, and dimples form on the final fracture zone of the low-cycle fatigue specimen.

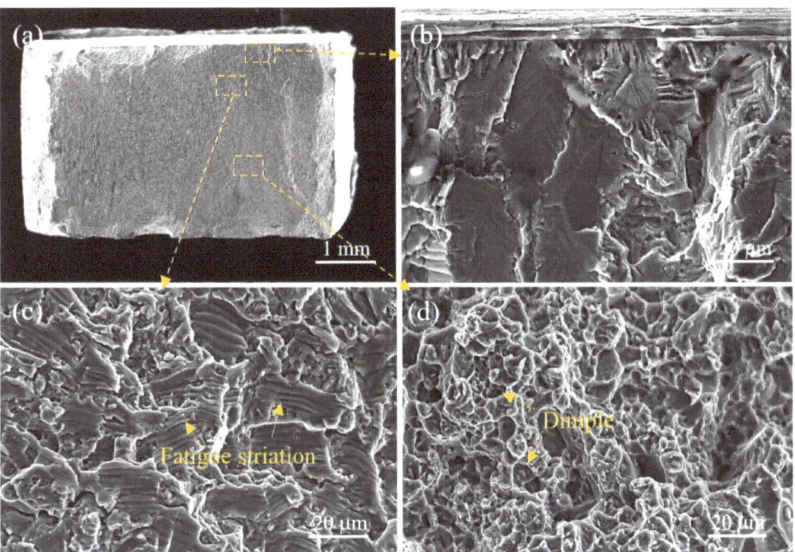

Figure 12. Fracture morphology of the low-cycle fatigue specimen of the Ti60 laser welded joint: (**a**) whole fracture surface; (**b**) crack initiation zone; (**c**) crack expansion zone; and (**d**) final fracture zone.

4. Conclusions

In this study, the microstructures and tensile and low-cycle fatigue properties of Ti60 welded joints with different laser heat inputs are investigated. The main conclusions are as follows:

(1) With the increase in the laser heat input, the macro morphology of the WZ changes from Y type to X type. Welding pores are mainly formed at the lower part of the X-type WZ.

(2) From the BM to the WZ, the microhardness increases gradually. With the increase in the laser heat input, the microhardness of the WZ increases first and then decreases, which is mainly due to the changes in the size and number of the acicular α′ phase.

(3) Although there are lots of pore defects at the bottom of the X-type WZ, the tensile and low-cycle fatigue specimens all fracture at the BM.

(4) Although the porosity defect in the WZ does not cause the tensile and low-cycle fatigue fracture of the welded joint at the WZ, it may damage the joint's high-cycle fatigue and persistent creep properties. Therefore, the welding parameters should be carefully controlled during the welding process to obtain a Y-type WZ to avoid porosity defects in the WZ, thereby improving the long-term service performance of the welded joints.

Author Contributions: Conceptualization, Q.Z., L.R. and X.L.; methodology, Q.Z.; validation, Q.Z. and K.Z.; formal analysis, Q.Z. and L.R.; investigation, Q.Z.; resources, J.Y. and J.Z.; data curation, Q.Z.; writing—original draft preparation, Q.Z. and L.R.; writing—review and editing, X.L. All authors have read and agreed to the published version of the manuscript.

Funding: This research was funded by Natural Science Foundation of Shaanxi Province, grant numbers 2019JQ-915 and 2019JQ-224, and Xi'an Scientific and Technological Project, grant number 2023JH-GXRC-0232.

Data Availability Statement: The original contributions presented in the study are included in the article; further inquiries can be directed to the corresponding author.

Acknowledgments: Thanks to Xi'an Aeronautical University, Xi'an Jiaotong University, and Northwestern Polytechnical University.

Conflicts of Interest: Author Lina Ren was employed by the company Western Titanium Technologies. Author Jiadian Yang was employed by the company Gui Zhou Aviation Technical Development. The remaining authors declare that the research was conducted in the absence of any commercial or financial relationships that could be construed as a potential conflict of interest.

References

1. Chen, W.; Li, R.; Liu, L.; Liu, R.; Cui, Y.; Chen, Z.; Wang, Q.; Wang, F. Effect of NaCl-rich environment on internal corrosion for Ti60 alloy at 600 °C. *Corros. Sci.* **2023**, *220*, 111307. [CrossRef]
2. Shao, L.; Li, W.; Li, D.; Xie, G.; Zhang, C.; Zhang, C.; Huang, J. A review on combustion behavior and mechanism of Ti alloys for advanced aero-engine. *J. Alloys Compd.* **2023**, *960*, 170584. [CrossRef]
3. Wu, J.; Yuan, S.; Wang, X.; Chen, H.; Huang, F.; Yu, C.; He, Y.; Yin, A. The Microstructure Characterization of a Titanium Alloy Based on a Laser Ultrasonic Random Forest Regression. *Crystals* **2024**, *14*, 607. [CrossRef]
4. Ai, Y.; Wang, Y.; Yan, Y.; Han, S.; Huang, Y. The evolution characteristics of solidification microstructure in laser welding of Ti-6Al-4V titanium alloy by considering transient flow field. *Opt. Laser Technol.* **2024**, *170*, 110195. [CrossRef]
5. Kotlarski, G.; Kaisheva, D.; Ormanova, M.; Stoyanov, B.; Dunchev, V.; Anchev, A.; Valkov, S. Improved Joint Formation and Ductility during Electron-Beam Welding of Ti6Al4V and Al6082-T6 Dissimilar Alloys. *Crystals* **2024**, *14*, 373. [CrossRef]
6. Fernandes, F.A.O.; Gonçalves, J.J.M.; Pereira, A.B. Evaluation of Laser Lap Weldability between the Titanium Alloy Ti-6Al-4V and Aluminum Alloy 6060-T6. *Crystals* **2023**, *13*, 1448. [CrossRef]
7. Wang, B.; Zeng, W.; Zhao, Z.; Jia, R.; Xu, J.; Wang, Q. Effect of micro-texture and orientation incompatibility on the mechanical properties of Ti60 alloy. *Mater. Sci. Eng. A* **2023**, *881*, 145419. [CrossRef]
8. Li, L.; Xie, F.; Wu, X.; He, J.; Li, G.; Zhang, T. Effect of the APS YAG coating on the fretting wear properties of Ti60 titanium alloy. *Surf. Coat. Technol.* **2024**, *489*, 131061. [CrossRef]
9. Li, H.; Chen, M.; Wang, N.; Wu, T. Microstructure Evolution and Mechanical Properties of TA15/Ti60 Tailor Welded Blank by Electron Beam Welding. *J. Mater. Eng. Perform.* **2024**. [CrossRef]
10. Song, D.; Wang, T.; Jiang, S.; Xie, Z. Influence of welding parameters on microstructure and mechanical properties of electron beam welded Ti60 to GH3128 joint with a Cu interlayer. *Chin. J. Aeronaut.* **2021**, *34*, 39–46. [CrossRef]
11. Song, D.; Wang, T.; Zhu, J.; Jiang, S.; Xie, Z. Influence of welding sequences on the microstructure and mechanical properties of dual-pass electron beam welded Ti60/V/Cu/GH3128 joints. *J. Mater. Res. Technol.* **2020**, *9*, 14168–14177. [CrossRef]
12. Guo, Z.; Ma, T.; Yang, X.; Chen, X.; Tao, J.; Li, J.; Li, W.; Vairis, A. Linear friction welding of Ti60 near-α titanium alloy: Investigating phase transformations and dynamic recrystallization mechanisms. *Mater. Charact.* **2022**, *194*, 112424. [CrossRef]
13. Guo, Z.; Ma, T.; Chen, X.; Yang, X.; Tao, J.; Li, J.; Li, W.; Vairis, A. Interfacial bonding mechanism of linear friction welded dissimilar Ti$_2$AlNb–Ti60 joint: Grain intergrowth induced by combined effects of dynamic recrystallization, phase transformation and elemental diffusion. *J. Mater. Res. Technol.* **2023**, *24*, 5660–5668. [CrossRef]
14. Liu, Y.-Y.; Tian, W.-T.; Yang, Q.-H.; Yang, J.; Wang, K.-S. Inertia radial friction welding of Ti60(near-α)/TC18(near-β) bimetallic components: Interfacial bonding mechanism, heterogenous microstructure and mechanical properties. *Mater. Charact.* **2024**, *208*, 113598. [CrossRef]
15. Zhao, K.; Liu, D.; Song, Y.; Hou, Z.; Song, X. Joining C/C–SiC composite and Ti60 alloy using a semi-solid TiNiCuNb filler. *J. Mater. Res. Technol.* **2023**, *27*, 8073–8083. [CrossRef]
16. Wang, P.; Zhao, S.; Nai, X.; Chen, H.; Wang, P.; Song, X.; Li, W. Homogenization treatment-induced reduction of brittle intermetallic compounds in Ti$_2$AlNb/Ti60 brazed joints: Effect on microstructure and mechanical properties. *Mater. Sci. Eng. A* **2024**, *909*, 146856. [CrossRef]
17. Park, C.; Hwang, T.; Kim, G.-D.; Nam, H.; Kang, N. Effect of the Initial Grain Size on Laser Beam Weldability for High-Entropy Alloys. *Crystals* **2022**, *13*, 65. [CrossRef]
18. Lassila, A.A.; Lönn, D.; Andersson, T.; Wang, W.; Ghasemi, R. Effects of different laser welding parameters on the joint quality for dissimilar material joints for battery applications. *Opt. Laser Technol.* **2024**, *177*, 111155. [CrossRef]
19. Sheng, J.; Kong, F.; Tong, W. Experimentally-guided finite element modeling on global tensile responses of AA6061-T6 aluminum alloy joints by laser welding. *J. Adv. Join. Process.* **2024**, *9*, 100229. [CrossRef]
20. Li, G.; Wang, Y.; Liang, Y.; Gao, P.; Liu, X.; Xu, W.; Yang, D. Microstructure and mechanical properties of laser welded Ti-6Al-4V (TC4) titanium alloy joints. *Opt. Laser Technol.* **2024**, *170*, 110320. [CrossRef]

21. Zhu, Y.; Lu, L.; Zhang, C.; Yuan, J.; Fu, C.; Wang, L. Microstructure, Variant Selection, and Mechanical Properties of Laser-Welded Ti-4Al-2V Joints. *Metals* **2024**, *14*, 405. [CrossRef]
22. Zhu, Y.; Zhang, Y.; Li, C.; Zhu, J.; Wang, L.; Fu, C. Investigation of Microstructure, Oxides, Cracks, and Mechanical Properties of Ti-4Al-2V Joints Prepared Using Underwater Wet Laser Welding. *Materials* **2024**, *17*, 1778. [CrossRef]
23. Wang, C.; Mi, G.; Zhang, X. Welding stability and fatigue performance of laser welded low alloy high strength steel with 20 mm thickness. *Opt. Laser Technol.* **2021**, *139*, 106941. [CrossRef]
24. Cai, D.; Luo, Z.; Han, L.; Han, S.; Yi, Y. Porosity and joint property of laser-MIG hybrid welding joints for 304 stainless steel. *J. Laser Appl.* **2020**, *32*, 022056. [CrossRef]

Disclaimer/Publisher's Note: The statements, opinions and data contained in all publications are solely those of the individual author(s) and contributor(s) and not of MDPI and/or the editor(s). MDPI and/or the editor(s) disclaim responsibility for any injury to people or property resulting from any ideas, methods, instructions or products referred to in the content.

Article

Effect of Exothermic Additions in Core Filler on Arc Stability and Microstructure during Self-Shielded, Flux-Cored Arc Welding

Vasyl Lozynskyi [1], Bohdan Trembach [2,*], Egidijus Katinas [3], Kostiantyn Sadovyi [4], Michal Krbata [5], Oleksii Balenko [6], Ihor Krasnoshapka [4], Olena Rebrova [7], Sergey Knyazev [7], Oleksii Kabatskyi [8], Hanna Kniazieva [7] and Liubomyr Ropyak [9]

1. Belt and Road Initiative Center for Chinese-European Studies (BRICCES), Guangdong University of Petrochemical Technology, Maoming 525000, China; lvg.nmu@gmail.com
2. Private Joint Stock Company «Novokramatorsky Mashinostroitelny Zavod», 04070 Kiev, Ukraine
3. Department of Electrical Engineering and Automation, Czech University of Life Sciences Prague, 165 00 Praha, Czech Republic; katinas@tf.czu.cz
4. Department of Radio-Technical and Special Troops, Ivan Kozhedub Kharkiv National Air Force University, 61045 Kharkiv, Ukraine; 971sadovyi@gmail.com (K.S.); igor-krasnij@ukr.net (I.K.)
5. Faculty of Special Technology, Alexander Dubcek University of Trenčín, 911 06 Trenčín, Slovakia; michal.krbata@tnuni.sk
6. Department of Computer Engineering and Programming, National Technical University "Kharkiv Polytechnic Institute", 61000 Kharkiv, Ukraine; alexibalenko@gmail.com
7. Department of Materials Science, National Technical University "Kharkiv Polytechnic Institute", 61000 Kharkiv, Ukraine; rebrovaem0512@gmail.com (O.R.); obmeninfoserg@ukr.net (S.K.); annapostelnik@ukr.net (H.K.)
8. Department of Fundamentals of Machine Design, Donbass State Engineering Academy, 84313 Kramatorsk, Ukraine; uncle.i.72@gmail.com
9. Department of Computerized Mechanical Engineering, Ivano-Frankivsk National Technical University of Oil and Gas, 15 Karpatska Str., 76019 Ivano-Frankivsk, Ukraine; l_ropjak@ukr.net
* Correspondence: btrembach89@gmail.com; Tel.: +380-994500811

Citation: Lozynskyi, V.; Trembach, B.; Katinas, E.; Sadovyi, K.; Krbata, M.; Balenko, O.; Krasnoshapka, I.; Rebrova, O.; Knyazev, S.; Kabatskyi, O.; et al. Effect of Exothermic Additions in Core Filler on Arc Stability and Microstructure during Self-Shielded, Flux-Cored Arc Welding. *Crystals* **2024**, *14*, 335. https://doi.org/10.3390/cryst14040335

Academic Editor: Hongbin Bei

Received: 22 February 2024
Revised: 24 March 2024
Accepted: 29 March 2024
Published: 31 March 2024

Copyright: © 2024 by the authors. Licensee MDPI, Basel, Switzerland. This article is an open access article distributed under the terms and conditions of the Creative Commons Attribution (CC BY) license (https://creativecommons.org/licenses/by/4.0/).

Abstract: In the conditions of an energy crisis, an important issue is the increase in energy efficiency and productivity of welding and hardfacing processes. The article substantiates the perspective of using exothermic additives introduced into core filler for flux-cored wire arc welding processes as a relatively cheap additional heat source, reducing energy consumption when melting filler materials, and increasing the deposition rate. The mixture design (MD) was selected as the design method to optimize the average values of current and voltage, as well as arc stability parameters depending on core filler composition. This article studies the influence of the introduction of exothermic addition (EA), as well as the ratios CuO/C and CuO/Al on arc stability for the FCAW S process. Parameters characterizing arc stability were determined using an oscillograph, and from the obtained oscillograms, an analysis was conducted on arc voltage and welding current signals during flux-cored arc welding. It was determined that various methods can be used to evaluate arc stability, which can be divided into two groups: graphical (current and voltage cyclograms, box plots with frequency histograms, ellipse parameters plotted on current, and voltage cyclograms) and statistical (standard variation and coefficients of variation for welding current and arc voltage). In this paper, a comprehensive evaluation of arc stability depending on the composition of the cored wire filler was carried out. It was determined that the most stable current parameters were observed for the flux-cored wire electrode with an average exothermic addition content at the level of EA = 26.5–28.58 wt.% and a high carbon content (low values of CuO/C = 3.75). Conversely, the lowest values of arc stability ($CV(U)$ and $Std(U)$) were observed during hardfacing with a flux-cored wire electrode with a high CuO/Al ratio $\geq$ 4.5 and a content of exothermic addition in the core filler below the average EA < 29 wt.%. Mathematical models of mean values, standard deviation, coefficient of variation for welding current, and arc voltage were developed. The results indicated that the response surface prediction models had good accuracy and prediction ability. The developed mathematical models showed that the ratio of oxidizing agent to reducing agent in the composition of exothermic addition (CuO/Al) had the greatest influence on the welding current and arc voltage characteristics under investigation. The percentage of exothermic mixture in the core filler (EA) only affected the average welding current (I_{aw}) and the average arc voltage (U_{aw}). The graphite content expressed through the CuO/C ratio had a significant impact on welding current parameters as well as the coefficient of variation of arc voltage ($CV(U)$). Two welding parameters were selected for optimization: the mean welding current (I_{aw}) and the standard deviation of arc voltage ($Std(U)$). The best arc stability when using exothermic addition CuO-Al in the core filler was observed at CuO/Al = 3.6–3.9, CuO/C = 3.5–4.26, and at an average EA content of 29–38 wt.%. The significant influence of the CuO/Al and CuO/C ratios on arc voltage parameters can also be explained by their impact on the elemental composition of the welding arc (copper, cupric oxide (CuO), and Al_2O_3). The more complete this reaction, the higher the amount of easily vaporized copper (Cu) in the arc plasma, enhancing arc stability. The influence of core filler composition on the microstructure of deposited metal of the Fe-Cr-Cu-Ti alloy system was investigated.

Keywords: hardfacing; flux-cored arc welding; arc stability; arc voltage; welding current; heat input; microstructure

1. Introduction

The role and significance of materials for the advanced economies of the world remain substantial. Materials form the basis of many sectors of the economy, among which construction [1,2], automobile manufacturing [3–7], and mechanical engineering [8–11] hold particular importance. Among materials, welding materials hold special significance. Chemical–thermal methods [12–14], functional coatings [15–17], electrospark deposition (ESD) [18–22], thermal spraying [23], and laser hardfacing [24] are some methods utilized. However, the most widespread application is found in the welding process [25]. Based on type, this application is classified into stick electrodes, solid wires, flux-cored wires, and submerged arc welding (SAW) wires with fluxes [26]. The flux-cored arc welding process is increasingly used in many industries, including construction, automobile production, agriculture, mining, etc. [27–31]. The growing demand for flux-cored wires is justified by the advantages of welding methods.

Flux-cored wires are widely used for hardfacing and the restoration of worn surfaces [32–36]. The hardfacing process consists forming a hardening layer of the deposited metal on surfaces subject to intense abrasive wear [37,38], impact abrasion [39–42], and other types of wear [43–46]. For the flux-cored arc welding process of filler materials, melting is mainly carried out by arc welding [47,48]. Additionally, an important aspect is the study of the mechanical properties of the deposited metal [49–52]. The final microstructure and mechanical properties depend on the cooling conditions of the deposited metal [53,54] and heat input. These determines the wear resistance of the deposited metal to various types of wear [55–59]. An unstable arc would deteriorate weld appearances and increase the amount of imperfections [60]. In this case, the flux-cored wire arc welding process has a specific character [61–63]. Melting is carried out due to the wandering of the arc spot around the perimeter of the metal sheath [64,65]. In most cases, the surfacing process using a flux-cored wire electrode is characterized by the short-circuiting, globular, or drop mode of metal transfer, due to the use of low welding currents. This reduces hardfacing productivity, but ensures a low dilution of variation. In this case, it is of great interest to improve the energy efficiency of the welding process and the quality of the deposited metal. One of the most promising ways to improve the energy efficiency of the welding and hardfacing processes is the introduction of exothermic addition to the core filler [26,66]. Additional heat is generated at the electrode tip or in the welding arc by an exothermic reaction. Many scientists have investigated the introduction of exothermic addition in filler materials [67–73]. These additives bring about changes in the thermodynamics and physics of the processes occurring both during the filler material heating stage and during the material transfer and arc burning stage, in the context of significant and purposeful changes in entropy, enthalpy, and internal energy [74,75]. Their studies show the positive effect of the introduction of exothermic addition on the uniformity of melting, productivity, and efficiency of hardfacing, arc stability, etc.

In self-shielded flux-cored wire arc welding with exothermic addition, the welding current is determined not only by the wire feed speed (WFS) but also by the exothermic addition used and its amount (EA) in the core filler [26,61], as well as the contact tip to work distance (CTWD) and set arc welding (U_{set}) [61,76]. The first parameter is determined by the characteristics of the power supply. The influence of the last two parameters can be explained by the cost of the supplied power for heating the flux-cored wire electrode (Joule heat heating), i.e., the preheating of the filler materials. This preheating enables the electrode to burn off faster and increases the deposition rate [77]. It has been concluded that the welding consumables, welding parameters, and equipment have a key role to play in the stability of the welding process [78–81].

Research by Allen et al. [67] demonstrated the effectiveness of introducing exothermic additions from the Fe_2O_3-Al system compared to the Fe_2O_3-Mg system. These observations indicated that the magnesium exothermic additions react ahead of the welding arc, delocalizing the exothermic source and its heating ability. Park et al. [65] conducted studies on the influence of introducing exothermic additions to the core filler, including simple exothermic systems Fe_2O_3-Al and Fe_2O_3-Mg, as well as a combined reducer Fe_2O_3-(50%Mg/50%Al). The results showed greater efficiency with the combined reducer Fe_2O_3-(50%Mg/50%Al). It is noteworthy that exothermic additions with Fe_2O_3-Al and Fe_2O_3-(50%Mg/50%Al) showed the lowest dilution values. The research also examined weld bead morphology and melting efficiency. FCAW with exothermic addition Fe_2O_3-Al exhibited the lowest dilution when the content of exothermic addition was <30 wt.% of the core filler. However, when the content of exothermic addition was $\geq$30 wt.% of the core filler, FCAW with exothermic addition Fe_2O_3-Mg showed the lowest dilution. However, most of the research on flux-cored wires with exothermic additions to the core filler pertains to the welding process. Meanwhile, the area of filler materials for hardfacing processes is insufficiently studied due to the demands placed on deposited alloys, particularly regarding their high degree of alloying. As a result, the composition of the core filler is limited, which is especially crucial for self-shielded flux-cored wire electrodes (FCAW SS) due to the mandatory presence of gas and slag-forming components [82]. In such cases, it is advisable to consider alternative exothermic addition systems [61]. For hardfacing, alloys from the Fe-C-B-Cr-Cg-Cu system [26,27,83] are of particular interest, especially when additionally alloyed with a significant amount of copper. The high degree of copper alloying improves the microstructure of the deposited metal, enhances its mechanical properties and resistance to abrasive wear [26], and significantly increases corrosion resistance [27]. Trembach et al. [64] first proposed the use of exothermic addition CuO-Al for hardfacing processes. Their research results demonstrated a high degree of alloying in the deposited metal due to the recovery of copper from CuO in the exothermic addition. Trembach et al. [61] optimized hardfacing conditions with the introduction of exothermic addition CuO-Al into the core filler.

Trembach et al. [79] investigated the influence of introducing exothermic addition CuO-Al to the core filler, the ratio of graphite to EA oxidizer (CuO/C), and the ratio of oxidizer to aluminum powder in the composition of the EA (CuO/Al) on core filler density and the fill factor. Experimental values are provided in Table 1. Mathematical models of these indicators and their response surfaces were constructed (Figure 1). Subsequently, Trembach et al. [84] studied the impact of these variables on melting characteristics such as the melt-off rate (MOR), deposited rate (DR), spattering factor (SF), and deposition efficiency (De). Experimental values are presented in Table 1. Response surface plots of the melting characteristics' deposited rate (DR) and spattering factor (SF) are shown in Figure 2. The next stage of the work involved investigating the element transition factor, the copper recovery factor η(Cu) [64], and the overall transition element factor η(SS) [85]. The obtained data are shown in Table 1 and Figure 3. The results of these studies showed that the ratio of components CuO/Al in the exothermic mixture and the ratio of graphite to oxidizing agent of the exothermic mixture (CuO/C) had the greatest effect. The research findings also indicated that increasing the amount of EA above 35% leads to intense combustion of filler components.

Table 1. Values of some characteristics of experimental self-shielded, flux-cored wire electrodes.

Run	Filler Rate C_{WF} [79]	Core Filler Density ρ_f (g·cm^{-3}) [79]	Deposited Rate DR (kg·h^{-1}) [84]	Spattering Factor SF (%) [84]	Overall Transition Element Factor η(SS) (%) [85]	Carbon Transition Factor η(C) [64]	Copper Recovery Factor η(Cu) [64]
1	0.334	0.334	4.19	14.35	74.705	0.422	0.863
2	0.341	0.341	4.69	9.90	81.582	0.536	0.938
3	0.341	0.341	4.65	14.95	76.828	0.477	0.904
4	0.337	0.337	4.59	13.62	81.252	0.457	0.971
5	0.341	0.341	3.97	20.84	75.435	0.446	0.959
6	0.345	0.345	4.81	9.74	75.802	0.456	0.909
7	0.327	0.327	4.77	10.75	81.233	0.520	0.939
8	0.361	0.361	4.65	10.85	81.228	0.521	0.932
9	0.34	0.34	4.23	20.27	80.240	0.554	0.899

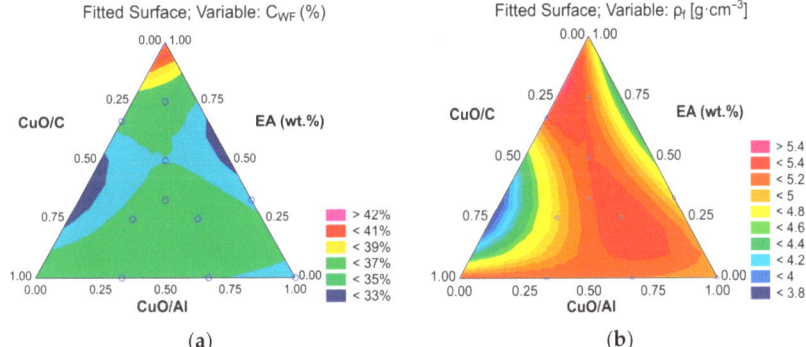

Figure 1. Contour surface graphs for (a) filler rate C_{WF} and (b) core filler density (ρ_f) [79].

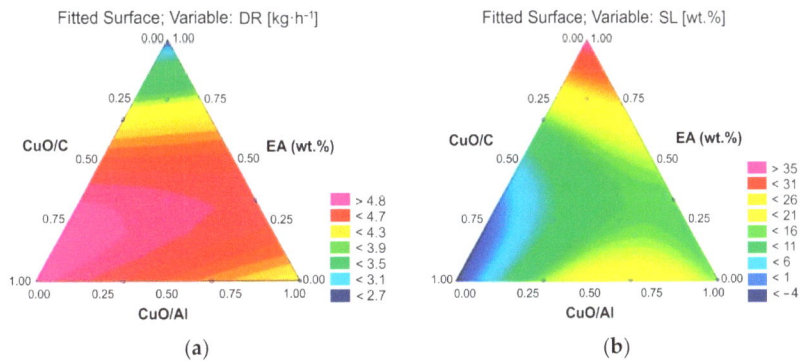

Figure 2. Contour surface graphs for (a) deposited rate (DR) and (b) spattering factor (SF) [84].

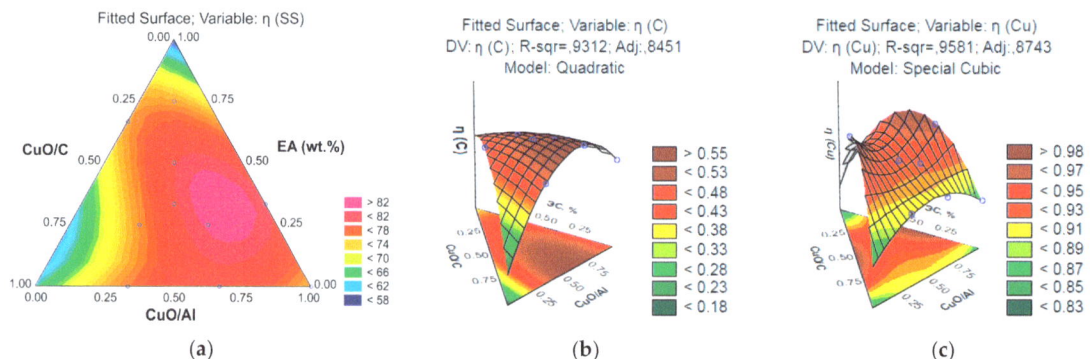

Figure 3. Contour surface graphs for (**a**) the overall transition element factor η(SS) [85], and 3D graphs for (**b**) the transition recovery factor η(C) [64] and (**c**) the copper recovery factor η(Cu) [64].

The mathematical dependencies of the received response surfaces are as follows [64,79,84,85]:

$$C_{wf} = 0.36 \cdot x_1 + 0.333 \cdot x_2 + 0.4276 \cdot x_3 - 0.2676 \cdot x_1 \cdot x_3 - 0.696 \cdot x_2 \cdot x_3 + \\ +0.696 \cdot x_1 \cdot x_2 \cdot x_3 + 0.272 \cdot x_2 \cdot x_3 \cdot (x_2 - x_3) \quad (1)$$

$$\rho_f = 3.2 \cdot x_1 + 2.95 \cdot x_2 + 3.8354 \cdot x_3 - 7.321 \cdot x_1 \cdot x_3 - 5.15 \cdot x_2 \cdot x_3 + \\ +30.105 \cdot x_1 \cdot x_2 \cdot x_3 + 6.705 \cdot x_2 \cdot x_3 \cdot (x_2 - x_3) \quad (2)$$

$$DR = 4.847 \cdot x_1 + 4.097 \cdot x_2 + 2.646 \cdot x_3 + 3.106 \cdot x_1 \cdot x_3 + 4.747 \cdot x_2 \cdot x_3 \quad (3)$$

$$SL = -4.23 \cdot x_1 + 15.19 \cdot x_2 + 38.635 \cdot x_3 + 64.03 \cdot x_1 \cdot x_2 - 41.3 \cdot x_1 \cdot x_3 - 39.79 \cdot x_2 \cdot x_3 \quad (4)$$

$$\eta(SS) = 64.8 \cdot x_1 + 74.62 \cdot x_2 + 57.43 \cdot x_3 + 39.35 \cdot x_1 \cdot x_2 + 40.52 \cdot x_1 \cdot x_3 + \\ +55.57 \cdot x_1 \cdot x_3 - 91.28 \cdot x_1 \cdot x_3 \cdot (x_1 - x_3). \quad (5)$$

$$Y(\eta(C)) = 0.17835 \cdot x_1 + 0.425 \cdot x_2 - 0.3747 \cdot x_3 + 0.94002 \cdot x_1 \cdot x_3 + \\ +0.5764 \cdot x_1 \cdot x_3 + 0.200058 \cdot x_2 \cdot x_3 \quad (6)$$

$$Y(\eta(Cu)) = 0.84814 \cdot x_1 + 0.86236 \cdot x_2 + 0.82342 \cdot x_3 + 0.208 \cdot x_1 \cdot x_3 + 0.556 \cdot x_1 \cdot x_3 + \\ +0.55714 \cdot x_2 \cdot x_3 - 1.67756 \cdot x_1 \cdot x_2 \cdot x_3 \quad (7)$$

However, the above-mentioned papers devote little attention to the issue of arc stability and its impact on weld bead morphology.

The purpose of this article is to study the influence of exothermic addition (CuO–Al) to core filler and core filler composition on the stability of the hardfacing process, to develop mathematical relationships for predicting current and voltage parameters, and to optimize core filler composition.

2. Materials and Methods

2.1. Design Experiment

Mixture design is known to be widely used to optimize composition. The main stages are the selection of a response variable (dependent variable) and a design experiment, as shown in Table 2 [82]. As variables, the following were chosen: x_1—the ratio of oxidizing agent and reducing agent of the exothermic mixture (CuO/Al); x_2—the ratio of exothermic mixture oxidizing agent to graphite content (CuO/C) with a lower level of 3 and an upper level of 6; and x_3—content of the exothermic mixture (EA, wt.%) in the core filler with a lower level of 20 wt.% and an upper level of 46 wt.%. When conducting an experiment on the optimization of processes and searching for mathematical models to predict indicators or parameters of optimization, the following block diagram, shown in general form in Figure 4, is used.

Table 2. Three-factor experiment design simplex trellis planning [64,79,84,85].

Run	Coded Values			Actual Values		
	x_1	x_2	x_3	CuO/C	CuO/Al	EA, (wt.%)
1	0	1	0	3/1	6/1	20
2	0.333	0.33	0.33	4/1	4/1	28.58
3	0.67	0.33	0	5/1	4/1	20
4	0	0.67	0.33	3/1	5/1	28.58
5	0.33	0	0.67	4/1	3/1	37.42
6	0.5	0.25	0.25	4.5/1	3.75/1	26.5
7	0.25	0.25	0.5	3.75/1	3.75/1	33
8	0.25	0.5	0.25	3.75/1	4.5/1	26.5
9	0.33	0.67	0	4/1	5/1	20

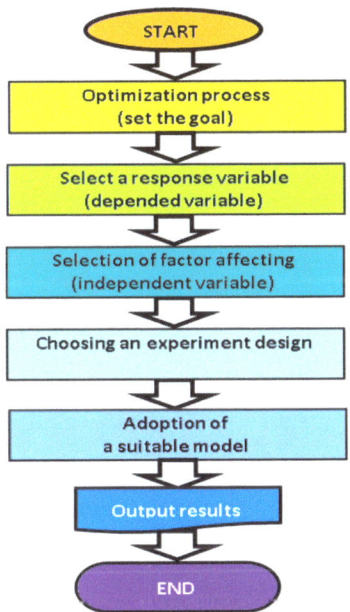

Figure 4. An overall block diagram elaborating the algorithm for conducting a design plan.

The full cubic model is represented in Equation (8):

$$y = \sum_{i=1}^{K} \beta_i \cdot x_i + \sum_{j>i}^{k} \beta_{ij} \cdot x_i \cdot x_j + \sum_{j>i}^{k} \delta_{ij} \cdot x_i \cdot x_j \cdot (x_i - x_j) + \sum_{k>j>i}^{K} \beta_{ijk} \cdot x_i \cdot x_j \cdot x_k + \varepsilon \quad (8)$$

where y—the predicted response variable; i and j are the number of ingredients in the mixture; K—components given by the equation; β_i—expected response at the top; β_{ij}—coefficients indicate the amount of quadratic curvature along the edge of the simplex region consisting of binary mixtures of x_i and x_j; and δ_{ij}—account for ternary blending among three separate components in the interior of the design. β_{ijk}—the coefficient of the regression coefficient of the product terms of two or three variables. x_i, x_j, and x_k represent three different design variables.

For modeling in this study, all possible regressions were used in combination with the stepwise regression method. The reason for using the two mentioned methods is to evaluate the suitability and appropriateness of the final model, using methods such as a residual analysis, a lack of fit test, and an evaluation of the effect of outliers.

2.2. Materials

The filler material was a TiO_2–CaF_2–CaO–SiO_2 slag system self-shielded, flux-cored wire with a diameter of 4 mm. The core filler composition of the experimental FCAW-SS are shown in Table 3.

Table 3. Composition of core filler FCAW-SS, wt.%.

Component	Marking of Experimental Flux-Cored Wires								
	FCAW-SS-E1	FCAW-SS-E2	FCAW-SS-E3	FCAW-SS-E4	FCAW-SS-E5	FCAW-SS-E6	FCAW-SS-E7	FCAW-SS-E8	FCAW-SS-E9
Fluorite concentrate GOST 4421–73	11	11	11	11	11	11	11	11	11
Rutile concentrate GOST 22938–78	6	6	6	6	6	6	6	6	6
Calcium carbonate GOST 8252–79	3	3	3	3	3	3	3	3	3
Ferromanganese FMN-88A GOST 4755–91	7	7	7	7	7	7	7	7	7
Ferrosilicon FS-92 GOST 1415–78	4	4	4	4	4	4	4	4	4
Ferrovanadium FVd–40 GOST 27130–94	4	4	4	4	4	4	4	4	4
Metal Chrome X99 GOST 5905–79	14	14	14	14	12	14	12	14	14
Titanium powder PTM-3 TU 14–22–57–92	3	3	3	3	3	3	3	3	3
Graphite is silver	5.6	5.8	3.75	7.6	7.1	4.9	7	5.8	4.3
Oxide of copper powder-like GOST 16539–79	16.7	23.3	17.2	22.8	28.4	21.8	26.1	21.8	17.2
Aluminum powder PA1 GOST 6058–73	2.8	5.8	4.3	4.6	9.5	5.8	7	4.9	3.5
Iron powder PZhR–1 GOST 9849–86	22.9	13.1	22.7	13	5	15.5	9.9	15.5	23

The base metal was S 235 J2G2 EN 10025–2 (St3ps), and its dimensions were 200 mm × 100 mm × 15 mm. The welding process was carried out using a polarity-reversed welding power source with a rigid external characteristic. The following modes were used for hardfacing: wire feed speed WFS = 111 m/h; set arc voltage U_s = 28 V; travel speed TS = 0.3 m/min; and contact tip to work distance CTWD = 40 mm. Hardfacing was carried out with reverse polarity by a self-powered welding head ABC automatic machine from a power source with a rigid external characteristic (Figure 5).

Figure 5. Experimental setup for hardfacing.

2.3. Investigation Parameters

Every researcher studying the stability process understands the process in a different way and applies different indicators of arc stability [77,86]. The arc stability indexes were the average voltage (I_{aw}) and current and their standard deviations ($Std(I)$ and $Std(U)$), the voltage and current square mean, the current coefficient of variation ($CV(I)$), and the voltage coefficient of variation (CV(U)) [87]. The current and voltage were measured using a welding waveform monitoring system (oscilloscope OWON SDS6062E) installed to the ABC welding machine.

The coefficient of variation (CV) is a statistical measure of the dispersion of data points in a data series around the mean [88]. The coefficient of variation represents the ratio of the standard deviation to the mean [89]. When comparing data sets, the coefficient of variation is utilized. The CV is a useful statistical measure for comparing the degree of variation of one data series with another, even if the averages differ sharply from each other [89].

$$Std = \sqrt{\frac{\sum_{i-1}^{n}(x - \overline{x})^2}{n - 1}} \qquad (9)$$

$$CV = \frac{Std}{\frac{1}{n}\sum_{i=1}^{n}|\overline{x_i}|} \qquad (10)$$

where $\overline{x}$—the sample mean; n—the sample size.

2.4. Welding Heat Indicate

Heat input (Q_{in}) is the quantity of energy introduced from the arc per unit length of weld. The heat input was calculated by the means equation [11,90].

$$Q_{in} = \frac{\eta_{FCAW} \cdot U_a \cdot I \cdot 60}{TS \cdot 1000} \qquad (11)$$

Q_{in}—heat input, (kJ/mm); U_a—arc voltage, V; I—arc current, A; TS—travel speed, mm/min; H—coefficient of efficiency of the process; η_{FCAW} = 0.8 [90].

2.5. Investigation of Composition, Phase Composition, and Microstructure of Deposited Metal

The average chemical composition of the alloys was determined by the atomic absorption spectroscopy method using a Spectrolab LA VFC01A device. The alloys were examined by a Neophot light optical microscope. Quantitative metallography was carried out with an Epiquant structural analyzer. An X-ray diffraction analysis was conducted to identify the existing phases in the produced samples on a DRON-UM-1 X-ray diffractometer with a CuK$_\alpha$ source. The phase transformations were investigated by means of a differential thermal analysis (DTA).

3. Results

3.1. Dynamic Characteristics

Figure 6 shows the dynamic characteristics of the welding voltage and current, which are shown with instantaneous values of current and voltage over a randomly chosen period of 0.5 s. The arc voltage and welding current fluctuate significantly, reflecting the poor stability of the hardfacing process.

The average values of the average welding current (I_{aw}), as well as the corresponding values of the standard deviation of the welding current ($Std(I)$) and the coefficient of variation of the welding current ($CV(I)$), are shown in Table 4.

The average arc voltage (U_{aw}), the standard deviation of the arc voltage ($Std(U)$), and the arc voltage coefficient of variation ($CV(U)$) are shown in Table 5.

Observations and analyses of the change in the welding electrical signal indicated that the waveform of the arc voltage waveform had significant fluctuation even with the same welding parameters, which in turn influenced the process of arc burning in flux-cored arc welding.

An analysis of the average arc voltage values showed (Figure 7) that the minimum value was observed when hardfacing a flux-cored wire electrode with index FCAW-SS-E1, whereas the maximum value was observed when hardfacing with FCAW-SS-E9. These flux-cored wire electrodes had low exothermic additive content. At the same time, FCAW-

SS-E1 was characterized by a high CuO/Al = 6 ratio (excess oxidant in exothermic addition) and a high CuO/C = 3 ratio (high proportion of carbon-containing component in the filler introduced as graphite).

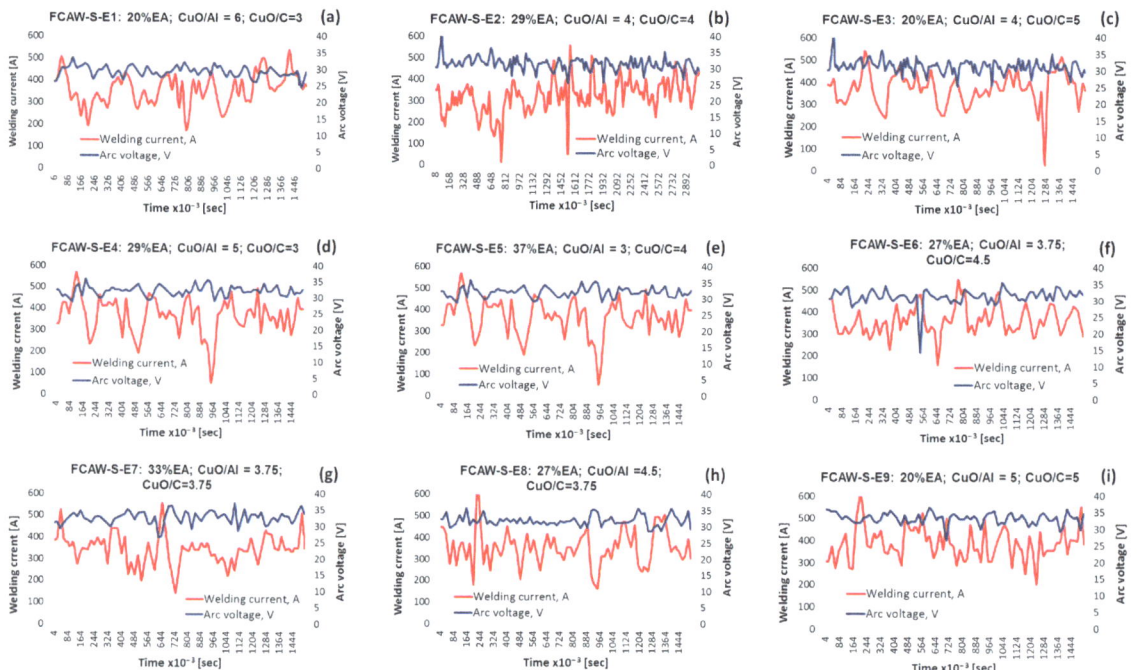

Figure 6. Dynamic characteristics of voltage and current with varying composition of core filler: (**a**) FCAW-SS-E1; (**b**) FCAW-SS-E2; (**c**) FCAW-SS-E3; (**d**) FCAW-SS-E4; (**e**) FCAW-SS-E5; (**f**) FCAW-SS-E6; (**g**) FCAW-SS-E7; (**h**) FCAW-SS-E8; and (**i**) FCAW-SS-E9.

Table 4. Welding current parameters during hardfacing with experimental flux-cored wires with different core filler compositions.

Run	Average Welding Current				Standard Deviation of Welding				Coefficient of Variation of Welding Current, (%)			
	Actual Value, (A)	Predicted Value, (A)	Diff. (A)	Dev., (%)	Actual Value, (A)	Predicted Value, (A)	Diff. (A)	Dev., (%)	Actual Value	Predicted Value	Diff.	Dev.
FCAW-SS-E1	365.0	362.4	2.57	0.70	66.3	64.8	1.48	2.23	18.2	17.46	0.74	4.09
FCAW-SS-E2	322.3	339.6	17.27	5.36	75.2	69.2	5.97	7.94	23.3	21.14	2.16	9.28
FCAW-SS-E3	372.7	373.9	1.22	0.33	70.5	69.6	0.92	1.30	18.9	18.25	0.65	3.44
FCAW-SS-E4	365.2	368.9	3.69	1.01	79.1	79.2	0.11	0.14	21.7	21.68	0.02	0.10
FCAW-SS-E5	365.2	364.7	0.51	0.14	79.1	78.9	0.24	0.30	21.7	21.41	0.29	1.35
FCAW-SS-E6	364.0	356.3	7.72	2.12	60.6	61.8	1.22	2.02	16.7	17.10	0.40	2.39
FCAW-SS-E7	345.2	337.8	7.36	2.13	64.2	66.5	2.32	3.61	18.6	18.75	0.15	0.78
FCAW-SS-E8	356.2	348.8	7.40	2.08	74.8	77.6	2.77	3.71	21.0	22.47	1.47	6.99
FCAW-SS-E9	388.4	389.7	1.32	0.34	65.1	67.2	2.07	3.17	16.8	17.85	1.05	6.23

Table 5. Arc voltage parameters during hardfacing with experimental flux-cored wires with different core filler compositions.

Run	Average Arc Voltage				Standard Deviation of Arc Voltage				Coefficient of Variation of Welding Current, %			
	Actual Value, (V)	Predicted Value, (V)	Diff. (V)	Dev., (%)	Actual Value, (V)	Predicted Value, (V)	Diff. (V)	Dev., (%)	Actual Value	Predicted Value	Diff.	Dev.
FCAW-SS-E1	29.0	28.97	0.03	0.11	1.36	1.37	0.01	0.43	4.7	4.70	0.00	0.00
FCAW-SS-E2	31.5	31.33	0.17	0.55	1.81	1.77	0.04	2.08	5.7	5.63	0.07	1.25
FCAW-SS-E3	31.5	31.55	0.05	0.16	1.81	1.81	0.01	0.28	5.7	5.75	0.05	0.89
FCAW-SS-E4	32.2	32.14	0.06	0.19	1.39	1.39	0.00	0.11	4.3	4.31	0.01	0.19
FCAW-SS-E5	32.2	35.42	3.22	10.0	1.38	1.36	0.03	2.36	4.2	4.23	0.07	1.58
FCAW-SS-E6	31.5	30.15	1.35	4.27	2.15	2.15	0.00	0.04	6.8	6.85	0.05	0.68
FCAW-SS-E7	32.6	34.00	1.40	4.31	1.70	1.70	0.00	0.12	5.2	5.22	0.02	0.44
FCAW-SS-E8	31.9	31.94	0.04	0.12	1.43	1.43	0.00	0.11	4.5	4.51	0.01	0.20
FCAW-SS-E9	33.3	33.39	0.09	0.28	1,46	1.46	0.00	0.16	4.4	4.36	0.04	0.93

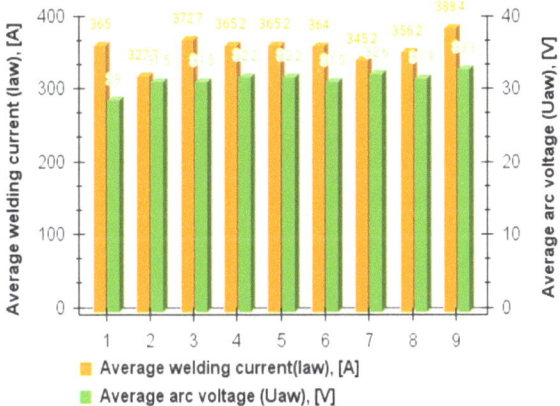

Figure 7. The effect of the composition of core filler on the average welding current and average arc voltage.

The lowest values of the average welding current could be observed for flux-cored wire with index FCAW-SS-E2 (Figure 7). A slightly higher value was observed for flux-cored wire with index FCAW-SS-E7. These filler materials were characterized by the average content of exothermic addition in the core filler (at EA = 28.5–33 wt.%) and the CuO/Al and CuO/C ratios (3.75–4). Whereas the highest values of average welding current were observed for flux-cored wire with index FCAW-SS-E9 and FCAW-SS-E3 (Figure 7). These filler materials were characterized by a low content of exothermic addition in the core filler (EA = 20 wt.%). It could be concluded that the amount of heat generated from the exothermic reaction was insufficient. Whereas for the parameter average arc voltage, the lowest values were observed for FCAW-SS-E2 and the highest for FCAW-SS-E9.

Filler materials were characterized by exothermic addition at a low level (EA = 20 wt.%), but the first flux-cored wire electrode was characterized by an excess of carbon containing material (low CuO/C = 3 ratio) and an excess of oxidants (high CuO/Al = 6 ratio). Whereas FCAW-SS-E9 was characterized by an average CuO/C = 4 ratio and a stoichiometric ratio of oxidizing agent and reducing agent of exothermic mixture CuO/Al = 5.

3.2. Arc Stability Analysis

The arc stability is accessed qualitatively using the probability distribution and current and voltage cyclograms, while the coefficients of variation of voltage and current can quantificationally evaluate arc stability [91]. Figure 8 shows the dependence of the coefficients of variation of arc voltage and welding current on different arc voltages. It is known that the smaller the coefficient of variation, the more stable the welding process and vice versa [91,92].

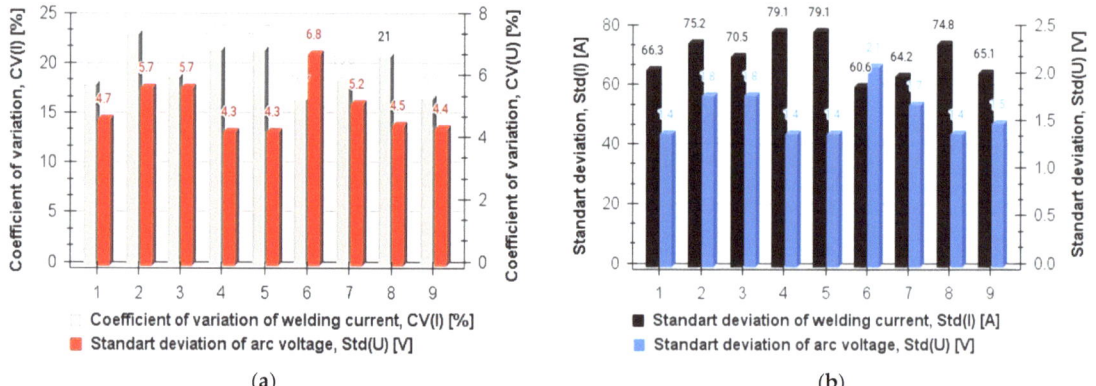

Figure 8. The effect of the composition of core filler on (**a**,**b**) the coefficient of variation and the standard variation of the arc voltage and welding current, respectively.

Analyzing the results shown in Figure 8a, it can be concluded that the lowest values of coefficients of variation of the welding current were observed for the flux-cored wire electrodes with indexes FCAW-SS-6 and FCAW-SS-9. The CV(I) values were 16.7% and 16.8%, respectively (Figure 8a). Somewhat higher values of coefficients of variation of the welding current were observed in flux-cored wire electrodes with indexes FCAW-SS-E1, FCAW-SS-E7, and FCAW-SS-E3 with corresponding values in an ascending order of 18.2%, 18.5%, and 18.9% (Figure 8a). Whereas the highest values corresponded to filler materials with indexes FCAW-SS-E2 → FCAW-SS-E4. FCAW-SS-E5 → FCAW-SS-E8 in descending order of values of coefficients of variation of the welding current, respectively: 23.3% → 21.7% → 21%. It should be noted that the highest values corresponded to flux-cored wire electrodes with an average content of exothermic additive at the level EA = 26.5–28.58 wt.% (x_3 = 0.25–0.33) and a high carbon content (low values of CuO/C = 33.75). Similarly, observations could be made regarding the standard variation of the welding current at the FCAW-SS location.

As for the arrangement (Figure 8a,b) of experimental self-shielded, flux-cored wire electrodes for coefficients of variation of the arc voltage $CV(U)$ and the standard variation of arc voltage $Std(U)$, the character of influence of the variable was different. The lowest values of the $CV(U)$ in the order of their value (Figure 8a) increasing had FCAW-SS with indices FCAW-SS-E4, FCAW-SS-E5, FCAW-SS-E8, FCAW-SS-E9, and FCAW-SS-E1. They corresponded to the following values of coefficients of variation of arc voltage: 4.3% → 4.4% → 4.5% → 4.7%. Whereas the highest values of the $CV(U)$ corresponded to flux-cored wire electrodes with indexes FCAW-SS-E6. FCAW-SS-E2, and FCAW-SS-E3, with corresponding values of 6.8%, 5.7%, and 5.7%. The matrix plan analysis showed that the lowest values of $CV(U)$ and $Std(U)$ were characterized by hardfacing processes with a high CuO/Al ratio $\geq$ 4.5 ($x_2 \geq$ 0.5) and a content of exothermic mixture in the core filler (EA) below the average EA < 29 wt.% (x_3 < 0.33). Whereas high values of coefficients of variation of the arc voltage and the standard variation of arc voltage were characterized by a hardfacing process carried out by self-shielded, flux-cored wire electrodes with a low ratio of oxidizing agent to reducing agent in the composition of the exothermic mixture CuO/Al = 3.75–4 (x_2 = 0.25–0.33) and a low content of exothermic mixture in the core filler EA < 29 wt.% (x_3 < 0.33).

Box plots (Figures 9a and 10a) were constructed to compare each process, and frequency histograms for each process were separately generated (Figures 9b–j and 10b–j) to better understand the distribution of welding current and arc voltage data. Box plots work best when a comparison in distributions needs to be performed between groups. For each flux-cored wire electrode, a box plot was made showing the difference between the minimum and maximum values with a dotted line. The colored box (rectangle) describes the values of the 25 and 75% quantiles, and the line inside the box shows the average value. Lines extend from each box to capture the range of the remaining data, with dots placed past the line edges to indicate outliers. This allowed us to compare the stability results of the current and arc voltage by medians through the box and whisker markings' positions. Displacement of the median from the middle indicated the asymmetry of the data. Vice versa, the closer the median was to the middle of the rectangle, the more symmetrical the data. The obtained box plots for the welding current are shown in Figure 9.

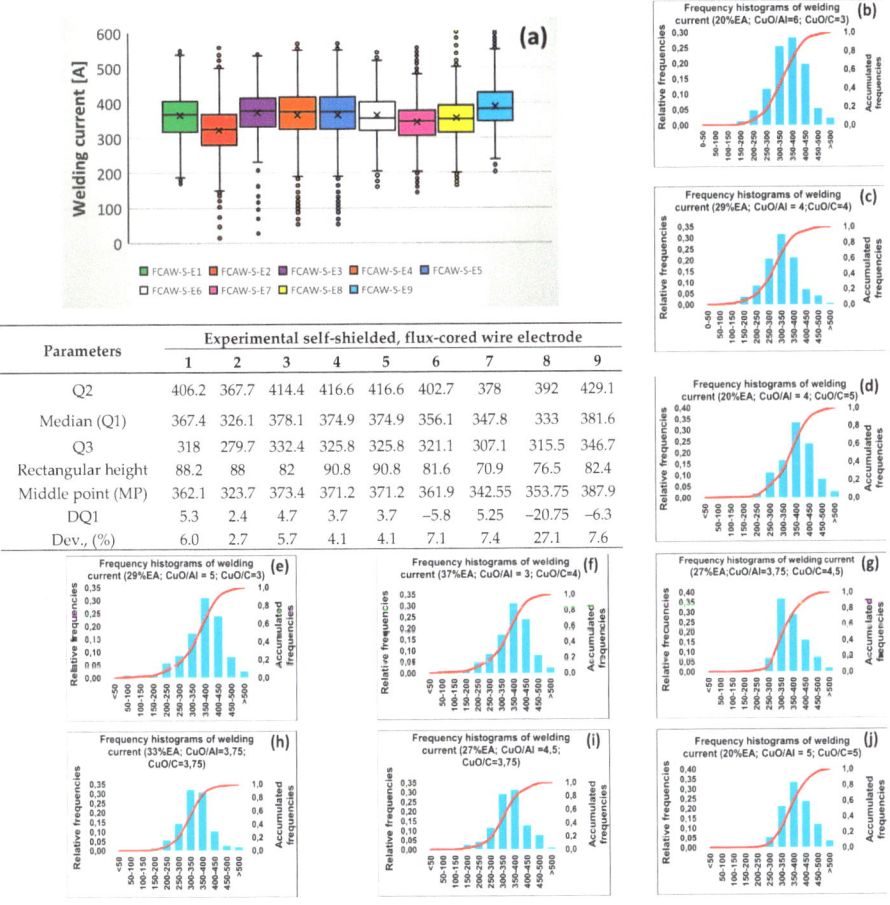

Parameters	Experimental self-shielded, flux-cored wire electrode								
	1	2	3	4	5	6	7	8	9
Q2	406.2	367.7	414.4	416.6	416.6	402.7	378	392	429.1
Median (Q1)	367.4	326.1	378.1	374.9	374.9	356.1	347.8	333	381.6
Q3	318	279.7	332.4	325.8	325.8	321.1	307.1	315.5	346.7
Rectangular height	88.2	88	82	90.8	90.8	81.6	70.9	76.5	82.4
Middle point (MP)	362.1	323.7	373.4	371.2	371.2	361.9	342.55	353.75	387.9
DQ1	5.3	2.4	4.7	3.7	3.7	−5.8	5.25	−20.75	−6.3
Dev., (%)	6.0	2.7	5.7	4.1	4.1	7.1	7.4	27.1	7.6

Figure 9. Box plots for the medians, quartiles, and scatter of the (**a**) welding current depending on the composition of the core filler of flux-cored wire electrodes and histograms (**b**–**j**) of the welding current distribution for all FCAW-S, where DQ1 = Q1–MP—difference of the median (Q1) from the midpoint (MP).

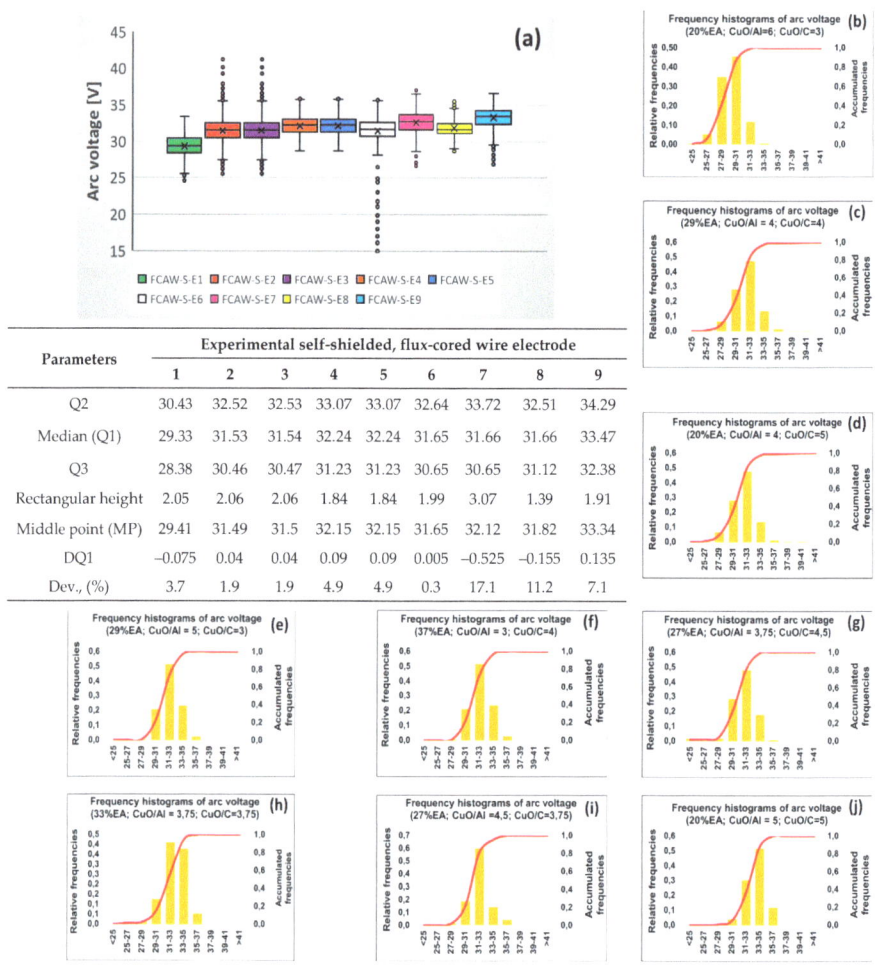

Figure 10. Box plots for the medians, quartiles, and scatter of the (**a**) arc voltage depending on the composition of the core filler of flux-cored wire electrodes and histograms (**b**–**j**) of the arc voltage distribution for all FCAW-SS, where DQ1 = Q1–MP—difference of the median (Q1) from the midpoint (MP).

The analysis of data for the welding current showed that the best symmetry was in the welding current values for processes hardfaced with flux-cored wire electrodes with indices FCAW-SS-E2 and FCAW-SS-E7 (Figure 9a). The analysis of frequency histograms (Figure 9c) and (Figure 9h) confirmed the uniformity of distribution of welding current values for these filler materials. The diagrams had an almost symmetrical appearance. Whereas a clear asymmetry of the welding current data was shown by processes using FCAW-SS labeled FCAW-SS-E9, less than FCAW-SS-E1 (Figure 9a) which was shown by their frequency histograms (Figure 9b,j). At the same time, their data shift towards higher values was noticeable. This was confirmed by the values of the average welding current for the hardfacing process. The analysis of filler compositions showed that symmetric frequency histograms corresponded to filler materials with an average content of exothermic addition (EA = 29–33 wt.%) to the core filler and average CuO/Al = 3.75–4 and CuO/C = 3.75–4 ratios. Whereas asymmetric frequency histograms corresponded to flux-cored wire electrodes with high values of the CuO/Al ratio ≥ 5 and exothermic additive content at the lower level EA = 20 wt.%.

Figure 10 shows the summary box plots and frequency histograms for arc voltage. The character of distribution of the arc voltage values was somewhat different from the welding current. According to Figure 10a, the smallest size of the quantile rectangle (colored field) was observed for FCAW-SS with indices FCAW-SS-E4 and FCAW-SS-E5, whereas the largest size was observed for the hardfacing process using FCAW-SS-E7. It is noteworthy that the most symmetrical arrangement of the median was observed for the hardfacing process performed by FCAW-SS-E6, FCAW-SS-E2, and FCAW-SS-E3. Whereas an asymmetric location of the median in the quantile rectangle was observed with FCAW-SS-E7 and FCAW-SS-E8, which was confirmed by the corresponding frequency histograms shown in Figure 10h,j.

The analysis of the core filler composition showed that the first group (with symmetrical median arrangement) differed from the second group in the CuO/C ratio.

Based on the above data, it can be stated that more stable values of the arc voltage were observed for flux-cored wire electrodes in which the ratio of the exothermic mixture oxidizing agent to graphite content was CuO/C = 4–5.

3.3. Current and Voltage Cyclograms

To describe the nature of the arc burning, its stability, as well as identifying characteristic zones during arc burning (AEA—arc extinction area; SCA—short-circuiting area; ABA—arc burning area), the use of current and voltage cyclograms has become widespread [86,91,93]. Current and voltage cyclograms show welding voltage as a function of the welding current. The cyclograms shown are also scatter density plots that show the number of dots in an area. The greater their number, the lighter the dots, gradually moving from purple for areas with the lowest density of dots to yellow with the maximum density of dots. Figure 11 displays current and voltage cyclograms for the conducted hardfacing processes.

The analysis of the obtained current and voltage cyclograms showed that the hardfacing process for the flux-cored wire with indexes FCAW-SS-E2 and FCAW-SS-E3 had an arc extinction area (AEA) (see Figure 11b,c). In the cyclogram (Figure 11f) for the flux-cored wire electrode with index FCAW-SS-E6, a short-circuiting SCA was observed. The analysis of the core filler compositions of FCAW—SS showed that the first ones (in which arc extinction was observed) were characterized by close to average values of CuO/Al = 4 and CuO/C = 4–5 ratios. Whereas the hardfacing process in which short-circuiting of the arc was observed was carried out with filler materials in which CuO/Al = 3.75 and CuO/C = 4.5, i.e., with an excess of aluminum in the core filler and a lack of graphite.

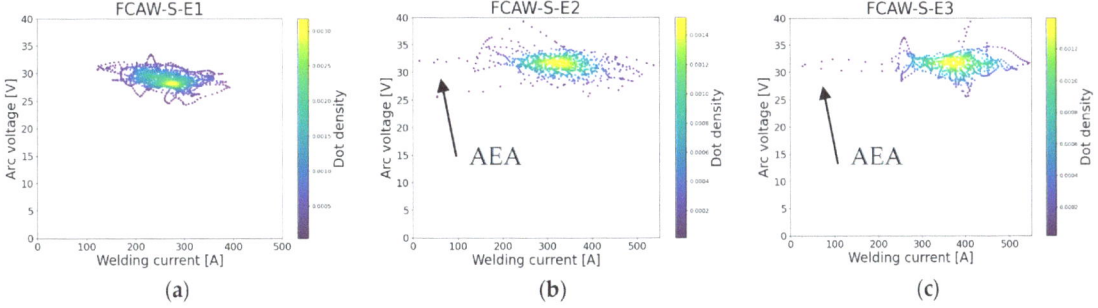

Figure 11. *Cont.*

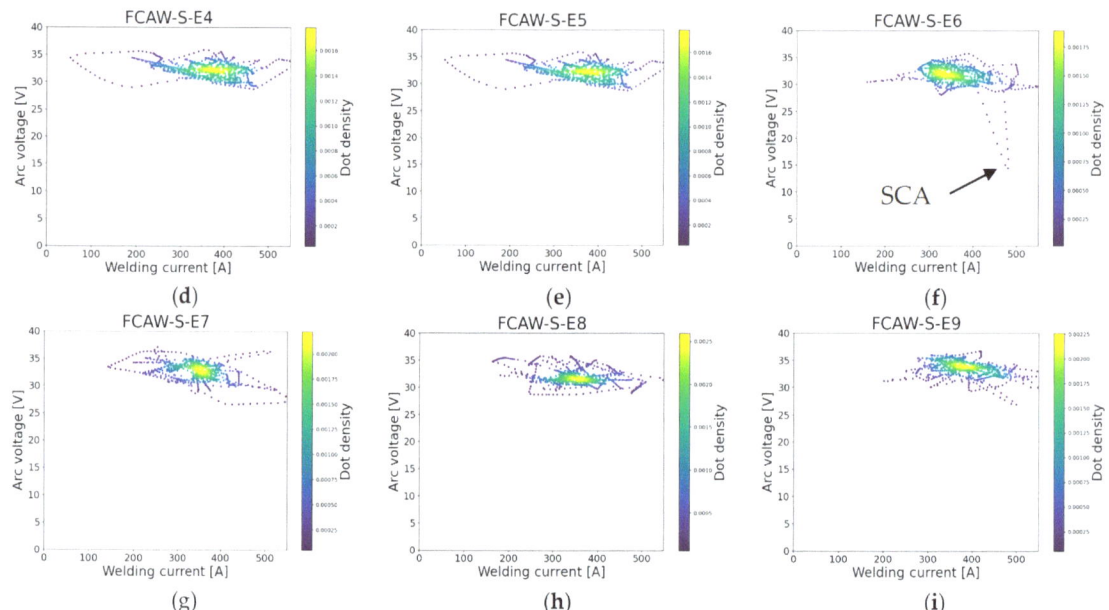

Figure 11. Current and voltage cyclograms (scatter density plot) with varying compositions of core filler: (**a**) FCAW-SS-E1; (**b**) FCAW-SS-E2; (**c**) FCAW-SS-E3; (**d**) FCAW-SS-E4; (**e**) FCAW-SS-E5; (**f**) FCAW-SS-E6; (**g**) FCAW-SS-E7; (**h**) FCAW-SS-E8; and (**i**) FCAW-SS-E9, where AEA—arc extinetion area; SCA—short-circuiting area.

3.4. Prediction of Mathematic Model of Welding Current and Arc Voltage Characteristics

For practical application, the use of mathematical models to predict the required dependent parameters of the hardfacing/welding process is of great importance. A mathematical model has been developed using the ANOVA technique and Response Surface Methodology (RSM). Results were used in the production of linear, quadratic special cubic and fully cubic regression models where the analysis of residue graphs and errors were generated to verify the quality of the models. The final choice of the mathematical model was made based on the values of the sum squared (SS) and the adjusted mean (MS). The higher their values, the more reliably the mathematical model describes the parameter under study [94]. Table 6 presents the regression coefficients for each response.

Table 6. An analysis of variance for the applied models.

Parameter	Source	Adjusted Sum of Square (SS)	Degree of Freedom (df)	Adjusted Mean Square (MS)	F-Value	*p*-Value
I_{aw}, (A)	Model	2165.891	5	433.1781	2.349205	0.256648
	Total Error	553.180	3	184.3935	–	–
	Total Adjusted	2719.071	8	339.8839	–	–
$CV(I)$, (%)	Model	37.00867	6	6.168112	1.641247	0.425755
	Total Error	7.51637	2	3.758186	–	–
	Total Adjusted	44.52504	8	5.565630	–	–
$Std(I)$, (A)	Model	322.4220	5	64.48441	3.895788	0.146137
	Total Error	49.6570	3	16.55234	–	–
	Total Adjusted	372.0791	8	46.50988	–	–

Table 6. Cont.

Parameter	Source	Adjusted Sum of Square (SS)	Degree of Freedom (df)	Adjusted Mean Square (MS)	F-Value	p-Value
$CV(U)$, (%)	Model	6.041557	7	0.863080	125.7341	0.068564
	Total Error	0.006864	1	0.006864	–	–
	Total Adjusted	6.048421	8	0.756053	–	–
$Std(U)$, (V)	Model	0.585934	7	0.083705	254.5907	0.048221
	Total Error	0.000329	1	0.000329	–	–
	Total Adjusted	0.586263	8	0.073283	–	–
U_{aw}, (V)	Model	11.66115	7	1.665878	13,508.76	0.006625
	Total Error	0.00012	1	0.000123	–	–
	Total Adjusted	11.66127	8	1.457659	–	–

The statistical characteristics of the obtained mathematical models indicate their reliability and adequacy. To assess the adequacy of the developed mathematical models, Fisher's criterion (F) was used. The significance of the obtained data was determined using Student's criterion (p). The developed mathematical models are considered adequate if the actual value of the Fisher's criterion exceeds the threshold table value $F_{act} > 3$ [64]. Therefore, the data obtained from the developed mathematical model is considered significant in the case that Student's criterion is $p > 0.05$. The analysis of variance for the applied models showed (Table 6) that the developed models meet the above requirements, except for the mathematical model for the $CV(I)$. All regression models (12)–(17) as mathematical functions have multivariate functions with bounded indexes in any direction [95] and bounded indexes in joint variables [96], and their indexes equal the corresponding maximum degree. The regression model of the fill coefficient has the following form:

$$I_{AW} = 401.21 \cdot x_1 + 362.43 \cdot x_2 + 382 \cdot x_3 - 1095.77 \cdot x_1 \cdot x_2 \cdot x_3 - \\ -192.73 \cdot x_1 \cdot x_2 \cdot (x_1 - x_2) + 314.56 \cdot x_1 \cdot x_3 \cdot (x_1 - x_3) \quad (12)$$

$$Std(I) = 71.93 \cdot x_1 + 64.82 \cdot x_2 + 48.74 \cdot x_3 + 217.9 \cdot x_1 \cdot x_2 \cdot x_3 - \\ -298.9 \cdot x_1 \cdot x_3 \cdot (x_1 - x_3) + 262 \cdot x_2 \cdot x_3 \cdot (x_2 - x_3) \quad (13)$$

$$CV(I) = 18.64 \cdot x_1 + 17.46 \cdot x_2 - 119 \cdot x_3 + 309.2 \cdot x_1 \cdot x_3 + 222.8 \cdot x_2 \cdot x_3 - \\ -266.9 \cdot x_1 \cdot x_2 \cdot x_3 - 354.16 \cdot x_1 \cdot x_3 \cdot (x_1 - x_3) \quad (14)$$

$$U_{AW} = 48.1 \cdot x_1 + 29 \cdot x_2 + 42.26 \cdot x_3 - 27.49 \cdot x_1 \cdot x_2 - 46.95 \cdot x_1 \cdot x_3 - 55.47 \cdot x_1 \cdot x_2 \cdot (x_1 - x_2) + \\ +21.5 \cdot x_1 \cdot x_3 \cdot (x_1 - x_3) - 16.44 \cdot x_2 \cdot x_3 \cdot (x_2 - x_3) \quad (15)$$

$$Std(U) = 2.41 \cdot x_1 + 1.362 \cdot x_2 - 3.47 \cdot x_3 - 1.134 \cdot x_1 \cdot x_2 + 13.07 \cdot x_1 \cdot x_3 + \\ +10.51 \cdot x_2 \cdot x_3 - 21.58 \cdot x_1 \cdot x_2 \cdot x_3 - 9.36 \cdot x_2 \cdot x_3 \cdot (x_2 - x_3) \quad (16)$$

$$CV(U) = 5.41 \cdot x_1 + 4.7 \cdot x_2 - 11 \cdot x_3 + 44.4 \cdot x_1 \cdot x_3 + 30.5 \cdot x_2 \cdot x_3 - \\ -63.34 \cdot x_1 \cdot x_2 \cdot x_3 + 7.65 \cdot x_1 \cdot x_2 \cdot (x_1 - x_2) - 26 \cdot x_2 \cdot x_3 \cdot (x_2 - x_3) \quad (17)$$

The evaluation of the accuracy of the prediction models and their values, predicted by these models, compared to the actual values is shown in Figure 12. These graphs represent an approximately linear trend. This indicates that the residuals follow a normal distribution and there is no significant correlation between them. The proximity of the data points to the $y = x$ line also indicates that there is minimal divergence. This allows you to make accurate predictions based on new and unknown data. The strong relationship between the predicted and actual values of the model indicates that it accurately captures the patterns and relationships in the data [79,97].

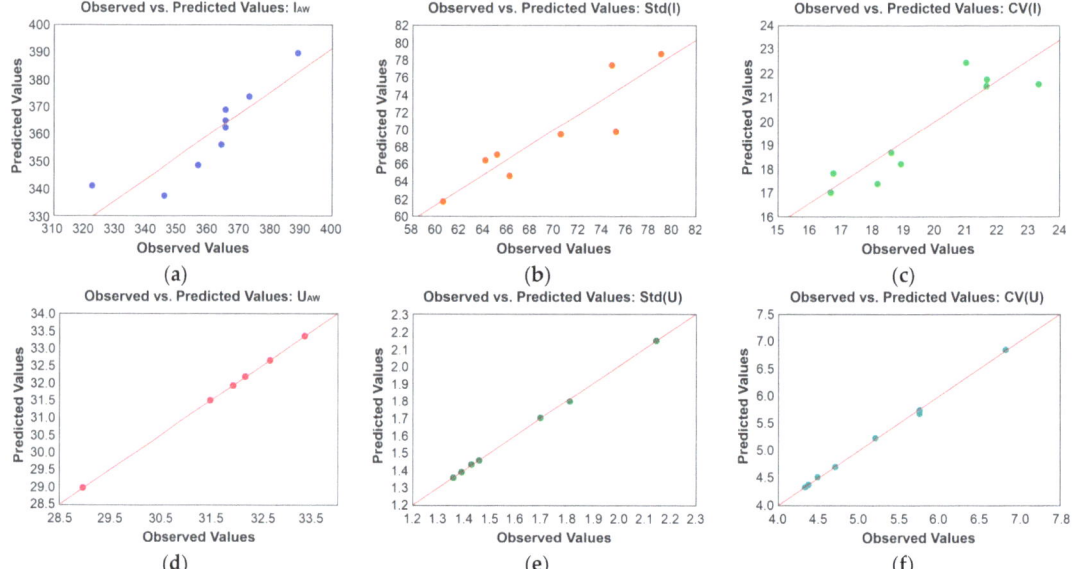

Figure 12. Diagram of predicted values versus actual values for (**a**) mean welding current, (**b**) coefficients of variation, (**c**) standard variation of welding current, (**d**) mean arcvoltage, (**e**) coefficients of variation, and (**f**) standard variation of arc voltage.

The graphs show that the developed models for the parameters characterizing the arc voltage (U_{aw}, $Std(U)$, and the $CV(U)$) have good prediction capabilities. This is indicated by the location of the points predicted by the developed model, which are practically located on a sloping straight line near the actual values. In this case, the model works well, and the errors are random. Whereas the location of the predicted points determined by using the developed mathematical models for the welding current parameters (I_{aw}, $Std(I)$, and $CV(I)$) are located away from the inclined straight red line. This indicates the scatter of predicted values and the limitations of the developed models.

Pareto diagrams were constructed (Figure 13) to better identify the influence of each variable in the resulting mathematical model.

The red line defines the threshold value, exceeding which indicates the importance of the effect for the optimization parameter. The analysis of the obtained Pareto diagrams (Figure 13) for the welding current indicators showed that the ratio of oxidizing agent to reducing agent in the composition of exothermic mixture (CuO/Al) and the ratio of graphite to oxidizing agent of exothermic mixture (CuO/C) had the greatest influence for all indicators (I_{aw}, $Std(I)$ and $CV(I)$). Whereas the percentage of exothermic mixture in the core filler (EA), which is CuO + Al, had a significant influence only on the average welding current (I_{aw}). Whereas the CuO/Al and CuO/C ratios had a significant influence on the process stability characterized by the indicators of standard deviation of the welding current ($Std(I)$) and the coefficient of variation of the welding current ($CV(I)$). The analysis of the Pareto diagram for the arc voltage parameters showed a similar influence of the studied variables. It should be noted that the influence of such variables as the ratio of oxidizing agent to reducing agent in the composition of exothermic mixture (CuO/Al) is more pronounced and significant. A significant influence on the parameter coefficient of variation of arc voltage ($CV(U)$) was from the ratio of graphite to oxidizing agent of exothermic mixture (CuO/C). Considering the influence of the studied equation effects on the average arc voltage (U_{aw}), the high influence of the effect of the percentage of exothermic mixture (EA) and the low influence of the effect of the ratio CuO/C can be noted.

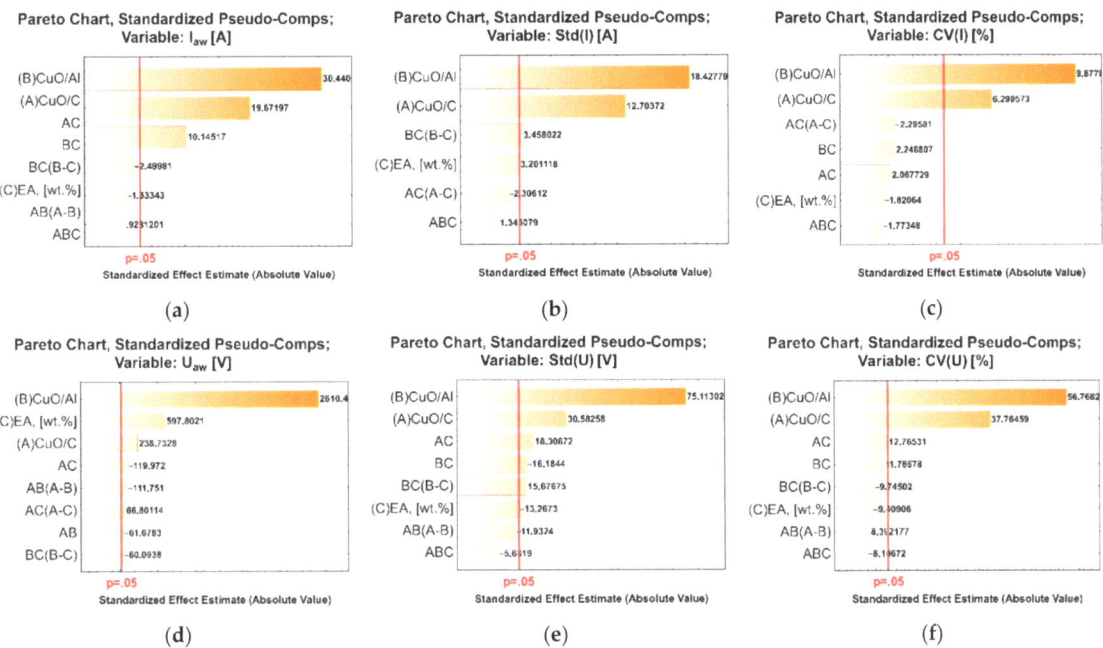

Figure 13. Pareto diagrams of effect of core filler composition on (**a**) mean welding current, (**b**) coefficients of variation, (**c**) standard variation of welding current, (**d**) meanarc voltage, (**e**) standard variation, and (**f**) coefficients of variation of arc voltage.

To visualize the relationships between the three variables and the output responses, and to determine the optimal values of the response parameters, contour graphs and 3D response surfaces are utilized. Figure 14 shows the 2D and 3D response representations of the optimized factors that lead to the efficiency of the average welding current (I_{aw}), the standard deviation of the welding current ($Std(I)$), and the coefficient of variation of the welding current ($CV(I)$).

A response surface plot (3D) and a contour plot (2D) were also constructed for indicators characterizing the arc voltage, shown as follows: average arc voltage U_{aw}, the standard deviation of arc voltage $Std(U)$, and the coefficient of variation of arc voltage $CV(U)$ (Figure 15).

Once the stationary point has been identified using a contour plot, it is important to characterize the response surface near the point to determine if it is a maximum, minimum, or saddle point. This information is useful for further optimizing the process and achieving the desired output response. For optimizing the core filler composition, the methodology described in [64] was utilized. This methodology enables the optimization of a research problem when there are many initial parameters to optimize. This technique allows us to determine the optimal values of the variables by overlaying the optimal areas of different initial parameters on one graph. Then, the intersection of these optimal areas is determined. If there is no intersection, then it is necessary to expand the threshold value for one of the selected dependent parameters to achieve this intersection. In our case, two welding parameters were selected for optimization: the mean welding current I_{aw} and the standard deviation of arc voltage $Std(U)$.

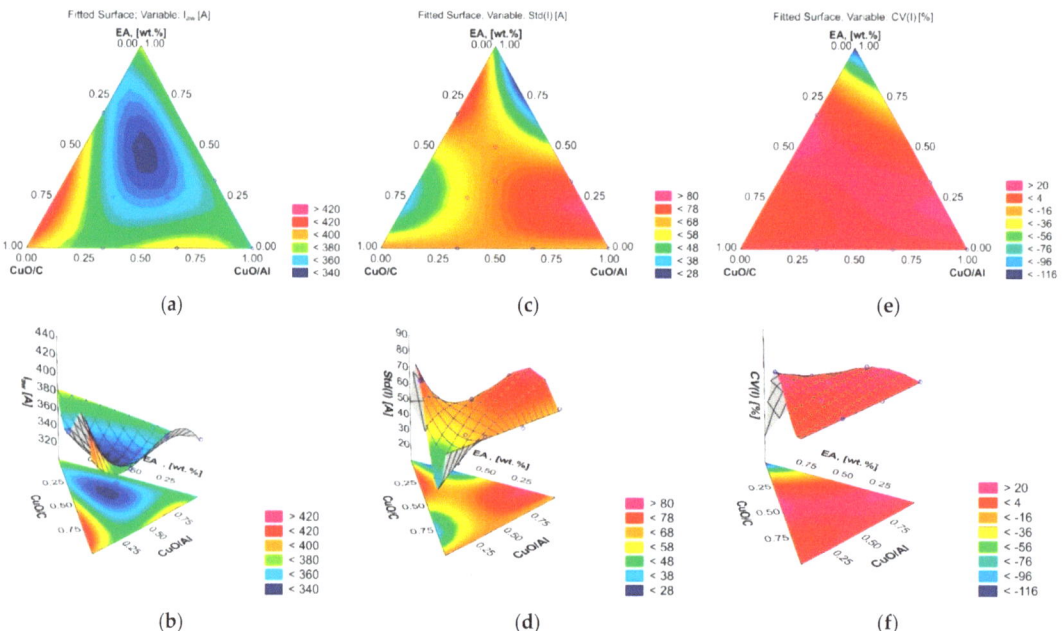

Figure 14. Response surface plot (3D) and contour plot (2D) showing the effects of variables on mean welding current (**a**) 2D and (**b**) 3D; standard variation of welding current (**c**) 2D and (**d**) 3D; and coefficients of variation of welding current (**e**) 2D and (**f**) 3D.

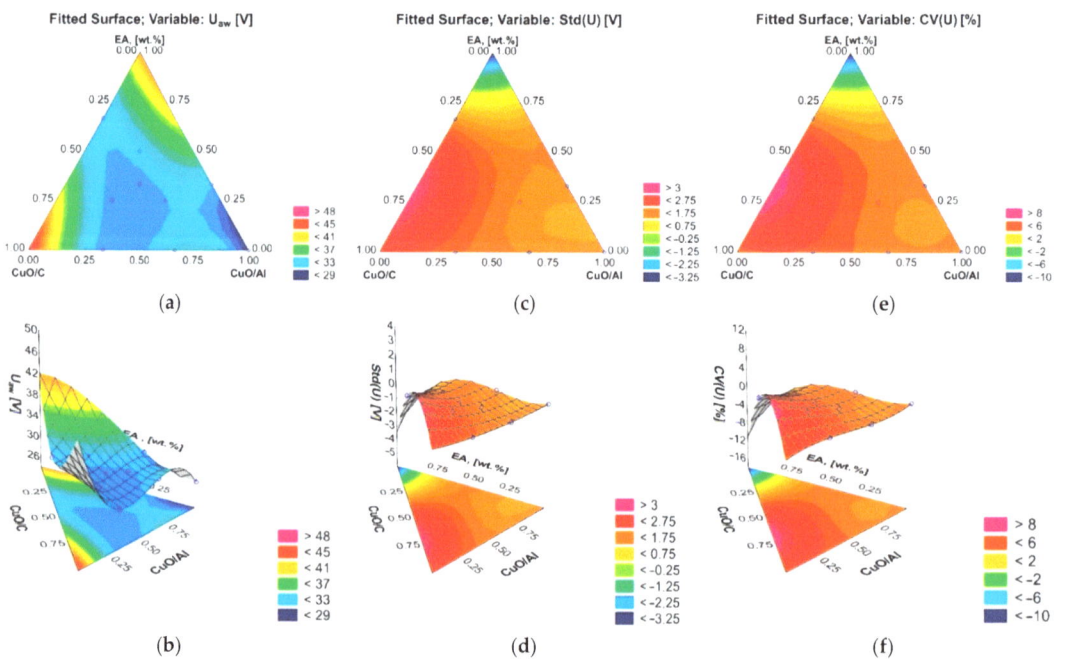

Figure 15. Response surface plot (3D) and contour plot (2D) showing the effects of variables on mean arc voltage (**a**) 2D and (**b**) 3D; standard variation of arc voltage (**c**) 2D and (**d**) 3D; and coefficients of variation of arc voltage (**e**) 2D and (**f**) 3D.

The first parameter was chosen because it can best characterize the probability of an exothermic reaction occurring before the arc, increasing the electrical resistance in the electrode extension area and improving the melting uniformity of the flux-cored wire electrode (the lag of melting of the core filler from the melting of the metal sheath [64]). The lower the observed mean welding current value, the greater the heat generated in the core filler from the exothermic reaction. The parameter of the standard deviation of arc voltages chosen to optimize the composition of the core filler because it can most informatively characterize the stability of the flux-cored arc welding process [86,87,98]. Based on experience and in accordance with the above methodology, the following threshold values were selected for optimization: $I_{aw} \leq 350$ A and $Std(U) \leq 2$ V.

Based on the obtained surfaces results (Figure 16), it can be seen that the optimum area with a maximum value observed is as follows: $x_1 = 0.17$–0.42 (CuO/C = 3.5–4.26), $x_2 = 0.18$–0.31 (CuO/Al = 3.6–3.9), $x_3 = 0.35$–0.7 (EA = 29–38 wt.%).

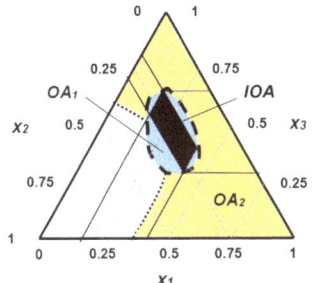

OA_1 – Optimal area of mean welding current
OA_2 – Optimal area of standard variation of arc voltage
IOA – Interaction of two optimal areas

Figure 16. Interpretation diagram of optimal value areas.

3.5. Prediction of Mathematic Model of Heat Input

To better understand the cooling conditions, heat input indicators were calculated for each experiment (Table 7). The values were calculated according to Equation (11).

Table 7. Heat input (Q_{in}) during hardfacing with experimental flux-cored wires with different core filler compositions.

Run	Actual Value, (kJ/mm)	Predicted Value, (kJ/mm)	Diff., (kJ/mm)	Dev., (%)
1	1586.1	1603.9	17.8	1.12
2	1522.4	1640.1	117.6	7.73
3	1760.8	1805.2	44.5	2.53
4	1762.5	1741.7	20.8	1.18
5	1762.5	1771.7	9.1	0.52
6	1718.9	1609.2	109.7	6.38
7	1690.7	1676.3	14.4	0.85
8	1705.9	1690.5	15.3	0.90
9	1942.7	1913.9	28.8	1.48
10	1586.1	1603.9	17.8	1.12

A mathematical model was constructed (Equation (18)). The mathematical model exhibits good accuracy ($R_{sqr} = 0.7345$), adequacy (Fisher's criterion $F_{act} = 3.766 > 3$), and significance (Student's criterion $p = 0.03740 > 0.05$). The equation of the obtained regression model for heat input is provided in Equation (18):

$$Q_{in} = 1278.47 \cdot x_1 + 1603.88 \cdot x_2 + 2017.89 \cdot x_3 + 1883 \cdot x_1 \cdot x_2 - 5471 \cdot x_1 \cdot x_2 \cdot x_3 \quad (18)$$

An analysis of the obtained mathematical model (18) showed that the linear terms of the equation (x_1, x_2, x_3) and their pairwise interactions increase the overall Q_{in}. A comparison of experimental values and predicted values showed good convergence, confirming the accuracy of the developed mathematical model for the heat input parameter.

Figure 17 presents a Pareto Chart, observed and predicted values, as well as a response surface and contour plot. An analysis of the Pareto diagram (Figure 17a) revealed that the two linear factors exert the greatest influence on Q_{in}, namely the oxidizer-to-reducer ratio (CuO/Al) and the amount of introduced exothermic addition (CuO/C) in the core filler. Experimental values and values predicted using the developed mathematical model (Equation (18)) showed good convergence, as evidenced by the proximity of points to the inclined line (Figure 17b). The 3D surface response plot (Figure 17c) and 2D contour plot (Figure 17d) of the obtained mathematical model demonstrate the complex interaction of components in the powder wire blend.

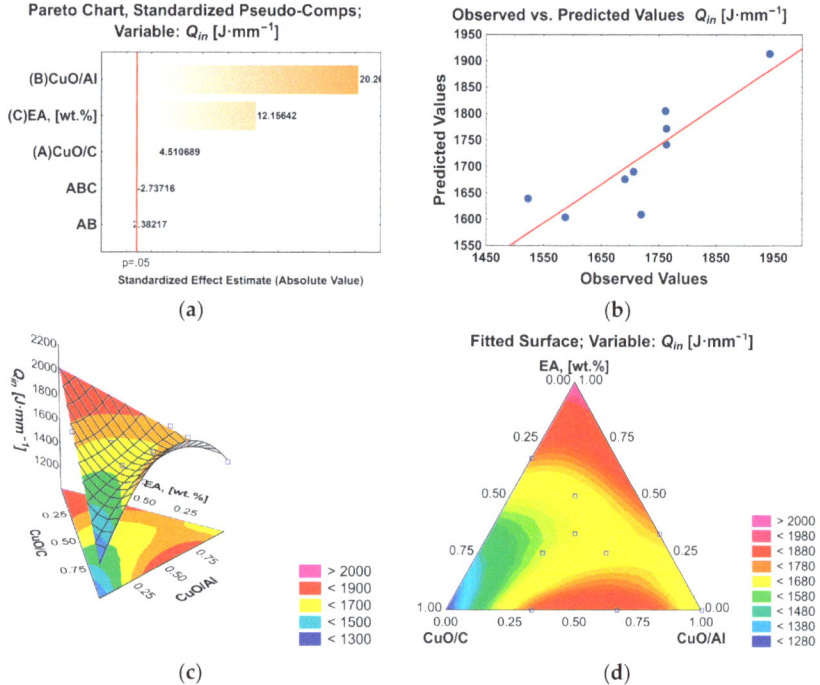

Figure 17. (**a**) Pareto chart, (**b**) plot of observed and predicted values, (**c**) response surface, and (**d**) contour surface graphs for heat input Q_{in}.

3.6. Microstructure Study

For structural investigations, two samples were selected based on the parameters with the highest and lowest values of the standard variation of the arc voltage $Std(U)$, which best assesses arc stability. These parameters (Table 5) correspond to samples of deposited metal applied by self-shielded, flux-cored wire electrodes indicated as FCAW-SS-E5 and FCAW-SS-E6. In this case, FCAW-SS-E5 contained EA = 37.42 wt%, while FCAW-SS-E6 contained EA = 26.5 wt%.

The chemical composition and mechanical properties were measured for the selected weld beads. Cross-sections of the weld beads, which were hardfaced by FCAW-SS-E5 and FCAW-SS-E6, are shown in Figure 18. Chemical composition of simple selected deposited metal is shown in Table 8.

(a)   (b)

Figure 18. Macrographic images of weld bead hardfacing using (**a**) FCAW-SS-E5; (**b**) FCAW-SS-E6.

Table 8. Chemical composition of simple selected deposited metal.

Item No.	Content of Alloying Element in Metal Deposit, wt.%							
	C	Cr	Si	Mn	Ti	V	Al	Cu
FCAW-SS-E5	0.73	3.95	0.99	1.77	0.52	0.32	0.16	7.15
FCAW-SS-E6	0.49	2.52	0.93	1.53	0.71	0.26	0.21	5.01

For a better understanding of the influence of the core filler composition, we decided to conduct a study and compare our experimental results with thermodynamic calculations performed using the JMatPro® API v7.0 software for the phase composition of the deposited metal [99,100]. Phase diagrams (Figure 5) and the phase composition of the deposited metal were constructed for the selected samples depending on the welding cycle and according to the obtained characteristics of the welding modes.

Figure 19a shows that the deposited metal produced using FCAW-SS-E5 exhibited an expansion of the austenitic zone and a decrease in temperature during the onset and completion of martensitic transformation compared to the sample produced using FCAW-SS-E6 (Figure 19b). The latter can be explained by differences in chemical composition, due to a higher content of carbon and copper in the FCAW-SS-E5 deposited sample. Such differences would lead to a change in the phase composition of the matrix, resulting in a decrease in the proportion of the ferritic phase in the deposited metal and an increase in the pearlitic phase (Figure 20). Structural-phase investigations were conducted to confirm the modeled phase composition of the obtained samples. The microstructure of the deposited metal is depicted in Figure 21, while the phase composition as shown in the XRD pattern is presented in Figure 22.

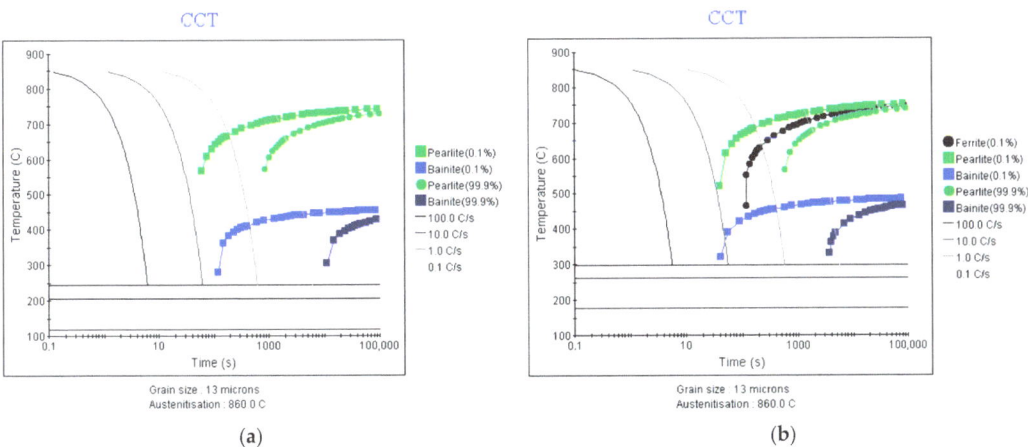

(a)   (b)

Figure 19. CCT diagram for deposited metals of hardfacing using (**a**) FCAW-SS-E5; (**b**) FCAW-SS-E6.

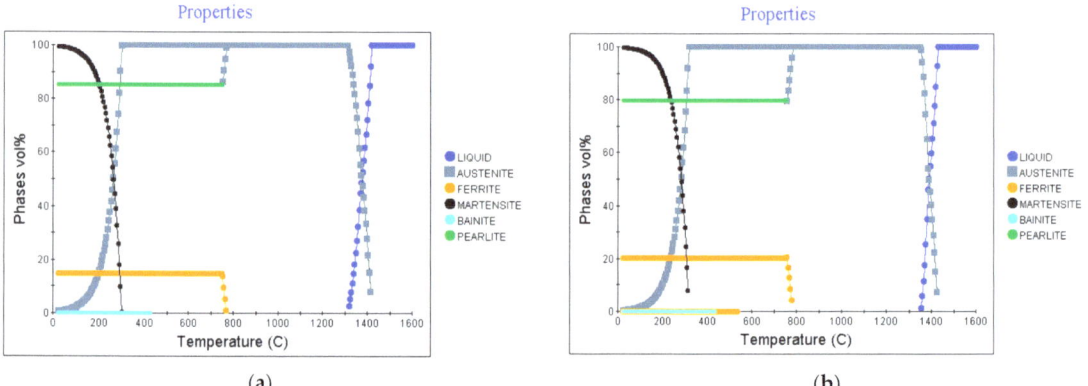

Figure 20. The phase composition of the deposited metals during the welding cycle (rate of heating $RH = 1000\ °C/s$, cooling rate $CR = 45\ °C/s$, graine size 13 μm) during hardfacing with a self-shielded, flux-cored wire electrode with indexes (**a**) FCAW-SS-E5; (**b**) FCAW-SS-E6.

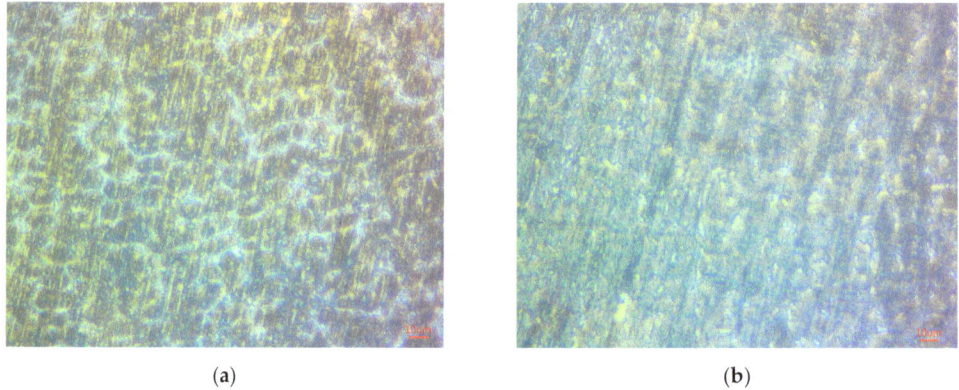

Figure 21. Microstructure of specimens of hardfacing using (**a**) FCAW-SS-E5; (**b**) FCAW-SS-E6.

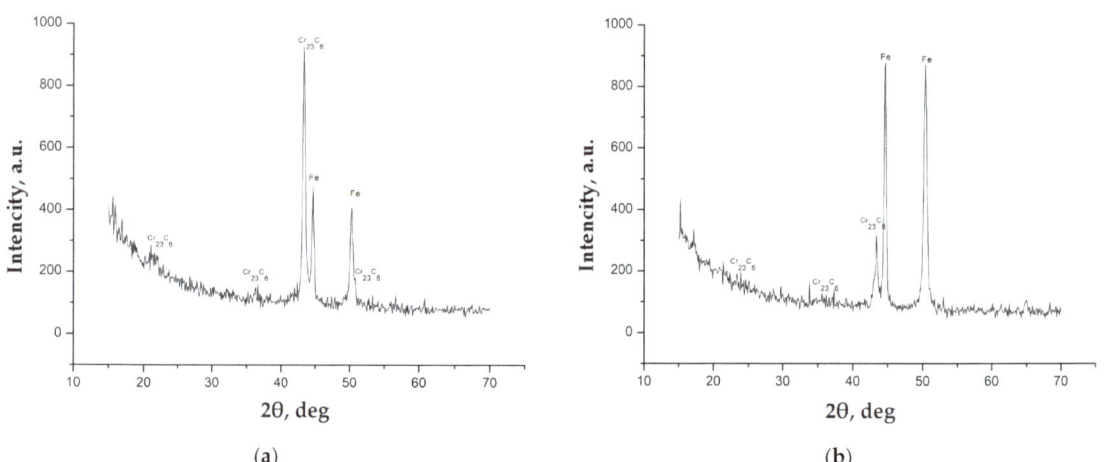

Figure 22. XRD pattern of specimens of hardfacing using (**a**) FCAW-SS-E5; (**b**) FCAW-SS-E6.

The microstructure of the sample deposited with a flux-cored wire electrode indexed as FCAW-SS-E5 consists of a ferrito-perlitic mixture with carbide eutectics (Figure 22a). The eutectic is located at the boundaries of dendrites in the form of a mesh (Figure 21a). The eutectic represents a solid solution based on ferrite and chromium carbides $Cr_{23}C_6$. The presence of eutectics increases the hardness of the deposited metal but also leads to a decrease in ductility (increased susceptibility to brittle fracture). Figures 21b and 22b show that the structure of the sample made with FCAW-SS-E6 consists of a ferrito-perlitic mixture with isolated chromium carbides ($Cr_{23}C_6$). That is, a ferritic base with uniformly distributed carbides. This structure possesses lower hardness.

4. Discussion

Our research has shown that the introduction of exothermic additive components into the core filler will have a significant impact on the stability of the flux-cored arc welding process. In this case, attention should be paid to the significant influence of variables including the ratio of oxidizing agent to reducing agent in the composition of the exothermic mixture (CuO/Al) on the studied parameters characterizing the welding current and arc voltage (Figure 13). Also, the ratio of graphite to oxidizing agent in the exothermic mixture (CuO/C) will have a significant influence, except for the influence on average arc voltage (U_{aw}). The value of the U_{aw} is more influenced by exothermic addition. The explanation of the influence of variables on the investigated parameters of the welding current and arc voltage will be considered separately.

The high influence of CuO/Al and CuO/C ratios on the average welding current (I_{aw}) can be explained by the change in power consumption transferred from the power supply at the contact-tip-to-work distance. The constant section from the contact tip to the substrate materials, referred to as contact-tip-to-work distance, can be considered as a parallel-connected chain of individual sections of the electrode extension—arc column, which dynamically changes due to mutual compensation. Increasing the length of the electrode extension section leads to heating of the flux-cored wire electrode in front of the tip, due to Joule heat [101,102]. The presence of an additional heat source (exothermic reaction) introduces changes into the energy distribution. Depending on the welding parameters, the exothermic reaction can take place both in the electrode extension section and in the arc column. In the first case, the introduction of exothermic addition will lead to the melting of the core filler and the heating of the metal sheath. The latter will lead to an increase in the electrical resistance of the sheath and a drop in the average welding current (I_{aw}). However, the presence of carbon-containing components in the core filler in the form of graphite leads to the possibility of cupric oxide (CuO) reduction due to a carbothermal reaction: $2CuO + C \rightarrow 2Cu + CO_2$ [84]. This reaction is characterized by a significantly lower thermal effect (Q_{CuO+C} = 16.49 kJ/mol O_2 against Q_{CuO+Al} = 798.87 kJ/mol O_2 [64]) compared to the thermal effect of the exothermic reaction from the additive: $2CuO + 4/3Al \rightarrow 2/3Al_2O_3 + 2Cu$. In the case of an exothermic reaction in the arc column, the arc temperature will increase [65,103] by changing the electrical characteristics of the arc, which leads to an increase in the average welding current [72].

The significant influence of the CuO/Al ratio on the arc voltage parameters can also be explained by its influence on the elemental composition of the welding arc, i.e., which compounds will enter the arc column. When using exothermic addition in the core filler, we can assume three different variants in the behavior of their components: (1) an ingress in the form of cupric oxide (CuO) and pure aluminum (Al) in the case that the exothermic reaction has not started; (2) fully reduced copper (Cu) and aluminum oxide; and (3) a combination of elements (1) and (2) in case of an incomplete exothermic reaction. The second case (2) can be explained by the insufficient level of heating of the filler to the temperature of the exothermic reaction [77]. When metal vapor enters the arc column, the distribution of the current density and the energy transmitted to the workpiece changes [104,105]. The vapor additives cause the changes, and in turn, affect the energy transport as well as temperature and velocity distributions in the arc column [106,107]. Thus, research by Jia et al. [108]

has shown that during the FCAW-SS process in the arc column, it is possible to detect in addition to the compounds providing gas protection (mainly CO and CO_2 as well as F_2 and F) and gases inevitably coming from the surrounding atmosphere (O_2, H, N, and N_2), also ions or molecules of most of the core filler components such as Mg, Si, K, Ca, Ti, Cr, Mn, Fe, and Ni. Their fraction in the arc column will depend primarily on the thermal stability of their compounds, as well as on the amount of the introduced component in the core filler and the fraction of the introduced component. Better arc stability could be obtained by adding low-ionizable components [109]. The more complete this reaction is, the greater the amount of easily evaporated copper (Cu) in the arc plasma will be. For content with a large amount of graphite (which will correspond to low values of CuO/C ratio), as well as lack of aluminum powder (high CuO/Al ratio), it becomes more thermodynamically probable that a less calorific carbothermic reaction of cupric oxide reduction at the expense of graphite will take place. In this case, the aluminum that has not entered into an exothermic reaction will be partially spent on the deoxidation of the weld pool, but mainly deposited metal will be absorbed. However, it can also enter the arc plasma in the form of vapor, changing its characteristics. Thus, the ingress of vapor containing copper or aluminum will explain the change in arc stability. The studies of Li et al. [72] showed similar results of the effect of adding Al + Fe_2O_3 powder of low ionization voltage into coating, which caused the arc voltage to be decreased at a given arc length with increasing thermite levels. In addition, the presence of Al + Fe_2O_3 powder had low ionization, thereby improving the arc stability. Research by Zhang et al. [107] also reported a reduction in the welding voltage and an increased welding current when introducing Fe_2O_3 into filler materials. This explains that the decomposition and evaporation of the filler materials produced oxygen, which led to a narrowing of the arc [107].

Many authors point out the positive effects of different contents of copper vapor on the arc morphology, arc voltage, arc pressure, current density, and arc axial temperature [105,106,110]. In this case. due to the relatively low boiling temperature of copper ($T_{boil.}$(Cu) = 2860 K) [64], copper vapors can fall into the arc zone in large quantities. Thermodynamic calculations performed by Guo et al. [110] showed the significant influence of copper vapor on arc plasma parameters at Cu mole fractions above a 10% start [111]. The influence of copper vapor has been explained by the fact that the current tends to flow through the edge of the electrode, which expands the conductive path and makes the arc disperse [105]. The content of copper vapor arc voltage gradually increased (1.5 V) [105]. This is consistent with the data we obtained, in which with the increase in the amount of introduced exothermic addition (EA wt.%), an increase in the average arc voltage (U_{aw}) is observed. Mostaghimi-Tehrani and Pfender [106] associate this with the significant contribution of electrons obtained from copper vapor to the electrical conductivity of the arc in a temperature range of 6000 K to 10,000 K.

Also, the influence of exothermic addition components on the surface tension of droplets when transferring filler materials from the electrode to the substrate should be taken into account. The authors of the paper point to a decrease in droplet size due to a decrease in the surface tension of the molten droplet [72].

The characteristics of the arc and plasma composition [112] will affect weld bead morphology [61,113], cooling conditions [114], as well as the thermo-stressed state [115–117], its microstructure [118], mechanical properties [119–125], and operational properties [126–130].

Future research plans to investigate the influence of changing the composition of the core filler with an exothermic addition on weld bead morphology and mechanical properties.

5. Conclusions

At present, there is great experience in analyzing and using parameters for research and comparison of arc burning stability in the fusion arc welding process. As parameters characterizing arc stability, both graphical (current and voltage cyclograms, box plots with frequency histograms) and statistical parameters such as standard variation and coefficients of variation for welding current and arc voltage have been used. In this paper, an experimental study of the influence of the introduction of exothermic addition (CuO-Al) into the core filler of a self-shielded, flux-cored wire electrode on the average of the welding current and arc voltage as well as arc stability, was carried out, with conclusions as follows:

1. It has been determined that the introduction of exothermic addition components into the core filler will have a significant effect on the stability of the flux-cored arc welding process. This can be attributed to changes in the chemical composition of the arc column, resulting from enrichment, depending on the composition of the core filler, with easily ionizable elements or compounds (such as Cu) or with less readily ionizable compounds (such as Al_2O_3);
2. The results of research have shown that the greatest influence on the parameters characterizing the welding current and arc voltage was exerted by the parameter of the ratio of oxidizing agent to reducing agent in the composition of exothermic mixture (CuO-Al), which had a significant influence on the completeness of the exothermic reaction in the core filler. To a lesser extent the ratio of exothermic mixture oxidizing agent to graphite content CuO/C had an influence, except for the parameter of average arc voltage;
3. The lowest values of coefficients of variation of the arc voltage $CV(U)$ and standard variation of the arc voltage $Std(U)$ are characterized by a hardfacing process with a high ratio of $CuO/Al \geq 4.5$ ($x_2 \geq 0.5$) and the content of exothermic mixture in the core filler (EA) being below average $EA = 26$–29 wt.% ($x_3 < 0.33$).
4. The investigation of microstructures revealed that weld beads deposited by self-shielded, flux-cored wire electrodes with a high content of exothermic additive (EA = 38 wt.%) and moderate graphite content (CuO/C = 4) exhibited higher hardness and are preferable for the reinforcement of surfaces subjected to abrasive wear in combination with metal-to-metal friction. Additionally, they showed high arc stability.

Author Contributions: Conceptualization, B.T.; methodology, B.T., S.K., H.K. and L.R.; software, S.K., M.K. and O.K.; validation, V.L. and B.T.; formal analysis, E.K., O.B. and K.S.; investigation, B.T.; resources, B.T.; data curation, B.T.; writing—original draft preparation, V.L., B.T. and L.R.; writing—review and editing, V.L., B.T. and O.R.; visualization, H.K. and I.K.; supervision, V.L.; project administration, B.T. and V.L.; funding acquisition, B.T. All authors have read and agreed to the published version of the manuscript.

Funding: This study was carried out as part of the project "Belt and Road Initiative Centre for Chinese-European studies (BRICES)" and was funded by the Guangdong University of Petrochemical Technology. This research was funded by the Ministry of Education and Science of Ukraine for the grant to implement projects 0123U101858 and 0124U000668.

Data Availability Statement: The data are contained within this article.

Acknowledgments: The authors express their sincere gratitude and respect to the Armed Forces of Ukraine, who made it possible to complete the preparation of this article for publication. The authors are also grateful to the editor and reviewers for their comments that helped improve the content of this paper.

Conflicts of Interest: The authors declare no conflicts of interest.

References

1. Kotov, M.A.; Konoplianyk, O.Y.; Volchuk, V.M.; Plakhtii, Y.G.; Plakhtii, A.O. Light Structurally Thermal Insulating Concrete with a Wide Range of Applications from Recycled Waste Polypropylene Container. *Adv. Transdiscipl. Eng.* **2023**, *43*, 515–521. [CrossRef]
2. Bausk, E.A.; Volchuk, V.M.; Uzlov, O.V. Remaining Service Life Evaluation of Nuclear Power Plants Construction Steel Elements. *J. Phys. Conf. Ser.* **2021**, *1926*, 012050. [CrossRef]
3. Bondarenko, I.; Keršys, R.; Neduzha, L. Studying of Dynamic Parameters Impulse Impact of the Vehicle Taking into Account the Track Stiffness Variations. In Proceedings of the Transport Means 2021: Proceedings of the 25th International Conference, Kaunas, Lithuania, 6–8 October 2021; pp. 684–689.
4. Ulbrich, D.; Kowalczyk, J.; Jósko, M.; Sawczuk, W.; Chudyk, P. Assessment of Selected Properties of Varnish Coating of Motor Vehicles. *Coatings* **2021**, *11*, 1320. [CrossRef]

5. Bondarenko, I.; Severino, A.; Olayode, I.O.; Campisi, T.; Neduzha, L. Dynamic Sustainable Processes Simulation to Study Transport Object Efficiency. *Infrastructures* **2022**, *7*, 124. [CrossRef]
6. Viňáš, J.; Brezinová, J.; Sailer, H.; Brezina, J.; Sahul, M.; Maruschak, P.; Prentkovskis, O. Properties Evaluation of the Welded Joints Made by Disk Laser. *Materials* **2021**, *14*, 2002. [CrossRef]
7. Kalivoda, J.; Neduzha, L. Running Dynamics of Rail Vehicles. *Energies* **2022**, *15*, 5843. [CrossRef]
8. Hlushkova, D.B.; Bagrov, V.A.; Saenko, V.A.; Volchuk, V.M.; Kalinin, A.V.; Kalinina, N.E. Study of wear of the building-up zone of martensite-austenitite-austenitic and secondary hardening steels of the Cr–Mn–Ti system. *Probl. At. Sci. Technol.* **2023**, *2*, 105–109. [CrossRef]
9. Selech, J.; Majchrzycki, W.; Ulbrich, D. Field and Laboratory Wear Tests of Machine Components Used for Renovation of Dirt Roads—A Case Study. *Materials* **2023**, *16*, 6180. [CrossRef]
10. Romek, D.; Ulbrich, D.; Selech, J.; Kowalczyk, J.; Wlad, R. Assessment of Padding Elements Wear of Belt Conveyors Working in Combination of Rubber–Quartz–Metal Condition. *Materials* **2021**, *14*, 4323. [CrossRef] [PubMed]
11. Bondarenko, I.; Campisi, T.; Tesoriere, G.; Neduzha, L. Using Detailing Concept to Assess Railway Functional Safety. *Sustainability* **2023**, *15*, 18. [CrossRef]
12. Dubei, O.Y.; Tutko, T.F.; Ropyak Lya Shovkoplias, M.V. Development of Analytical Model of Threaded Connection of Tubular Parts of Chrome-Plated Metal Structures. *Metallofiz. Noveishie Tekhnologii* **2022**, *44*, 251–272. [CrossRef]
13. Pokhmurskii, V.I.; Zin, I.M.; Bily, L.M.; Vynar, V.A.; Zin, Y.I. Aluminium alloy corrosion inhibition by chromate-free composition of zinc phosphate and ion-exchanged zeolite. *Surf. Interface Anal.* **2013**, *45*, 1474–1478. [CrossRef]
14. Volkov, O.; Knyazev, S.; Vasilchenko, A.; Doronin, E. Alternative Strengthening of Jewelry Tools Using Chemical-Thermal and Local Surface Treatments. *Mater. Sci. Forum* **2021**, *1038*, 68–76. [CrossRef]
15. Shatskyi, I.; Makoviichuk, M.; Ropyak, L.; Velychkovych, A. Analytical Model of Deformation of a Functionally Graded Ceramic Coating under Local Load. *Ceramics* **2023**, *6*, 1879–1893. [CrossRef]
16. Shatskyi, I.P.; Makoviichuk, M.V.; Ropyak, L.Y. Equilibrium of Laminated Cu/Ni/Cr Coating Under Local Load. *Nanosistemi Nanomater. Nanotehnologii* **2023**, *21*, 379–389. [CrossRef]
17. Bembenek, M.; Makoviichuk, M.; Shatskyi, I.; Ropyak, L.; Pritula, I.; Gryn, L.; Belyakovskyi, V. Optical and Mechanical Properties of Layered Infrared Interference Filters. *Sensors* **2022**, *22*, 8105. [CrossRef]
18. Martsynkovskyy, V.; Tarelnyk, V.; Konoplianchenko, I.; Gaponova, O.; Dumanchuk, M. Technology support for protecting contacting surfaces of half-coupling—Shaft press joints against fretting wear. In *Design, Simulation, Manufacturing: The Innovation Exchange*; Springer International Publishing: Cham, Switzerland, 2020; pp. 216–225. [CrossRef]
19. Gaponova, O.P.; Antoszewski, B.; Tarelnyk, V.B.; Kurp, P.; Myslyvchenko, O.M.; Tarelnyk, N.V. Analysis of the Quality of Sulfomolybdenum Coatings Obtained by Electrospark Alloying Methods. *Materials* **2021**, *14*, 6332. [CrossRef]
20. Tarelnyk, V.B.; Gaponova, O.P.; Konoplianchenko, I.e.V.; Tarelnyk, N.V.; Dumanchuk, M.Y.; Pirogov, V.O.; Voloshko, T.P.; Hlushkova, D.B. Development the Directed Choice System of the Most Efficient Technology for Improving the Sliding Bearings Babbitt Covers Quality. Pt. 2. Mathematical Model of Babbitt Coatings Wear. Criteria for Choosing the Babbitt Coating Formation Technology. *Metallofiz. Noveishie Tekhnol.* **2022**, *44*, 1643–1659. [CrossRef]
21. Umanskyi, O.P.; Storozhenko, M.S.; Tarelnyk, V.B.; Koval, O.Y.; Gubin, Y.V.; Tarelnyk, N.V.; Kurinna, T.V. Electrospark Deposition of Fenicrbsic–Meb2 Coatings on Steel. *Powder Metall. Met. Ceram.* **2020**, *59*, 57–67. [CrossRef]
22. Kotsyubynsky, V.; Shyyko, L.; Shihab, T.; Prysyazhnyuk, P.; Aulin, V.; Boichuk, V. Multilayered MoS_2/C nanospheres as high performance additives to lubricating oils. *Mater. Today Proc.* **2019**, *35*, 538–541. [CrossRef]
23. Student, M.; Gvozdetsky, V.; Student, O.; Prentkovskis, O.; Maruschak, P.; Olenyuk, O.; Titova, L. The Effect of Increasing the Air Flow Pressure on the Properties of Coatings during the Arc Spraying of Cored Wires. *Stroj. Časopis-J. Mech. Eng.* **2019**, *69*, 133–146. [CrossRef]
24. Tarelnyk, V.B.; Konoplianchenko, I.V.; Gaponova, O.P.; Tarelnyk, N.V.; Martsynkovskyy, V.S.; Sarzhanov, B.O.; Sarzhanov, O.A.; Antoszewski, B. Effect of Laser Processing on the Qualitative Parameters of Protective Abrasion-Resistant Coatings. *Powder Metall. Met. Ceram.* **2020**, *58*, 703–713. [CrossRef]
25. Prysyazhnyuk, P.; Ivanov, O.; Matvienkiv, O.; Marynenko, S.; Korol, O.; Koval, I. Impact and abrasion wear resistance of the hardfacings based on high-manganese steel reinforced with multicomponent carbides of Ti–Nb–Mo–V–C system. *Procedia Struct. Integr.* **2022**, *36*, 130–136. [CrossRef]
26. Lozynskyi, V.; Trembach, B.; Hossain, M.M.; Kabir, M.H.; Silchenko, Y.; Krbata, M.; Sadovyi, K.; Kolomiitse, O.; Ropyak, L. Prediction of phase composition and mechanical properties Fe–Cr–C–B–Ti–Cu hardfacing alloys: Modeling and experimental Validations. *Heliyon* **2024**, *10*, e25199. [CrossRef]
27. Trembach, B.O.; Sukov, M.G.; Vynar, V.A.; Trembach, I.O.; Subbotina, V.V.; Rebrov, O.Y.; Rebrova, O.M.; Zakiev, V.I. Effect of incomplete replacement of Cr for Cu in the deposited alloy of Fe–C–Cr–B–Ti alloying system with a medium boron content (0.5% wt.) on its corrosion resistance. *Metallofiz. Noveishie Tekhno* **2022**, *144*, 493–513. [CrossRef]
28. Wolski, A.; Świerczyńska, A.; Lentka, G.; Fydrych, D. Storage of high-strength steel flux-cored welding wires in urbanized areas. *Int. J. Precis. Eng. Manuf.-Green Technol.* **2023**, *11*, 55–70. [CrossRef]
29. Vynnykov, Y.; Kharchenko, M.; Manhura, S.; Muhlis, H.; Aniskin, A.; Manhura, A. Analysis of corrosion fatigue steel strength of pump rods for oil wells. *Min. Miner. Depos.* **2022**, *16*, 31–37. [CrossRef]

30. Student, M.M.; Pokhmurs'ka, H.V.; Zadorozhna, K.R.; Veselivs'ka, H.H.; Hvozdets'kyi, V.M.; Sirak, Y.Y. Corrosion resistance of VC–FeCr and VC–FeCrCo coatings obtained by supersonic gas-flame spraying. *Mater. Sci.* **2019**, *54*, 535–541. [CrossRef]
31. Świerczyńska, A.; Varbai, B.; Pandey, C.; Fydrych, D. Exploring the trends in flux-cored arc welding: Scientometric analysis approach. *Int. J. Adv. Manuf. Technol.* **2024**, *130*, 87–110. [CrossRef]
32. Bembenek, M.; Prysyazhnyuk, P.; Shihab, T.; Machnik, R.; Ivanov, O.; Ropyak, L. Microstructure and wear characterization of the Fe–Mo–B–C–based hardfacing alloys deposited by flux-cored arc welding. *Materials* **2022**, *15*, 5074. [CrossRef]
33. Vahrusheva, V.S.; Hlushkova, D.B.; Volchuk, V.M.; Nosova, T.V.; Mamhur, S.I.; Tsokur, N.I.; Bagrov, V.A.; Demchenko, S.V.; Ryzhkov, Y.V.; Scrypnikov, V.O. Increasing the corrosion resistance of heat-resistant alloys for parts of power equipment. *Probl. At. Sci. Technol.* **2022**, *4*, 140. [CrossRef]
34. Hlushkova, D.B.; Volchuk, V.M. Multifractal analysis of the influence of chromium-nickel cast iron structure on its quality criteria. *Probl. At. Sci. Technol.* **2024**, *2024*, 145–153. [CrossRef]
35. Boiko, I.A.; Grin', A.G. Effect of the surface condition of fluxed-cored wires on the stability of the arc process. *Weld. Int.* **2015**, *29*, 543–547. [CrossRef]
36. Spiridonova, I.M.; Sukhovaya, E.V.; Pilyaeva, S.B. Wear-resistant composite coatings with fillers of Fe-BC system. *Paton Weld. J.* **2002**, *2003*, 30–33.
37. Selech, J.; Ulbrich, D.; Romek, D.; Kowalczyk, J.; Wlodarczyk, K.; Nadolny, K. Experimental Study of Abrasive, Mechanical and Corrosion Effects in Ring-on-Ring Sliding Contact. *Materials* **2020**, *13*, 4950. [CrossRef]
38. Guzanova, A.; Džupon, M.; Draganovská, D.; Brezinová, J.; Viňáš, J.; Cmorej, D.; Janoško, E.; Maruschak, P. The corrosion and wear resistance of laser and MAG weld deposits. *Acta Met. Slovaca* **2020**, *26*, 37–41. [CrossRef]
39. Shihab, S.T.A.; Prysyazhnyuk, P.; Andrusyshyn, R.; Lutsak, L.; Ivanov, O.; Tsap, I. Forming the structure and the properties of electric arc coatings based on high manganese steel alloyed with titanium and niobium carbides. *East-Eur. J. Enterp. Technol.* **2020**, *1*, 38–44. [CrossRef]
40. Ivanov, O.; Prysyazhnyuk, P.; Lutsak, D.; Matviienkiv, O.; Aulin, V. Improvement of Abrasion Resistance of Production Equipment Wear Parts by Hardfacing with Flux-Cored Wires Containing Boron Carbide/Metal Powder Reaction Mixtures. *Manag. Syst. Prod. Eng.* **2020**, *28*, 178–183. [CrossRef]
41. Viňáš, J.; Brezinová, J.; Brezina, J.; Hermel, P. Innovation of Biomass Crusher by Application of Hardfacing Layers. *Metals* **2021**, *11*, 1283. [CrossRef]
42. Hlushkova, D.B.; Bagrov, V.A.; Volchuk, V.M.; Murzakhmetova, U.A. Influence of structure and phase composition on wear resistance of sparingly alloyed alloys. *Funct. Mater.* **2023**, *30*, 74–78.
43. Viňáš, J.; Kaščák, L. Possibilities of using welding-on technologies 1 in crane wheel renovation. *Bull. Mater. Sci.* **2008**, *31*, 125–131. [CrossRef]
44. Viňáš, J.; Brezinová, J.; Brezina, J.; Maruschak, P.O. Structural and mechanical features of laser-welded joints of zinc-coated advanced steel sheets. *Mater. Sci.* **2019**, *55*, 46–51. [CrossRef]
45. Sawczuk, W.; Merkisz-Guranowska, A.; Ulbrich, D.; Kowalczyk, J.; Cañás, A.-M.R. Investigation and Modelling of the Weight Wear of Friction Pads of a Railway Disc Brake. *Materials* **2022**, *15*, 6312. [CrossRef]
46. Ulbrich, D.; Stachowiak, A.; Kowalczyk, J.; Wieczorek, D.; Matysiak, W. Tribocorrosion and Abrasive Wear Test of 22MnCrB5 Hot-Formed Steel. *Materials* **2022**, *15*, 3892. [CrossRef] [PubMed]
47. Erohin, A.A. Fundamentals of fusion welding. In *Physical and Chemical Laws*; Mashinostroyeniye: Moscow, Russia, 1973.
48. Świerczyńska, A.; Fydrych, D.; Rogalski, G. Diffusible hydrogen management in underwater wet self-shielded flux cored arc welding. *Int. J. Hydrog. Energy* **2017**, *42*, 24532–24540. [CrossRef]
49. Sukhova, O.V. Solubility of Cu, Ni, Mn in boron-rich Fe–B–C alloys. *J. Phys. Chem. Solids* **2021**, *22*, 110–116. [CrossRef]
50. Sukhova, O.V. Effect of Ti, Al, Si on the structure and mechanical properties of boron-rich Fe–B–C alloys. *East. Eur. J. Phys.* **2021**, *2021*, 115–121. [CrossRef]
51. Sukhova, O.V. Formation of structure and properties of boron-rich Fe–B–C alloys alloyed with Cr, V, Nb or/and mo. *Metallofiz. Noveishie Tekhnologii* **2021**, *43*, 355–365. [CrossRef]
52. Sukhova, O.V. Structure and properties of Fe–B–C powders alloyed with Cr, V, Mo or Nb for plasma-sprayed coatings. *Probl. At. Sci. Technol.* **2020**, *4*, 77–83. Available online: http://dspace.nbuv.gov.ua/handle/123456789/194417 (accessed on 18 December 2023). [CrossRef]
53. Samardžć, I.; Žagar, D.; Klarić, Š. Application and Upgrading of On-line Monitoring System for Measurement and Processing of Electric Signals at Arc Stud Welding Process. *Stroj. Časopis Za Teor. I Praksu U Stroj.* **2009**, *51*, 355–363.
54. Krbata, M.; Krizan, D.; Eckert, M.; Kaar, S.; Dubec, A.; Ciger, R. Austenite Decomposition of a Lean Medium Mn Steel Suitable for Quenching and Partitioning Process: Comparison of CCT and DCCT Diagram and Their Microstructural Changes. *Materials* **2022**, *15*, 1753. [CrossRef]
55. Krbata, M.; Ciger, R.; Kohutiar, M.; Eckert, M.; Barenyi, I.; Trembach, B.; Dubec, A.; Escherova, J.; Gavalec, M.; Beronská, N. Microstructural Changes and Determination of a Continuous Cooling Transformation (CCT) Diagram Using Dilatometric Analysis of M398 High-Alloy Tool Steel Produced by Microclean Powder Metallurgy. *Materials* **2023**, *16*, 4473. [CrossRef]
56. Gramajo, J.; Gualco, A.; Svoboda, H. Effect of welding parameters on nanostructured Fe–(C.B)–(Cr.Nb) alloys. *Mater. Res.* **2019**, *22*, 1–8. [CrossRef]

57. Gualco, A.; Svoboda, H.G.; Surian, E.S. Effect of welding parameters on microstructure of Fe-based nanostructured weld overlay deposited through FCAW–S. *Weld. Int.* **2016**, *30*, 573–580. [CrossRef]
58. Hlushkova, D.B.; Volchuk, V.M.; Polyansky, P.M.; Saenko, V.A.; Efimenko, A.A. Fractal modeling the mechanical properties of the metal surface after ion-plasma chrome plating. *Funct. Mater.* **2023**, *30*, 275–281. [CrossRef]
59. Spiridonova, I.M.; Sukhovaya, E.V.; Butenko, V.F.; Zhudra, A.P.; Litvinenko, A.I.; Belyj, A.I. Structure and properties of boron-bearing iron granules for composites. *Powder Metall. Met. Ceram.* **1993**, *32*, 45–49. [CrossRef]
60. Krbaťa, M.; Eckert, M.; Cíger, R.; Kohutiar, M. Physical modeling of CCT diagram of tool steel 1.2343. *Procedia Struct. Integr.* **2023**, *43*, 270–275. [CrossRef]
61. Trembach, B.; Grin, A.; Turchanin, M.; Makarenko, N.; Markov, O.; Trembach, I. Application of Taguchi method and ANOVA analysis for optimization of process parameters and exothermic addition (CuO–Al) introduction in the core filler during self-shielded flux-cored arc welding. *Int. J. Adv. Manuf. Technol.* **2021**, *114*, 1099–1118. [CrossRef]
62. Júnior, J.G.F.; Cardoso, A.H.C.; Bracarense, A.Q. Effects of TiC formation in situ by applying titanium chips and other ingredients as a flux of tubular wire. *J. Braz. Soc. Mech. Sci. Eng.* **2020**, *42*, 375. [CrossRef]
63. Júnior, J.G.F.; Clementecardoso, A.H.; Bracarense, A.Q. Addition of TiO$_2$, CaCo$_3$ and CaF$_2$ as a flux of tubular wire applied for TiC in-situ reaction. *Soldag. Insp.* **2020**, *25*, e2511. [CrossRef]
64. Trembach, B.; Grin, A.; Makarenko, N.; Zharikov, S.; Trembach, I.; Markov, O. Influence of the core filler composition on the recovery of alloying elements during the self-shielded flux-cored arc welding. *J. Mater. Res. Technol.* **2020**, *9*, 10520–10528. [CrossRef]
65. Park, Y.D.; Kang, N.; Malene, S.H.; Olson, L. Effect of exothermic additions on heat generation and arc process efficiency in flux-cored arc welding. *Metals Mater. Int.* **2007**, *13*, 501–509. [CrossRef]
66. Malene, S.H.; Park, Y.D.; Olson, D.L. Response of exothermic additions to the flux cored arc welding electrode-Part 1 Effectiveness of exothermically reacting magnesium-type flux additions was investigated with the flux-cored arc welding process. *Weld. J.* **2007**, *86*, 293–302.
67. Allen, J.W.; Olson, D.L.; Frost, R.H. Exothermically assisted shielded metal arc welding. *Weld. J.* **1998**, *77*, 277–285.
68. Zharikov, S.V.; Grin, A.G. Investigation of slags in surfacing with exothermic flux-cored wires. *Weld. Int.* **2015**, *29*, 386–389. [CrossRef]
69. Vlasov, A.F.; Makarenko, N.A. Special features of heating and melting electrodes with an exothermic mixture in the coating. *Weld. Int.* **2016**, *30*, 717–722. [CrossRef]
70. Chigarev, V.V.; Zarechenskii, D.A.; Belik, A.G. Optimisation of the composition and melting parameters of powder strips with the exothermic mixture in the filler. *Weld. Int.* **2016**, *30*, 557–559. [CrossRef]
71. Vlasov, A.F.; Makarenko, N.A.; Kushchiy, A.M. Using exothermic mixtures in manual arc welding and electroslag processes. *Weld. Int.* **2017**, *31*, 565–570. [CrossRef]
72. Li, H.; Liu, D.; Guo, N.; Chen, H.; Du, Y.; Feng, J. The effect of alumino-thermic addition on underwater wet welding process stability. *J. Mater. Process Technol.* **2017**, *245*, 149–156. [CrossRef]
73. Wang, J.; Li, H.; Hu, C.; Wang, Z.; Han, K.; Liu, D.; Wang, J.; Zhu, Q. The Efficiency of Thermite-Assisted Underwater Wet Flux-Cored Arc Welding Process: Electrical Dependence, Microstructural Changes, and Mechanical Properties. *Metals* **2023**, *13*, 831. [CrossRef]
74. Levchuk, K.H.; Radchenko, T.M.; Tatarenko, V.A. High-temperature entropy effects in tetragonality of the ordering interstitial–substitutional solution based on body-centred tetragonal metal. *Metallofiz. Noveishie Tekhnol.* **2021**, *43*, 1–26. [CrossRef]
75. Solomenko, A.G.; Balabai, R.M.; Radchenko, T.M.; Tatarenko, V.A. Functionalization of quasi-two-dimensional materials: Chemical and strain-induced modifications. *Prog. Phys. Met.* **2022**, *23*, 147–238. [CrossRef]
76. Henckell, P.; Gierth, M.; Ali, Y.; Reimann, J.; Bergmann, J.P. Reduction of Energy Input in Wire Arc Additive Manufacturing (WAAM) with Gas Metal Arc Welding (GMAW). *Materials* **2020**, *13*, 2491. [CrossRef]
77. Lankin, Y.N. Indicators of stability of the GMAW process. *Paton Weld. J.* **2011**, *1*, 6–13.
78. Kah, P.; Edigbe, G.O.; Ndiwe, B.; Kubicek, R. Assessment of arc stability features for selected gas metal arc welding conditions. *SN Appl. Sci.* **2022**, *4*, 268. [CrossRef]
79. Trembach, B.O.; Hlushkova, D.V.; Hvozdetskyi, V.M.; Vynar, V.A.; Zakiev, V.I.; Kabatskyi, O.V.; Savenok, D.V.; Zakavorotnyi, O.Y. Prediction of Fill Factor and Charge Density of Self-Shielding Flux-Cored Wire with Variable Composition. *Mater. Sci.* **2023**, *59*, 18–25. [CrossRef]
80. Jartovsky, O.V.; Larichkin, O.V. Pressure welding through a layer of hydrocarbon substance: Physical processes of a diffusion joint formation. *Prog. Phys. Met.* **2021**, *22*, 440–460. [CrossRef]
81. Tarelnyk, V.B.; Gaponova, O.P.; Tarelnyk, N.V.; Myslyvchenko, O.M. Aluminizing of metal surfaces by electric-spark alloying. *Prog. Phys. Met.* **2023**, *24*, 282–318. [CrossRef]
82. Trembach, B.; Starikov, V.; Sukov, M.G.; Zharikov, S.; Kabatskyi, O.; Ivanova, Y. Application of Mixture design in optimization of physical properties of slag during self-shielded flux-cored wire arc welding process. In Proceedings of the IEEE 5th International Conference on Modern Electrical and Energy System, MEES 2023, Kremenchuk, Ukraine, 27–30 September 2023; pp. 1–5. [CrossRef]
83. Trembach, B. Comparative studies of the three-body abrasion wear resistance of hardfacing Fe-Cr-C-B-Ti alloy. *Proc. IOP Conf. Ser. Mater. Sci. Eng.* **2023**, *1277*, 012016. [CrossRef]

84. Trembach, B.; Balenko, O.; Davydov, V.; Brechko, V.; Trembach, I.; Kabatskyi, O. Prediction the Melting Characteristics of Self-Shielded Flux Cored arc Welding (FCAW–S) with Exothermic Addition (CuO–Al). In Proceedings of the 2022 IEEE 4th International Conference on Modern Electrical and Energy System, MEES 2022, Kremenchuk, Ukraine, 20–23 October 2022; pp. 1–6. [CrossRef]
85. Trembach, B.O.; Silchenko, Y.u.A.; Sukov, M.G.; Ratska, N.B.; Duriagina, Z.A.; Krasnoshapka, I.V.; Kabatskyi, O.V.; Rebrova, O.M. Development of a model of assimilation of alloying elements self-protecting powder wire and optimization of its charge composition. *Mater. Sci.* 2024; in press.
86. Suban, M.; Tušek, J. Methods for the determination of arc stability. *J. Mater. Process Technol.* 2003, *143*, 430–437. [CrossRef]
87. Pessoa, E.; Ribeiro, L.F.; Bracarense, A.Q.; Tao, X.; Jin, P.; Feng, J. Arc stability indexes evaluation on underwater wet welding. *Int. Conf. Offshore Mech. Arct. Eng.* 2010, *49149*, 19–201. [CrossRef]
88. Joseph, A.; Harwig, D.; Farson, D.F.; Richardson, R. Measurement and calculation of arc power and heat transfer efficiency in pulsed gas metal arc welding. *Sci. Technol. Weld. Join.* 2003, *8*, 400–406. [CrossRef]
89. Cottis, R.A. Electrochemical noise for corrosion monitoring. In *Techniques for Corrosion Monitoring*, 2nd ed.; Yang, L.T., Ed.; Woodhead Publishing: Cambridge, UK, 2021; pp. 99–122. [CrossRef]
90. Świerczyńska, A.; Łabanowski, J.; Fydrych, D. The effect of welding conditions on mechanical properties of superduplex stainless steel welded joints. *Adv. Mater. Sci.* 2014, *14*, 14–23. [CrossRef]
91. Wang, J.; Sun, Q.; Jiang, Y.; Zhang, T.; Ma, J.; Feng, J. Analysis and improvement of underwater wet welding process stability with static mechanical constraint support. *J. Manuf. Process* 2018, *34*, 238–250. [CrossRef]
92. Mazzaferro, J.A.E.; Machado, I.G. Study of arc stability in underwater shielded metal arc welding at shallow depths. *Proc. Inst. Mech. Eng. Part C* 2009, *223*, 699–709. [CrossRef]
93. Chen, T.; Xue, S.; Wang, B.; Zhai, P.; Long, W. Study on short-circuiting GMAW pool behavior and microstructure of the weld with different waveform control methods. *Metals* 2019, *9*, 1326. [CrossRef]
94. Ricardo, P.C.; Serudo, R.L.; Ţălu, Ş.; Lamarão, C.V.; da Fonseca Filho, H.D.; de Araújo Bezerra, J.; Sanches, E.A.; Campelo, P.H. Encapsulation of bromelain in combined sodium alginate and amino acid carriers: Experimental design of simplex-centroid mixtures for digestibility evaluation. *Molecules* 2022, *27*, 6364. [CrossRef]
95. Bandura, A.I. Composition of entire functions and bounded L-index in direction. *Mat. Stud.* 2017, *47*, 179–184. [CrossRef]
96. Baksa, V.; Bandura, A.; Skaskiv, O. Growth estimates for analytic vector-valued functions in the unit ball having bounded L-index in joint variables. *Constr. Math. Anal.* 2020, *3*, 9–19. [CrossRef]
97. Kumar, J.; Kumar, G.; Mehdi, H.; Kumar, M. Optimization of FSW parameters on mechanical properties of different aluminum alloys of AA6082 and AA7050 by response surface methodology. *Int. J. Interact. Des. Manuf.* 2023, 1–13. [CrossRef]
98. Li, H.; Liu, D.; Yan, Y.; Guo, N.; Liu, Y.; Feng, J. Effects of heat input on arc stability and weld quality in underwater wet flux-cored arc welding of E40 steel. *J. Manuf. Process* 2018, *31*, 833–843. [CrossRef]
99. Halmešová, K.; Procházka, R.; Koukolíková, M.; Džugan, J.; Konopík, P.; Bucki, T. Extended Continuous Cooling Transformation (CCT) Diagrams Determination for Additive Manufacturing Deposited Steels. *Materials* 2022, *15*, 3076. [CrossRef]
100. Dykas, J.; Samek, L.; Grajcar, A.; Kozłowska, A. Modelling of Phase Diagrams and Continuous Cooling Transformation Diagrams of Medium Manganese Steels. *Symmetry* 2023, *15*, 381. [CrossRef]
101. Assunção, M.T.; Bracarense, A.Q. A novel strategy to improve melting efficiency and arc stability in underwater FCAW via contact tip air chamber. *J. Manuf. Process* 2023, *104*, 1–16. [CrossRef]
102. Mohamat, S.A.; Ibrahim, I.A.; Amir, A.; Ghalib, A. The effect of flux core arc welding (FCAW) processes on different parameters. *Procedia Eng.* 2012, *41*, 1497–1501. [CrossRef]
103. Bozhenko, B.L.; Shalimov, V.N.; Puchkin, P.A.; Kupar, R.Y. Oxidation kinetics of metal vapours in the arc in consumable electrode welding. *Weld. Int.* 1999, *13*, 894–896. [CrossRef]
104. Haidar, J. The dynamic effects of metal vapour in gas metal arc welding. *J. Phys. D* 2010, *43*, 165204. [CrossRef]
105. Ding, F.; Xinglong, Y.; Yingjie, H.; Jiankang, H.; Dequan, L. The study of arc behavior with different content of copper vapor in GTAW. *China Weld.* 2022, *31*, 1–14. [CrossRef]
106. Mostaghimi-Tehrani, J.; Pfender, E. Effects of metallic vapor on the properties of an argon arc plasma. *Plasma Chem. Plasma Process* 1984, *4*, 129–139. [CrossRef]
107. Zhang, S.; Wang, Y.; Xiong, Z.; Zhu, M.; Zhang, Z.; Li, Z. Mechanism and optimization of activating fluxes for process stability and weldability of hybrid laser-arc welded HSLA steel. *Weld. World* 2021, *65*, 753–766. [CrossRef]
108. Jia, C.; Zhang, T.; Maksimov, S.Y.; Yuan, X. Spectroscopic analysis of the arc plasma of underwater wet flux-cored arc welding. *J. Mater. Process Technol.* 2013, *213*, 1370–1377. [CrossRef]
109. Fattahi, M.; Nabhani, N.; Vaezi, M.R.; Rahimi, E. Improvement of impact toughness of AWS E6010 weld metal by adding TiO_2 nanoparticles to the electrode coating. *Mater. Sci. Eng. A* 2011, *528*, 8031–8039. [CrossRef]
110. Guo, X.; Li, X.; Murphy, A.B.; Zhao, H. Calculation of thermodynamic properties and transport coefficients of CO_2–O_2–Cu mixtures. *J. Phys. D* 2017, *50*, 345203. [CrossRef]
111. Li, D.; Fan, D.; Huang, J.; Yao, X. Effect of copper vapor on arc characteristics under DC magnetic field. *Trans. China Weld. Inst.* 2023, *44*, 71–76. [CrossRef]
112. Ropyak, L.; Shihab, T.; Velychkovych, A.; Bilinskyi, V.; Malinin, V.; Romaniv, M. Optimization of Plasma Electrolytic Oxidation Technological Parameters of Deformed Aluminum Alloy D16T in Flowing Electrolyte. *Ceramics* 2023, *6*, 146–167. [CrossRef]

113. Touileb, K.; Attia, E.; Djoudjou, R.; Hedhibi, A.C.; Benselama, A.; Ibrahim, A.; Ahmed, M.M.Z. Effect of Microchemistry Elements in Relation of Laser Welding Parameters on the Morphology 304 Stainless Steel Welds Using Response Surface Methodology. *Crystals* **2023**, *13*, 1138. [CrossRef]
114. Sukhova, O.V. Influence of the Structure and Cooling Rate of Fe–B–C Alloys on Mechanical Properties and Wear Resistance. *Metallofiz. Noveishie Tekhno* **2023**, *45*, 1337–1348. [CrossRef]
115. Nouira, M.; Oliveira, M.C.; Khalfallah, A.; Alves, J.L.; Menezes, L.F. Comparative fracture prediction study for two materials under a wide range of stress states using seven uncoupled models. *Eng. Fract. Mech.* **2023**, *279*, 108952. [CrossRef]
116. Viňáš, J.; Brezinová, J.; Pástor, M.; Šarga, P.; Džupon, M.; Brezina, J. Determination of the Effect of Heat Input during Laser Welding on the Magnitude of Residual Stresses in the Refurbishment of Al Alloy Casting. *Metals* **2023**, *13*, 2003. [CrossRef]
117. Bembenek, M.; Kopei, V.; Ropyak, L.Y.; Levchuk, K. Stressed State of Chrome Parts During Diamond Burnishing. *Metallofiz. Noveishie Tekhnol* **2023**, *45*, 239–250. [CrossRef]
118. Mabuwa, S.; Msomi, V. Investigation of the Microstructure and Mechanical Properties in Friction Stir Welded Dissimilar Aluminium Alloy Joints via Sampling Direction. *Crystals* **2023**, *13*, 1108. [CrossRef]
119. Ivanov, O.O.; Prysiazhniuk, P.M.; Bodrova, L.G.; Kramar, G.M.; Marynenko, S.Y.; Koval, I.V.; Guryk, O.Y. 3D Modeling of the Structure of Deposited Materials Based on Fe–Ti–Mo–B–C System. *Mater. Sci.* **2024**, *59*, 163–169. [CrossRef]
120. Beygi, R.; Zarezadeh Mehrizi, M.; Akhavan-Safar, A.; Mohammadi, S.; da Silva, L.F.M. A Parametric Study on the Effect of FSW Parameters and the Tool Geometry on the Tensile Strength of AA2024–AA7075 Joints: Microstructure and Fracture. *Lubricants* **2023**, *11*, 59. [CrossRef]
121. Boughrara, N.; Benzarti, Z.; Khalfallah, A.; Oliveira, J.C.; Evaristo, M.; Cavaleiro, A. Thickness-dependent physical and nanomechanical properties of $Al_xGa_{1-x}N$ thin films. *Mater. Sci. Semicond. Process.* **2022**, *151*, 107023. [CrossRef]
122. Veselivska, H.H.; Hvozdetskyi, V.M.; Student, M.M.; Zadorozhna, K.R.; Dzioba, Y.V. The Influence of the Electrolyte Composition for Hard Anodizing of Aluminum on Corrosion Resistance of Synthesized Coatings. *Mater. Sci.* **2024**, *59*, 228–233. [CrossRef]
123. Andreikiv, O.Y.; Dolinska, I.Y.; Zviahin, S.; Liubchak, M.O. Acoustic-emission method for determining residual life of power equipment with creep cracks under static load. *Mater. Sci.* **2023**, *59*, 103–111. [CrossRef]
124. Prysyazhnyuk, P.; Di Tommaso, D. The thermodynamic and mechanical properties of Earth-abundant metal ternary boride Mo2(Fe,Mn)B2 solid solutions for impact- and wear-resistant alloys. *Mater. Adv.* **2023**, *4*, 3822–3838. [CrossRef]
125. Luzan, S.A.; Bantkovskiy, V.A. Structure and Tribotechnical Properties of Deposited Composite Layers Based on PG 10N-01 Alloy Containing AL_2O_3. *Mater. Sci.* **2023**, *59*, 328–334. [CrossRef]
126. Katinas, E.; Antonov, M.; Jankauskas, V.; Goljandin, D. Effect of Local Remelting and Recycled WC-Co Composite Reinforcement Size on Abrasive and Erosive Wear of Manual Arc Welded Hardfacings. *Coatings* **2023**, *13*, 734. [CrossRef]
127. Pashechko, M.I.; Shyrokov, V.V.; Duryahina, Z.A.; Vasyliv, K.B. Structure and Corrosion-Mechanical Properties of the Surface Layers of Steels after Laser Alloying. *Mater. Sci.* **2003**, *39*, 108–117. [CrossRef]
128. Vynnykov, Y.; Kharchenko, M.; Manhura, S.; Aniskin, A.; Manhura, A. Degradation of the internal well equipment steel under continuous service in the corrosive and aggressive environments. *Min. Miner. Depos.* **2023**, *17*, 84–92. [CrossRef]
129. Allan, I.; Aknazarov, S.; Golovchenko, O.; Bairakova, O.; Ponomareva, Y.; Mutushev, A. Influence of fluxing additives on aluminothermic synthesis of aluminum borides. *Eng. J. Satbayev Univ.* **2022**, *144*, 19–25. [CrossRef]
130. Andreikiv, Y.; Dolinska, I.Y.; Zviahin, N.S. Acoustic-Emission Method of Determining Residual Life of Thin-Walled Structural Elements Under the Action of Force Loads and Corrosive Environments. *Mater. Sci.* **2023**, *58*, 693–700. [CrossRef]

Disclaimer/Publisher's Note: The statements, opinions and data contained in all publications are solely those of the individual author(s) and contributor(s) and not of MDPI and/or the editor(s). MDPI and/or the editor(s) disclaim responsibility for any injury to people or property resulting from any ideas, methods, instructions or products referred to in the content.

Article

Particle Swarm Method for Optimization of ATIG Welding Process to Joint Mild Steel to 316L Stainless Steel

Kamel Touileb [1], Rachid Djoudjou [1,*], Abousoufiane Ouis [1], Abdeljlil Chihaoui Hedhibi [2], Sahbi Boubaker [3] and Mohamed M. Z. Ahmed [1]

[1] Department of Mechanical Engineering, College of Engineering in Al-Kharj, Prince Sattam bin Abdulaziz University, P.O. Box 655, Al-Kharj 16273, Saudi Arabia; k.touileb@psau.edu.sa (K.T.); a.ouis@psau.edu.sa (A.O.); moh.ahmed@psau.edu.sa (M.M.Z.A.)
[2] Laboratory of Mechanics of Sousse (LMS), National Engineering School of Sousse, Erriadh City, Sousse P.O. Box 264, Tunisia; achedhibi@isetso.mu.to
[3] Department of computer and Network Engineering, College of Computer Science and Engineering, University of Jeddah, Jeddah 21959, Saudi Arabia; sboubaker@uj.edu.sa
* Correspondence: r.djoudjou@psau.edu.sa

Abstract: 316L stainless steel joined to mild steel is widespread in several applications to reach a requested good association of mechanical properties at a lower cost. The activating tungsten inert gas (ATIG) weld was carried out using a modified flux composed of 76.63% SiO_2 + 13.37% Cr_2O_3 + 10% NaF to meet standard recommendations in terms of limiting the root penetration. Modified optimal flux gave a depth of penetration 1.84 times greater than that of conventional tungsten inert gas (TIG) welds and a root penetration of up to 0.8 mm. The microstructure of the dissimilar joints was investigated using a scanning electron microscope and EDS analysis. The mechanical properties of the weld were not affected by the modified flux. The results show that the energy absorbed in the fusion zone in the case of ATIG weld (239 J/cm^2) is greater than that of TIG weld (216 J/cm^2). It was found that the weld bead obtained with the optimal flux combination in ATIG welding can better withstand sudden loads. The obtained UTS value (377 MPa) for ATIG welding was close to that of TIG welding (376 MPa). The average Vickers hardness readings for ATIG welds in the fusion zone are up to 277 HV, compared to 252 HV for conventional TIG welding.

Keywords: particle swarm method; ATIG weld; NaF flux; mechanical properties

1. Introduction

Stainless steel is a widespread material used in various industries such as automotive, construction, rolling, and chemical processing. Tungsten inert gas (TIG) is largely used for joining parts. TIG welding is one of the most widely used welding processes for stainless steel due to its excellent quality and sound weld [1,2]. However, the main drawback of TIG welding is its limitation to weld thicknesses less than 3 mm. Thus, it is highly required of the welders to use the filler rod, edge preparation, and multi-passes, and it is synonymous with less productivity and an increase in the product cost [3]. Activating tungsten inert gas (ATIG) welding is a genius and alternative technique allowing to achieve a full penetrated weld with only one pass, square edges, and without filler rod [4–6]. In the ATIG technique, the used equipment and welding condition parameters are the same as those of the TIG process, except that, prior to the welding process, a flux is deposited. This flux, in the form of a paste, is applied to the joints that will be welded by a brush, sprayed, or conveyed to the pieces to be joined. A shielding gas covers the coating with a density range between 5 and 6 mg/cm^2 [7].

Three main mechanisms have been proposed to explain the phenomenon occurring in the ATIG weld pool. The first mechanism, proposed by Heiple et al. [8], says that the surfactant elements such as O, Se, S, and Te present in the weld pool contribute to reversing the Marangoni circulation of molten metal, leading to inward convection and hence a deeper weld bead. The second mechanism, proposed by Howse et al. [9], suggests that the weld depth is ascribed to an arc constriction linked to the migration of elements with high electronegativity, such as halides such as fluorine and elements such as oxygen. These elements react with the outer arc electrons, leading to a constriction of the weld arc. The latter enhances the current density at the anode arc root, enabling deeper penetration in weld metal compared to that of TIG welding. The third mechanism stipulates that the contraction of the arc is ascribed to the insulation effect of the powders used, particularly silicon dioxide [10]. By means of the insulating effect of the high electrical resistivity of flux, the anode spot at the workpiece diminishes, and the heat density at this region increases.

Compared to popular TIG welding, ATIG welding can significantly enhance the welding efficiency and reduce welding costs without altering the mechanical properties or corrosion resistance under the same welding conditions [11–13]. Several dissimilar materials have been welded by ATIG welding, such as carbon steel, stainless steel, and light materials such as aluminum, magnesium alloys, titanium alloys, etc. The effects of single-component activating fluxes on the morphology and weld mechanical properties of ATIG welding were extensively investigated [14–16]. Austenitic stainless steel metals are considerably used in many industries, such as chemical processing, aerospace oil and petrol plants, pharmaceutical manufacturing, and food processing. The weldability of these steels is usually very good. Austenitic stainless steel 304 grade is extensively used in day-to-day life applications [17,18]. Mild steels, well known as low-carbon steel, contain mainly iron, ranging from 0.05 to 0.2% carbon (C) and 0.60 to 0.90% manganese (Mn), and are also mostly used in various structural applications for their strength, weldability, and formability properties [19]. Patel et al. [20] reported in their study on 316LN that the maximum depth of penetration was obtained with the use of flux Co_3O_4 and TiO_2. The authors noticed that the enhancement in depth of penetration is attributed to the reverse Marangoni effect and arc constriction. Meanwhile, dissimilar welding is becoming a more common technique in order to meet the requirements of many industries [21–23]. However, achieving efficient welding posed a major challenge compared to welding similar materials due to differences in the thermo-mechanical and chemical properties of the materials to be joined under common welding conditions [24–26].

Dissimilar welding of stainless steel and low carbon steel is a competitive alternative widely used in power plants, oil industries, and infrastructure structure building owing to its combination of structural performance and cost-efficient designs [27–29]. Sanjay et al. [30] investigated the effect of current, welding speed, joint gap, and electrode diameter on dissimilar ATIG welds between carbon steel (CS) and stainless steel (SS). They tested the mono-oxide flux, such as TiO_2, ZnO, and MnO_2. They observed the benefits of oxides in increasing penetration without altering the mechanical properties of the joint. They noticed the migration of carbon from CS to SS. On the other hand, many optimization methods were used to optimize the welding parameters to enhance the mechanical properties of the joint including the Taguchi approach [31,32], the Response Surface Method [33,34], the Jaya Algorithm Method [35], and even the Particle Swarm Method for Optimization (PSMO) [36]. Some other works were interested in improving weld bead geometry by using the Support Vector Machine (SVM) method [37] or the Artificial Neural Network (ANN) method [38]. Several works [39,40] were dedicated to investigating the effect of single powder oxide on the morphology, mechanical properties, and corrosion resistance of the weld bead. Other studies focused on optimizing the proportions of powders in the flux to be deposited prior to welding. Mixing method design is the main method applied when two or more mixed powders are used [41]. The flux can be a binary flux composed of two different types of powders [42,43] or a tri-component flux [44,45].

This study uses PSMO for dissimilar tri-component flux welding in order to optimize the best flux composition and achieve a penetrated sound weld without altering the mechanical properties of the joint consisting of 316L stainless steel (316L) and mild steel (MS). The weld line carried out with the ATIG technique is compared to another one performed with the conventional TIG process in terms of bead morphology, SEM-EDS analysis, and mechanical properties. The obtained results will enrich the database relating to fusion welding and will help manufacturers and researchers develop innovative methods and accomplish performant welds at a low cost. The present study demonstrates that the novel idea to associate oxides and fluorides ensures a sound weld that meets the standards recommendations.

2. Materials and Methods

2.1. Materials

The materials to be joined are austenitic stainless steel (316L grade) and mild steel. Table 1 shows the chemical compositions of both alloys.

Table 1. Chemical composition of 316L stainless steel and mild steel (weight %).

Elements	C	Mn	Si	P	S	Cr	Ni	Mo	N	Cu	Al	Fe
316L	0.026	1.47	0.42	0.034	0.0016	16.60	10.08	2.14	0.044	0.50	-	Balance
Mild steel	0.0521	0.165	0.009	0.0098	0.0137	0.0234	0.0277	0.00647	0.0053	0.0985	0.0245	Balance

2.2. Welding Procedure

Experiments consist of welding a line of about 20 cm on a rectangular plate of 6 mm thickness. Prior to welding, the plates were cleaned using acetone, and the powders were heated in a furnace at 100 °C for a period of one hour to eliminate humidity. After that, the powders were mixed with acetone in a (1 ÷ 1) ratio, and a layer of paste of about 0.3–0.4 mm was applied with a brush to the edges of the plates up to 10 mm wide, as shown in Figure 1. The joints were accomplished with a square butt weld design without any edge preparation on clamped plates with zero clearance distance. The deposited used powders and their characteristics are listed in Table 2, where the oxygen percentage in oxides was determined by XPS tool analysis. Table 3 represents the welding parameters after several trials to adjust the operational welding parameters.

Table 2. Powders, melting, and evaporation temperatures.

Powders	Melting Temperature (°C)	Evaporation Temperature (°C)	Free Enthalpy of Formation $|\Delta_d H°|$ (kJ/mol)	Oxygen Percentages in Oxide %
SiO_2	1722	2950	902	68.35
TiO_2	1830	2972	941	63.26
Fe_2O_3	1540	1987	826	68.35
Cr_2O_3	2435	300	1128	61.71
V_2O_5	681	1750	1550.6	49.05
MoO_3	802	1155	745.1	67.73
NaF	933	1704	573.6	-

Table 3. Welding parameters.

Parameters	Range
Welding speed	150 mm/min
Welding current	150 A
Arc Length	2 mm
Electrode tip angle	45°
Shielding gas on the workpiece	Argon with flow rate 8 L/min
Shielding gas on the backside	Argon with flow rate 4 L/min
Welding mode	Negative direct current electrode

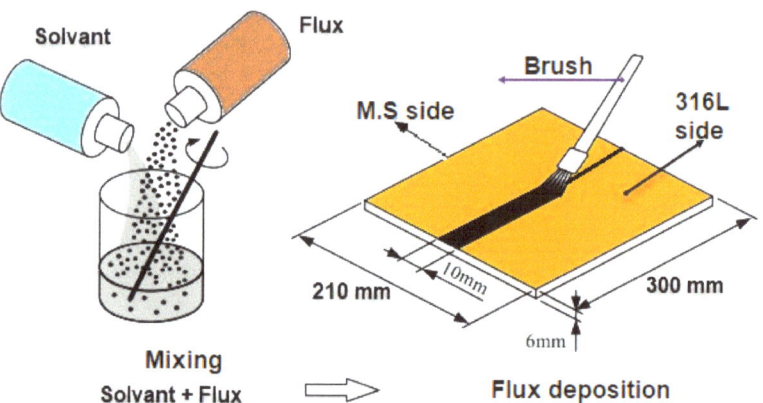

Figure 1. Mixing and deposition of flux on the workpiece.

After welding, the samples were cut far from the welding starting point to be sure that the arc welding was stabilized, as shown in Figure 2.

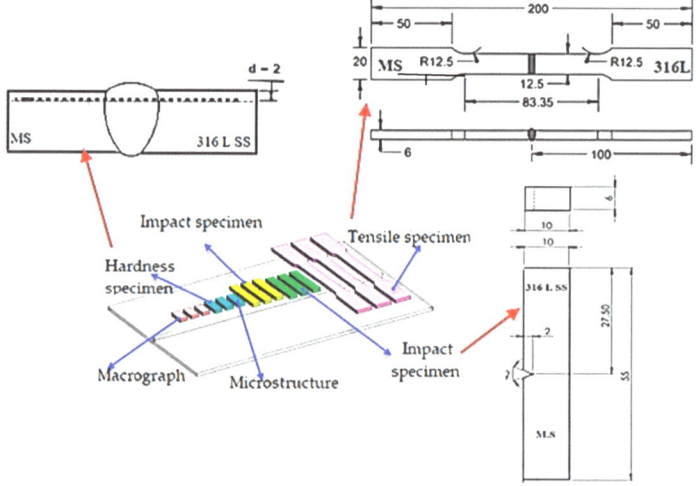

Figure 2. Locations of the test specimens taken for the various characterization techniques carried out in this study (units in mm).

The tensile tests were performed with a computerized universal testing machine. The samples were prepared in accordance with ASTM E8M-04 and shown in Figure 2. The tests were carried out on 3 samples for each TIG and ATIG welded using the modified optimal flux. Vickers hardness tests were carried out according to ASTM E-384-99. Figure 2 shows the hardness reading position and tracks micro-indentation. Impact tests were carried out using the Charpy "V" notch impact testing machine on 3 samples for each TIG and ATIG weld according to ASTM E23.

2.3. Design of the Experiment Methodology

The design of the experiment will be applied to the depth and ratio as shown in Figure 3. The optimal flux obtained from the highest depth penetration and weld aspect ratio resulting from macrograph analysis will be used to investigate the mechanical properties. Note that W_f is the face weld width, W_b is the back weld width (root width), D is the depth weld penetration till the plate thickness, b is the excess depth penetration, and R is the weld aspect ratio.

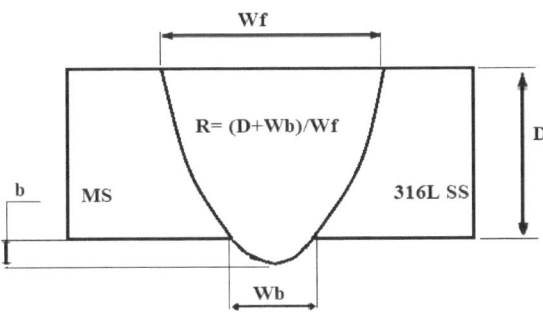

Figure 3. Ratio expression in a full penetration weld bead.

The fusion zone microstructure of both TIG and ATIG welds was analyzed. Micrographs, chemical composition, and elemental distribution (mapping) were characterized by a field-emission scanning electron microscope. To avoid sample charging under the electron beam, samples were coated with a layer of platinum with a 25 nm thickness for 35 s. An accelerating voltage of 20 kV was used, with a working distance of 8 mm for the best signal intensity.

A mathematical model was developed, where D (depth penetration) is written in terms of selected oxides' percentages, as will be explained in the discussion. In the second step, the optimal combination that maximizes D is determined. In the third step, an equation relating D to the proportions of the selected oxides was developed using the particle swarm optimization (PSO) method. Finally, the Matlab R2020 software module(Mathworks, Natick, Massachusetts · USA) was used to obtain the optimal combination of oxides that permits maximizing the depth of weld by a constrained optimization algorithm. Using the optimal flux obtained by the mathematical model, ATIG weld lines were performed to join mild steel to 316L.

3. Results and Discussions

3.1. Weld Bead Aspect

3.1.1. Selection of Appropriate Fluxes

Six oxide fluxes were used for dissimilar ATIG welding. The weld lines were performed using a square butt join design. Table 4 shows that the sample welded with the flux SiO_2 has the highest values of depth (5.96 mm) and ratio (0.49), followed by the samples welded with the fluxes Fe_2O_3 and Cr_2O_3. Based on these results, the oxides SiO_2, Fe_2O_3, and Cr_2O_3 were selected.

Table 4. Weld aspect of a single oxide flux of dissimilar ATIG welds.

Oxides	SiO_2	TiO_2	Fe_2O_3	Cr_2O_3	V_2O_5	MoO_3
Depth (mm)	5.96	3.28	4.59	4.33	3.26	3.95
Width (mm)	12.22	11.28	11.40	11.08	11.54	11.16
Ratio	0.49	0.29	0.40	0.39	0.28	0.35

3.1.2. Mathematical Modeling

In this study, the three chosen fluxes (Fe_2O_3, Cr_2O_3, and SiO_2) with different percentages, varying within a range of 100%, were mixed to form a ternary flux. In Table 5, depth penetration (D) is expressed as a function of stream mixes as follows: ($D = f(X_1, X_2, X_3)$), where X_1, X_2, and X_3 indicate, respectively, the proportions of the fluxes Fe_2O_3, Cr_2O_3, and SiO_2 (in terms of percentages, forming a variable mix of 100%).

Table 5. Actual (experimental) depth vs. predicted depth (in mm) for the different ternary mixtures.

Order	X_1 (en%)	X_2 (en%)	X_3 (en%)	D Actual	D Predicted
1	75	25	0	6.4625	6.2868
2	75	0	25	6.4575	6.0203
3	0	75	25	6.2875	6.0868
4	25	75	0	6.2375	6.0402
5	25	0	75	6.7733	7.3305
6	0	25	75	8.2467	7.6437
7	0	50	50	6.7867	7.1309
8	50	0	50	7.1200	6.7955
9	50	50	0	6.1275	6.632
10	50	25	25	7.5467	6.9794
11	25	50	25	7.4600	7.0016
12	25	25	50	7.4700	7.4317
13	33.33	33.33	33.33	6.4125	7.2325
14	66.67	16.67	16.67	6.7033	6.5238
15	16.67	66.67	16.67	6.6600	6.4711
16	16.67	16.67	66.67	6.6900	7.5636
17	100	0	0	4.5850	5.0048
18	0	100	0	4.3250	4.5114
19	0	0	100	7.9600	7.6253

The selected mathematical model that depicts the effect of mixing flux on the depth penetration (depth) of the weld bead is the coupling of a second-order model as proposed in [46] with classical linear regression as in [47]. This mathematical model involves three components: (i) the linear effect of the proportions, (ii) their quadratic effects, and (iii) the interactions between these proportions. The starting model formulation is given in Equation (1):

$$D = \alpha_1(X_1) + \alpha_2(X_2) + \alpha_3(X_3) + \alpha_4(X_1)^2 + \alpha_5(X_2)^2 + \alpha_6(X_3)^2 + \alpha_7(X_1)(X_2) + \alpha_8(X_1)(X_3) + \alpha_9(X_2)(X_3) \quad (1)$$

where: X_1 = % Fe_2O_3, X_2 = % Cr_2O_3, X_3 = % SiO_2.

3.1.3. First Step of the Modeling Process

The first step of the modeling process is to find out the optimal parameters that minimize the squared error between actual D (experimental) and predicted D using the proposed model.

Therefore, the modeling problem is transformed into an optimization problem comprising decision variables and quadratic error into an objective function to be minimized.

Knowing that the problem can have irregularities such as the non-convexity of the criterion as well as numerous feasibility constraints, the meta-heuristic method known as particle swarm optimization (PSO) was used. A set of coefficients was randomly initialized within the bounds of the search space to solve the problem in PSO, and gradually a suboptimal solution that combines three components was found. Three components consist of (i) tracking their speeds, (ii) returning to their best positions, and (iii) going to the position of the best neighbor. Four vectors were assigned to the swarm as follows:

- The position;
- A speed;
- The best personal position;
- The best common position.

To optimize the process, the equations of motion at the kth iteration were developed as follows [48,49]:

$$V_{k+1}^i = w_k V_k^i + c_1 r_1 \left(P^i - \alpha^i\right) + c_2 r_2 \left(G^i - \alpha^i\right) \tag{2}$$

$$\alpha_{k+1}^i = \alpha_k^i + V_{k+1}^i \tag{3}$$

$$w_k = w_{max} - \frac{w_{max} - w_{min}}{k_{max}} \times k \tag{4}$$

As reported in the literature [48,49], the inertial weight decreases linearly from {0.9 to 0.4}. Where $c_1 = c_2 = 0.75$ are the cognitive and social factors, and r_1 and r_2 are two random numbers generated between {0 and 1}. The search limits of the model parameters are fixed, respectively, at −1 and +1 for the global stability of the model [48]. After a certain number of iterations, the optimization process will stop (k_{max} = 5000), and then a local search will be performed to obtain better solutions around the sub-optimal global solution obtained by the algorithm of PSO. In order to select model D, the optimization process will be executed several times. This model is expressed by X_1, X_2, and X_3 in Equation (5). To evaluate the efficiency of the model, three performance measures were used as follows [48]:

$$\begin{aligned} D = &\; 5.8818(\%Fe_2O_3) + 6.2719(\%Cr_2O_3) + 6.4349(\%SiO_2) \\ &-0.877(\%Fe_2O_3)^2 - 1.7605(\%Cr_2O_3)^2 + 1.1904(\%SiO_2)^2 \\ &+4.858(\%Fe_2O_3)(\%Cr_2O_3) + 2.2355(\%Fe_2O_3)(\%SiO_2) \\ &+3.6801(\%Cr_2O_3)(\%SiO_2) \end{aligned} \tag{5}$$

$$MAPE = \frac{100}{19} \sum_{t=1}^{19} \frac{|D(t) - D\,predicted(t)|}{D\,mean} \tag{6}$$

$$R^2 = 100 \times \left(1 - \frac{\frac{1}{19}\sum_{t=1}^{19}(D(t) - D\,predicted(t))^2}{\frac{1}{19}\sum_{t=1}^{19}(D(t) - D\,mean)^2}\right) \tag{7}$$

$$RMSE = \sqrt{\frac{1}{19}\sum_{t=1}^{19}(D(t) - D\,predicted(t))^2} \tag{8}$$

The precision of this model is measured by the calculation indicators mentioned above as follows: MAPE = 5.8674%, R^2 = 71.15% and RMSE = 0.448 mm.

3.1.4. Second-Step Modeling Process

The second step to consider in the optimization process is to determine the optimal percentages of flux to obtain the maximum D.

To reach this target, a constrained optimization algorithm (Matlab 2020 Optimization Toolkit, Mathworks, Natick, MA, USA) was used. This algorithm permits to obtain the optimal combination presented by Equation 9 which allows to predict the value of (D = 7.7012 mm) with a root mean square error (RSME = 0.448 mm).

$$Flux\;Optimal = 0\%Fe_2O_3 + 13.368\%Cr_2O_3 + 86.632\%SiO_2 \tag{9}$$

3.1.5. Validation Test

The optimal flux is tested in the validation step by comparing the weld beads obtained by ATIG and conventional TIG. Table 6 shows the depth bead aspect data for ATIG and

TIG. It can be seen clearly that for ATIG, the penetration depth D increased by 2.2 times and the aspect ratio $(D + W_B)/W_F$ was enhanced by 5 times. Moreover, the obtained depth of weld (8.2 mm) was higher than the predicted one (7.7 mm).

Table 6. Depth, widths, and ratios for TIG and ATIG.

Welds	D	W_F	W_B	$(D + W_B)/W_F$
TIG Weld [50]	3.7	12.02	0	0.31
ATIG Weld (optimal flux)	8.2	9.94	7.3	1.56

Figure 4 shows macrographs of TIG and ATIG weld beads. It is clearly seen that the ATIG weld bead is fully penetrated, unlike the TIG weld bead, which is not.

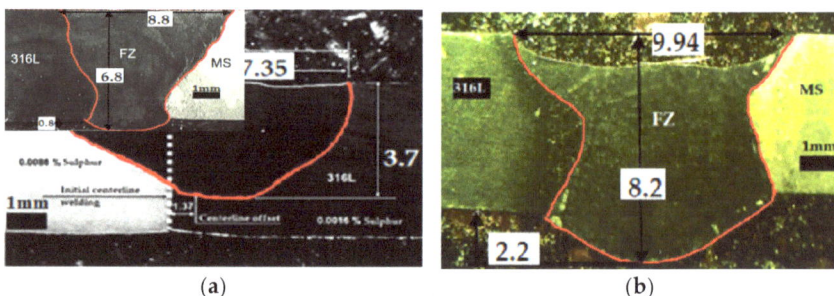

Figure 4. Dissimilar 316L-MS weld bead. (**a**) TIG process and (**b**) ATIG process with optimal flux. (All dimensions are in mm).

According to ISD 341-2 [51] Engineering Standards, excessive root penetration should be less than 25% of the nominal thickness of the base material of the thinnest component to be joined. In other words, the root penetration in our study should be less than 1.5 mm, knowing that the workpiece thickness is 6 mm. The full penetration of the ATIG weld depth, using the optimal flux, is 8.2 mm; consequently, the excess of penetration is 2.2 mm beyond the back side of the base material, which exceeds 1.5 mm, as shown in Figure 4.

A temporary copper support plate is usually used to support the molten weld metal (WM) and prevent excessive root penetration. Unfortunately, the latter can be a source of stress concentration in addition to the appearance of rust-promoting crevices and additional manufacturing costs. For these reasons, a support plate as a solution is discarded. To decrease the root weld penetration to an acceptable range to meet the requirements of BS EN ISO 15614-1:2017 industrials and standards [52], the optimal combination obtained was modified by adding 10% NaF. The modified optimal combination obtained is 13.368% Cr_2O_3 + 76.632% SiO_2 + 10% NaF. The addition of NaF to the optimal flux is aimed to constrict the arc, leading to an increase in the arc density and temperature, as reported by several studies [53,54].

Another weld line using modified optimal flux was carried out and compared to the welds performed with unmodified optimal flux and with conventional TIG welding.

Figure 5 shows the benefit of adding 10% NaF at the optimal flux to reduce root penetration to an acceptable range without weld bed collapse. The upper surface of the modified flux weld is almost flat and at the same level as the materials to be joined, as shown in Figure 5b.

The results depicted in Table 7 show that the modified optimal flux meets industrial requirements as long as penetration is 6.8 mm with an excess penetration of 0.8 mm. Ion fluoride is released from NaF and is characterized by a relatively low melting point (682 °C) and a low free enthalpy of formation (573.6 kJ/mol). The migration of fluoride ions into the arc react with the outer free electrons, leading to the constriction of the arc as mentioned above. In a further study of this work, we will focus on the microstructure and mechanical properties in comparison between double-sided TIG and the modified optimal flux. Also, the fluoride ions migration in the arc welding decreases the anode spot and contributes to increasing the energy density of both the heat source and the electromagnetic force in the weld pool. Consequently, the weld morphology is relatively narrow and deep [55].

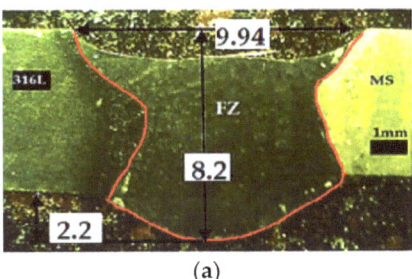
(a)

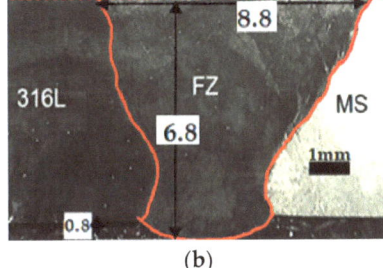
(b)

Figure 5. Micrographs of ATIG weld beads with optimal flux (**a**) and welds carried out with modified optimal flux (**b**).

Table 7. Morphology of TIG, ATIG with optimal flux, and weld with modified optimal flux.

Welds	D	W_F	W_B	$(D + W_B)/W_F$
Conventional TIG Weld [50]	3.7	12.02	0	0.31
ATIG Weld (optimal flux)	8.2	9.9	7.3	1.56
ATIG Weld (with modified optimal flux)	6.8	8.8	3.5	1.17

3.2. SEM-EDS investigation

3.2.1. ATIG Weld

In Table 8, the chemical composition obtained for dissimilar ATIG butt joints along a horizontal line from the MS side to the 316L side throughout the ATIG weld region is summarized. The weld chemistry composition analysis is performed using an EDS surface area scan from the MS side to the 316L side for the dissimilar ATIG weld presented in Table 9.

Table 8. EDS analyzes regions across the ATIG weldment from 316 L to MS base metal (weight %).

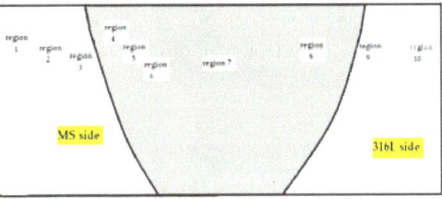

Elements	Reg.1	Reg.2	Reg.3	Reg.4	Reg.5	Reg.6	Reg.7	Reg.8	Reg.9	Reg.10
C	2.51	2.30	2.17	2.62	2.44	2.45	0	0	0	0
Cr	0	0	0	4.49	9.81	11.25	10.75	10.56	18.82	18.63
Fe	97.49	97.70	97.83	90.73	83.29	81.82	84.42	83.40	70.82	71.21
Ni	0	0	0	2.16	4.46	4.48	4.82	4.92	8.29	8.13

Table 9. EDS analyses across the ATIG weldment from 316L to MS base metal.

Regions	EDS Spectrum for Dissimilar 316-MS ATIG Weld	
Inside MS region 1	Element: C K 2.51; Fe K 97.49	Electron image (MS side)
Inside MS region 2	Element: C K 2.30; Fe K 97.70	Electron image
MS/WZ border (MS side) region 3	Element: C K 2.17; Fe K 97.83	Electron image (MS side / WZ side)
MS/WZ border (WZ side) region 4	Element: C K 2.62; Cr K 4.49; Fe K 90.73; Ni K 2.16	Electron image (MS side / WZ side)
MS/WZ border (WZ side) Region 5	Element: C K 2.44; Cr K 9.81; Fe K 83.28; Ni K 4.46	Electron image
MS/WZ border (WZ side) region 6	Element: C K 2.45; Cr K 11.25; Fe K 81.82; Ni K 4.48	Electron image

Table 9. Cont.

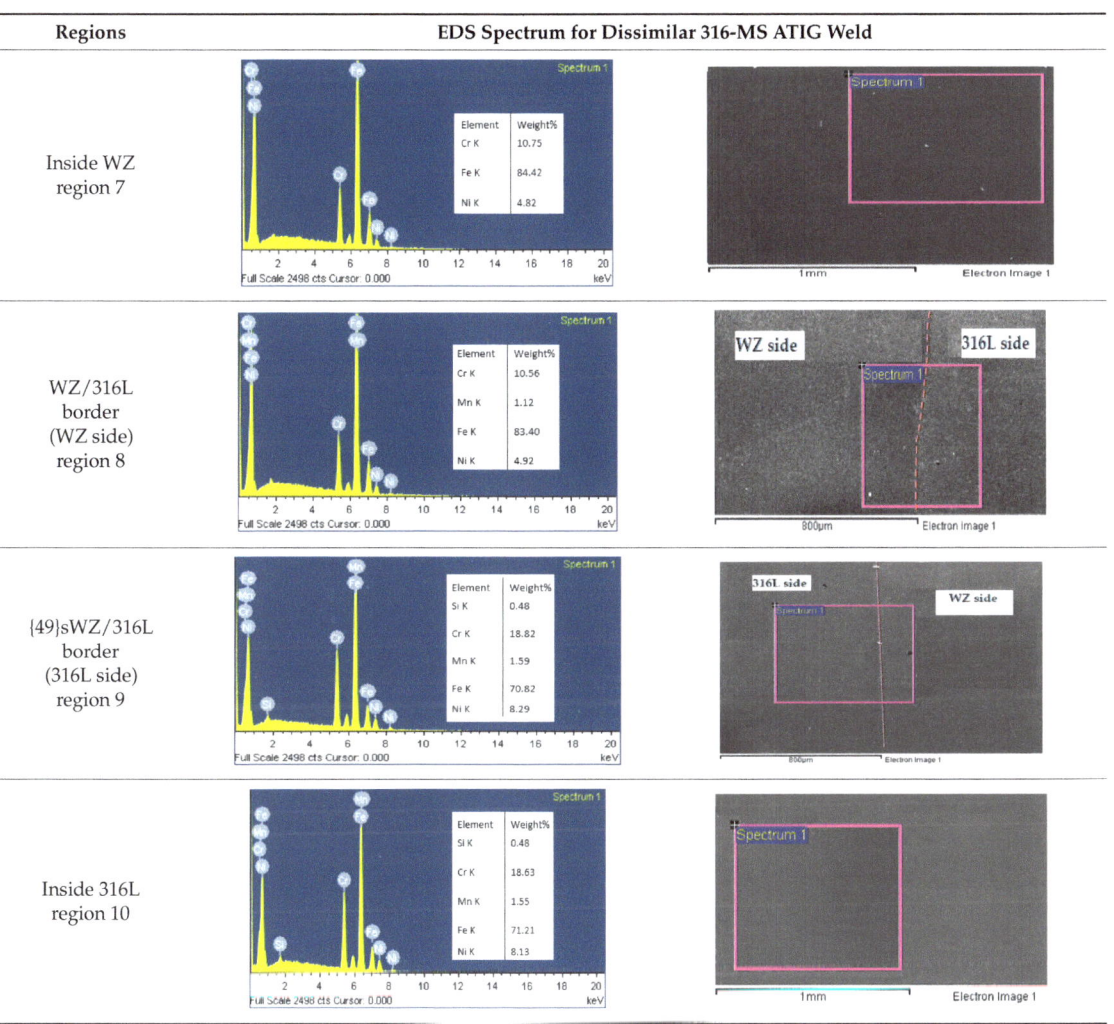

The analysis reveals a decrease in chromium and nickel content from the 316L side to the M.S. side throughout the weld zone. On the other hand, we notice an increase in iron content in the same direction.

Close to the boundary located between the MS side and the ATIG weld zone, the weight percentages of carbon are low, as revealed by Table 8 at region 3 (2.17%). In this region, the depletion of carbon is suspect. The percentage of carbon at the weld zone border is up to 2.62%. The migration of carbon from MS to the weld fusion zone is altered due to the fast cooling rate characterized by the ATIG weld.

Figure 6 depicts the mapping of the principal chemistry elements available at the MS base metal and the weld zone (WZ) border. The mapping of C content across the boundary fusion line shows a gradient of C elements, that is marked in favor of the MS region. However, the density of elements such as Cr, Ni, and Mn is more pronounced in WZ, and Fe content decreases from MS to WZ.

At the ATIG 316L-WZ border, the diffusion of elements such as Cr, Ni, and Mn from 316 L SS to WZ is pronounced. However, the migration of carbon is not obvious, as shown in Figures 6 and 7.

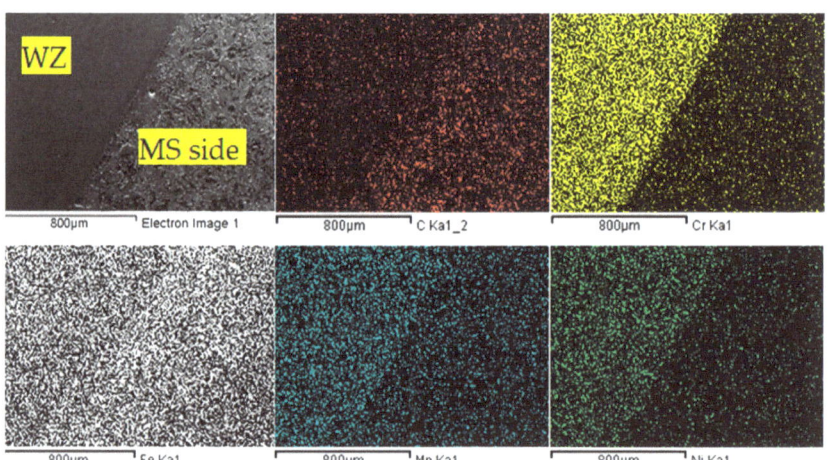

Figure 6. EDS map scan showing variation of alloying elements C, Cr, Fe, Mn, and Ni across the ATIG dissimilar weld of MS base metal and weld zone border.

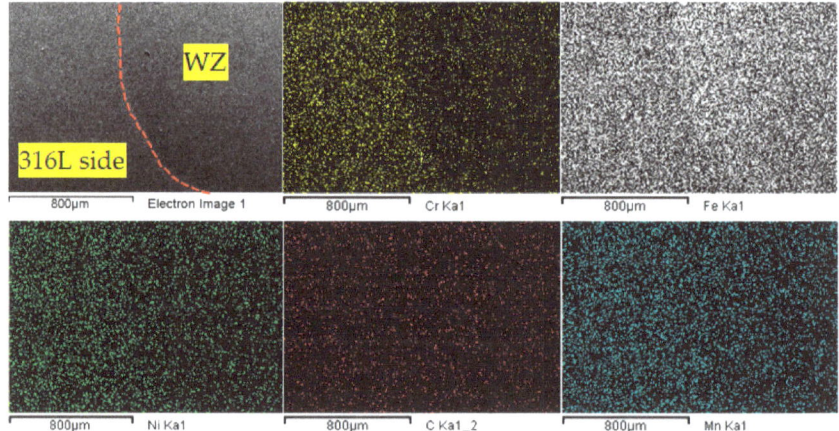

Figure 7. EDS map scan showing variation of alloying elements C, Cr, Fe, Mn, and Ni across the ATIG dissimilar weld of 316L and WZ borders.

3.2.2. TIG Weld

The results collected in Table 10 are extracted from the Table 11 and tell us about the evolution of the content of principal elements such as C, Cr, Fe, and Ni throughout the TIG dissimilar weld.

Table 10. EDS analyzes regions across the ATIG weldment from 316L to MS base metal (weight %).

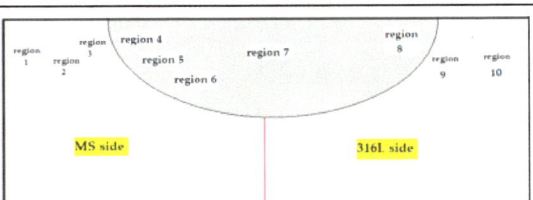

Elements	Reg.1	Reg.2	Reg.3	Reg.4	Reg.5	Reg.6	Reg.7	Reg.8	Reg.9	Reg.10
C	5.03	4.99	1.82	4.67	3.58	1.92	0	0	0	0
Cr	0	0	0	4.98	4.86	14.41	11.47	16.09	18.55	18.50
Fe	94.97	95.01	98.18	87.46	89.51	76.01	82.69	75.08	70.89	71.17
Ni	0	0	0	2.89	2.06	6.41	4.83	6.53	8.06	8.30

Table 11. EDS analyses across the TIG weldment from 316 L to MS base metal.

Regions	EDS Spectrum for Dissimilar 316-MS Weld	
Inside MS region 1	Spectrum 1: C K 5.03, Fe K 94.97	MS side
Inside MS region 2	Spectrum 1: C K 4.99, Fe K 95.01	WZ side / MS side
MS/WZ border (MS side) region 3	Spectrum 1: C K 1.82, Fe K 98.18	MS side / WZ side
MS/WZ border (WZ side) region 4	Spectrum 1: C K 4.67, Cr K 4.98, Fe K 87.46, Ni K 2.89	WZ side / MS side

Table 11. *Cont.*

Regions	EDS Spectrum for Dissimilar 316-MS Weld	
MS/WZ border (WZ side) Region 5	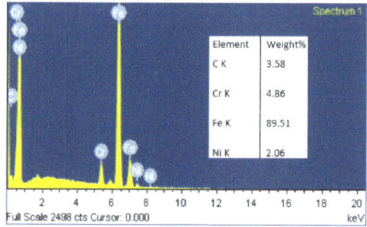	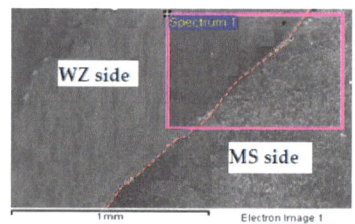
MS/WZ border (WZ side) region 6	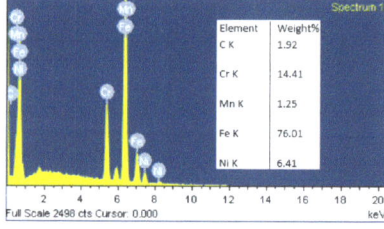	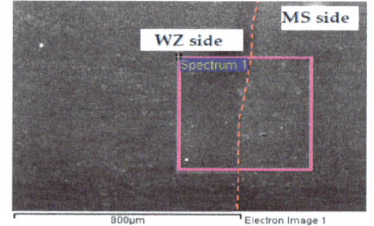
Inside WZ region 7	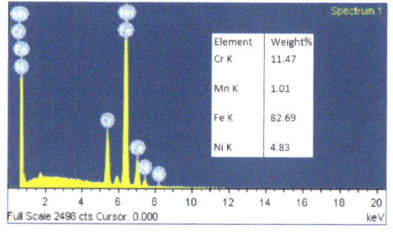	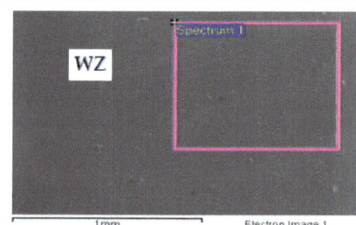
WZ/316L border (WZ side) region 8	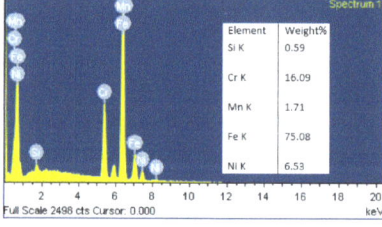	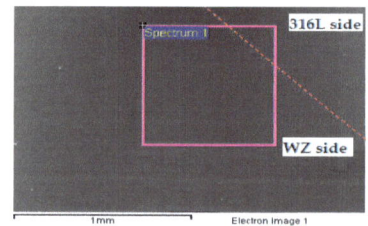
WZ/316L border (316L side) region 9	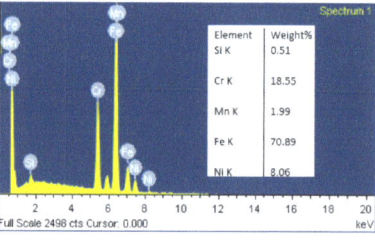	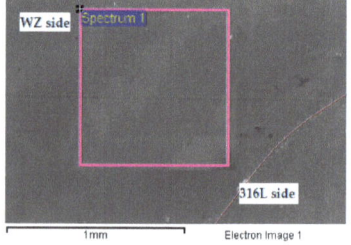

Table 11. *Cont.*

Regions	EDS Spectrum for Dissimilar 316-MS Weld	
Inside 316L region 10	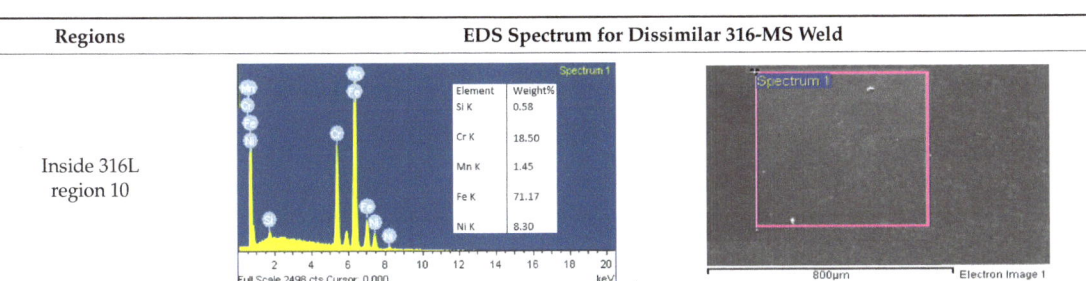	

Element	Weight%
Si K	0.58
Cr K	18.50
Mn K	1.45
Fe K	71.17
Ni K	8.30

The weight percentages of carbon elements revealed by EDS scan analysis at WZ border region 4 are up to 4.67% against 1.82% at region 3 at the closest location to the boundary MS/WZ. The carbon content in region 1, far from weld solidification, is up to 5.03 % and steeply decreases to 1.82% in region 3, near the border line, which reveals the migration of the carbon element towards the weld zone. The carbon migration from MS base metal to WZ leads to a carbon depletion zone at the heat-affected zone (HAZ)-MS side. On the other hand, the WZ zone bordering the MS side is enriched by carbon, which favors the formation of carbides, as revealed by Wenyong et al. [56]. Chromium and nickel contents decrease from the 316L side to the MS side throughout the weld zone. We also notice an increase in iron content in the same direction.

The mapping scan depicted in Figure 8 shows the decrease in Cr, Ni, and Mn content, whereas Fe content gradually increases from WZ to MS base metal across the WZ-MS interface. We can remark that the chromium content gained a gradual decrease across the WM-MS interface, while the ferrous content had no notable change. Regarding the carbon content, the scan map shows a decarburized layer in MS near the fusion line and a carburized layer in WM. This can be explained by the carbon diffusion from MS to WM, and the same phenomenon was reported in some studies [30,50].

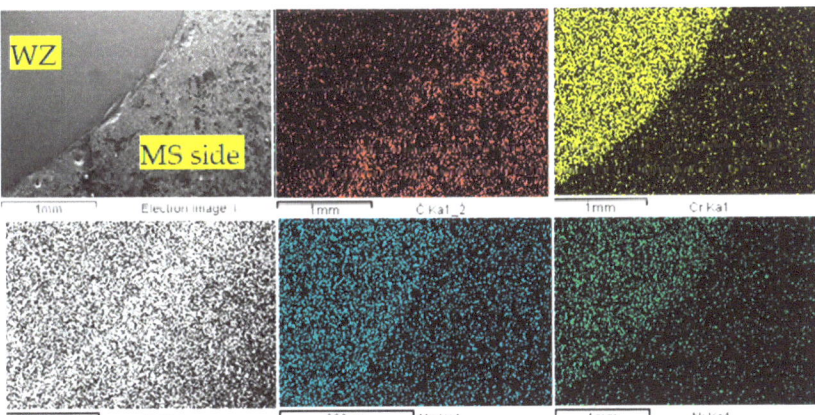

Figure 8. EDS map scan showing variation of alloying elements C, Cr, Fe, Mn, and Ni across the TIG dissimilar weld of MS base metal and weld zone border.

Figure 9 obviously shows a gradient in Cr, Ni, and Fe contents across the border 316L-WZ; an increase in Cr and Ni contents and a decrease in Fe content from 316L to WZ. However, the migration of carbon apparently has not occurred.

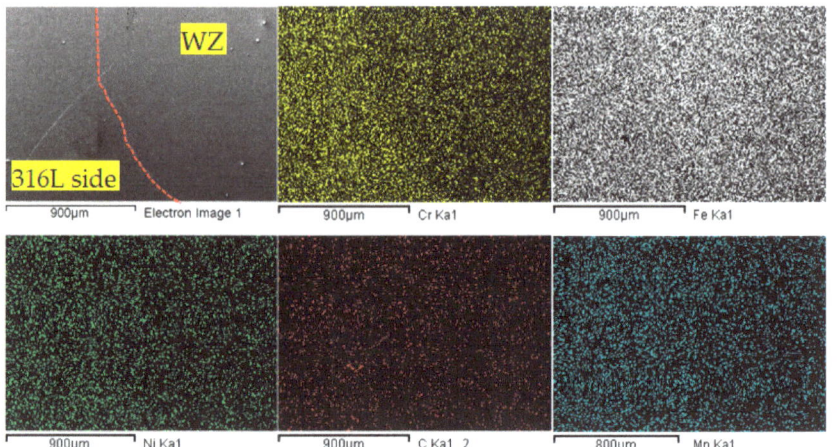

Figure 9. EDS map scan showing variation of alloying elements C, Cr, Fe, Mn, and Ni across the TIG dissimilar weld of 316L and WZ border.

3.3. Tensile Test

Table 12 shows that the average value of UTS for ATIG dissimilar welds is 377 MPa, which is almost equal to that of conventional TIG welding (376 MPa).

Table 12. Measurements of tensile strength and standard deviations of TIG and ATIG (modified optimal flux).

Sample	Number of Tests	UTS Max. (MPa)	UTS Min. (MPa)	UTS Average (MPa)	Standard Deviations σ
As received MS base metal	3	364	361	362	2.02
As received 316L base metal	3	626	623	624	1.4
TIG MS/316L [50]	3	380	372	376	4.00
ATIG MS/316L	3	382	370	377	4.04

We remark that the fracture for both TIG and ATIG dissimilar welds occurs on the mild steel side, far from the weld joint, which attests that this region is the weakest location in comparison to the entire specimen, as shown in Figure 10. This phenomenon shows, firstly, that the MS base metal is weaker than the weld metal and, secondly, that the weld joint quality is good. We also notice that the necking for ATIG is more pronounced than that of TIG welds, which confirms the enhanced ability of plastic deformation until fracture occurs.

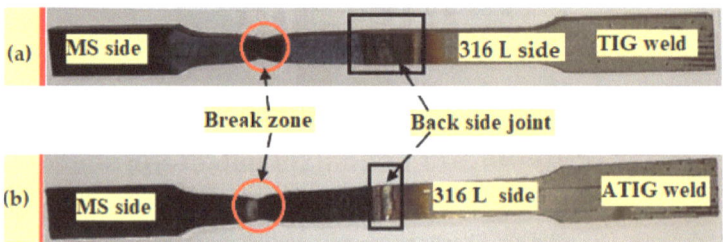

Figure 10. Break zones for TIG weld (**a**) and ATIG dissimilar weld (**b**) after tensile test.

3.4. Hardness Test

Figure 11 shows the variations of Vickers micro-hardness as compared to the distance from MS to 316 L SS. Table 13 shows the hardness measurements and standard deviations of ATIG and TIG at FZ. The hardness of TIG and ATIG weld regions is higher than that of both 316L and MS base metals, as reported in Figure 11.

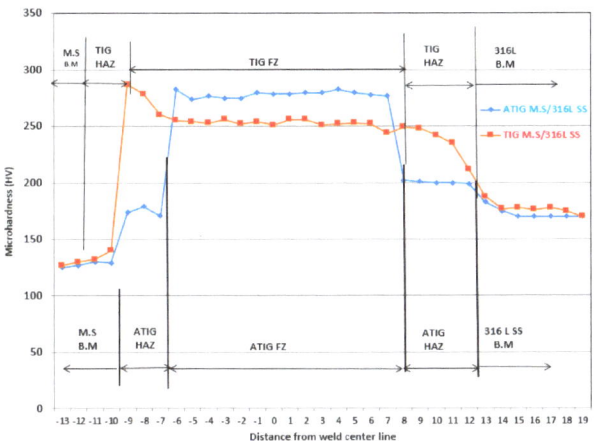

Figure 11. Microhardness profiles across the dissimilar TIG and ATIG welds.

Table 13. Measurements of hardness and standard deviations of TIG and ATIG (with modified optimal flux) at FZ.

Sample	Zone of Tests	HV Max.	HV Min.	HV Average	Standard Deviations σ
TIG [50]	FZ	287	235	252	12.75
ATIG	FZ	283	270	277	4.84

The highest hardness value in the TIG weld is situated in the weld zone, close to the fusion boundary beneath the mild steel side. This aspect is due to the formation of harder micro-constituents in this region by the migration of carbon into weld metal from the mild steel side, as previously shown in the SEM-EDS analyses as a decarbonized zone. On the other hand, there is no migration of carbon in the ATIG weld, which explains the homogenous values of hardness in the weld. The tendency of carbon diffusion from the MS side to the weld zone is due to the presence of elements like chromium in the weld zone and the high cooling rate characterized by ATIG weld decreases. The average hardness measurements in FZ are higher for ATIG than for TIG welds, which can also be explained by the rapid cooling rate in ATIG.

In the ATIG weld zone, as shown in Table 13, the obtained hardness values are uniform (hardness standard deviation is less than 5 HV), which means that there is a homogenization of hardness. Contrarily, in the TIG weld zone, the obtained hardness values are non-uniform (hardness standard deviation is more than 5 HV), as reported by Osoba et al. [57]. Moreover, we can observe a decrease in the hardness values from FZ to base metal in both MS and SS.

3.5. Impact Test

Table 14 presents the obtained results of the absorbed energy in impact tests and the standard deviations of TIG and ATIG at the fusion zone. We can remark that the average absorbed energy in ATIG weld (239 J/cm^2) is higher than that of TIG weld (216 J/cm^2) by 23 J/cm^2, which means that the ATIG welds withstand more sudden loads.

Table 14. Measurements of absorbed energy and standard deviations of TIG and ATIG (with optimal flux) at the fusion zone for dissimilar 316L/MS welds.

Sample	Number of Tests	Absorbed Energy (J/cm^2) Min	Absorbed Energy (J/cm^2) Max	Absorbed Energy (J/cm^2) Average	Standard Deviations σ
TIG—316L SS/MS [50]	3	215	238	216	15.56
ATIG—316L SS/MS	3	234	243	239	4.58

One set of fractured impact specimens obtained after V-notch testing at room temperature is illustrated in Figure 12.

Figure 12. The fractured impact specimen for ATIG weld (**a**) and for TIG weld (**b**).

Figure 13b shows the fractography of an ATIG impact test specimen. We can see small and deep dimples with spreading small voids, which indicate a fully ductile fracture mode. On the other hand, Figure 13a shows the morphology of the fractured surface of a TIG specimen. The fractography exhibits mixed fracture, which is composed of a large number of islets of fine dimples separated by the facies of quasi-cleavage and indicates less resistance to sudden impact loads.

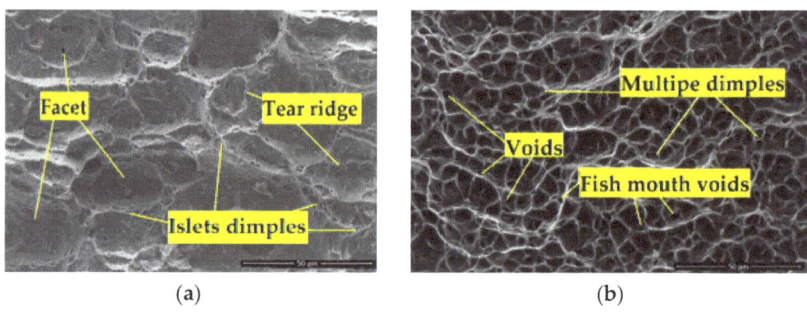

Figure 13. Fractography of TIG (**a**) and ATIG (**b**) impact tests for dissimilar 316L/MS welds (×2500).

4. Conclusions

PSMO was used to obtain the optimum combination of flux in order to weld dissimilar MS and austenitic 316L SS. The resultant optimal flux ensured full penetration of the weld bead with an excess root. A modification of optimal flux by adding fluorine (NaF fluorine) avoids excessive root penetration. The morphologies and mechanical properties of the weld beads were compared for ATIG with optimal flux and conventional TIG. The following main conclusions can be deduced:

- The results of this investigation show that 316L can be joined with MS using a tricomponent flux composed of 76.63% SiO_2, 13.37% Cr_2O_3, and 10% NaF.
- A fully penetrated bead in a single pass without edge preparation is achieved. The excess root penetration can reach 0.8 mm, which meets the requirements of industrial

standards. The obtained depth (D) is 6.8 mm, the bead face width (W_F) is 8.8 mm, and the back face width (W_B) is 3.5 mm, leading to an aspect ratio $(D + W_B)/WF$ of 1.17. Hence, compared to TIG welding, the depth and the ratio increased by 1.83 and 3.77 times, respectively.

- The fluoride in the form of ions present in the modified flux migrates to the welding arc and contributes to an arc constriction. Thus, a phenomenon of reduction of the weld bead is observed, leading to an increase in the penetration of the weld compared to conventional TIG. Moreover, the surfactant elements, such as oxygen liberated in the weld pool, contribute to the reversal of Marangoni convection, resulting in a fully penetrated weld.
- SEM-EDS analysis shows carbon depletion closest to the weld zone at the MS side border. The migration of carbon from MS to the weld zone is suspected to occur in cases of TIG weldment. However, this situation is very limited or inexistent in ATIG welding due to its high cooling rate.
- The tensile test reveals that the strength has almost the same value, and the fracture happens at MS base metal for both ATIG and TIG welding. This indicates that the welding zone is stronger than that of the MS parent metal.
- The hardness of ATIG welds exhibits homogenous values ($\approx$277 HV), and the average value is higher than that of TIG welds ($\approx$255 HV). The disparities in hardness between ATIG weld readings are negligible.
- The fractography in the impact test shows a fully ductile fracture with many fine dimples overall on the fractured surface in ATIG welding, compared to gatherings of dimples separated by quasi-cleavage facies in TIG welding. ATIG welding exhibits more resistance to sudden loads (239 J/cm^2) than TIG welding (216 J/cm^2).

Author Contributions: Conceptualization, K.T. and A.C.H.; methodology, K.T. and R.D.; software, K.T. and S.B.; validation, K.T., R.D., A.O. and A.C.H.; formal analysis, K.T.; investigation, K.T., R.D. and A.C.H.; resources, R.D.; data curation, K.T. and S.B.; writing—original draft preparation, K.T. and S.B.; writing—review and editing, K.T., R.D., A.O. and M.M.Z.A.; visualization, K.T.; supervision, K.T. All authors have read and agreed to the published version of the manuscript.

Funding: The authors extend their appreciation to the Deputyship for Research and Innovation, Ministry of Education in Saudi Arabia, for funding this research work through the project number (IF2/PSAU/2022/01/21943).

Data Availability Statement: The data used to support the findings of this study are included within the article.

Acknowledgments: The authors express their deep thanks and acknowledge the collaboration of Hany S. Abdo from the Center of Excellence for Research in Engineering Materials (CEREM), King Saud University, Saudi Arabia, for his help in performing EDS tests.

Conflicts of Interest: The authors declare no conflict of interest.

References

1. Dixit, P.; Suketu, J. Techniques to weld similar and dissimilar materials by ATIG welding—An overview. *Mater. Manuf. Process.* **2021**, *36*, 1–16.
2. Anbarasu, P.; Yokeswaran, R.; Godwin Antony, A.; Sivachandran, S. Investigation of filler material influence on hardness of TIG welded joints. *Mater. Today Proc.* **2020**, *21*, 964–967. [CrossRef]
3. Singh, A.K.; Dey, V.; Rai, R.N. Techniques to improve weld penetration in TIG welding (a review). *Mater. Today Proc.* **2017**, *4*, 1252–1259. [CrossRef]
4. Vijay, S.J.; Mohanasundaram, S.; Ramkumar, P.; Kim, H.G.; Tugirumubano, A.; Go, S.H. Experimental investigations on activated-TIG welding of Inconel 625 and AISI 304 alloys. In *Advances in Materials and Manufacturing Engineering, Proceedings of the ICAMME 2019, Bhubaneswar, India, 15–19 March 2019*; Springer: Singapore, 2020; pp. 311–317. [CrossRef]
5. Patel, D.; Jani, S. ATIG welding: A small step towards sustainable manufacturing. *Adv. Mater. Process. Technol.* **2021**, *7*, 514–536. [CrossRef]
6. Vidyarthy, R.S.; Sivateja, P. Influence of activating flux tungsten inert gas welding on mechanical and metallurgical properties of the mild steel. *Mater. Today Proc.* **2020**, *28*, 977–981. [CrossRef]

7. Chern, S.K.; Tseng, H.; Tsai, H.-L. Study of the characteristics of duplex stainless steel activated tungsten inert gas welds. *Mater. Des.* **2011**, *32*, 255–263. [CrossRef]
8. Heiple, C.R.; Roper, J.R. Mechanism for minor element effect on GTA fusion zone geometry. *Weld. J.* **1982**, *61*, 97–102.
9. Howse, D.S.; Lucas, W. An investigation into arc constriction by active fluxes for TIG (A-TIG). *Weld. Sci. Technol. Weld. Join.* **2000**, *5*, 189–193. [CrossRef]
10. Lowke, J.J.; Tanaka, M.; Ushio, M. Mechanisms giving increased weld depth due to a flux. *J. Phys. D Appl. Phys.* **2005**, *38*, 3438–3445. [CrossRef]
11. Kumar, M.S.; Sankarapandian, S.; Shanmugam, N.S. Investigations on mechanical properties and microstructural examination of activated TIG-welded Nuclear grade stainless steel. *J. Braz. Soc. Mech. Sci. Eng.* **2020**, *42*, 292. [CrossRef]
12. Singh, R.S.; Khanna, P. A-TIG (activated flux tungsten inert gas) welding:—A review. *Mater. Today Proc.* **2020**, *44*, 808–820. [CrossRef]
13. Baghel, A.; Sharma, C.; Rathee, S.; Srivastava, M. Activated flux TIG welding of dissimilar SS202 and SS304 alloys: Effect of oxide and chloride fluxes on microstructure and mechanical properties of joints. *Mater. Today Proc.* **2021**, *47*, 7189–7195. [CrossRef]
14. Niagaj, J. Influence of Activated Fluxes on the Bead Shape of A-TIG Welds on Carbon and Low-Alloy Steels in Comparison with Stainless Steel AISI 304L. *Metals* **2021**, *11*, 530. [CrossRef]
15. Bhanu, V.; Gupta, A.; Pandey, C. Role of A-TIG process in joining of martensitic and austenitic steels for ultra-supercritical power plants -a state of the art review. *Nucl. Eng. Technol.* **2022**, *54*, 2755–2770. [CrossRef]
16. Kumar, K.; Deheri, C.S.; Masanta, M. Effect of Activated Flux on TIG Welding of 304 Austenitic Stainless Steel. *Mater. Today Proc.* **2019**, *18*, 4792–4798. [CrossRef]
17. Prasadarao Pydi, H.; Prakash Pasupulla, A.; Vijayakumar, S.; Abebe Agisho, H. Study on microstructure, behavior and Al_2O_3 content flux A-TIG weldment of SS-316L steel. *Mater. Today Proc.* **2022**, *51*, 728–734. [CrossRef]
18. Paul, B.G.; Ramesh Kumar, K.C. Effect of single component and binary fluxes on the depth of penetration in a-TIG welding of Inconel alloy 800H austenitic stainless steel. *Int. J. Adv. Eng. Global Technol.* **2017**, *5*, 1791–1795.
19. Rajesh Kumar, G.; Chandra Sekhara Reddy, M. Stainless steel effect of flux activated TIG tungsten inert gas welding on weld bead morphology of austenitic stainless steel 304. *JETIR* **2019**, *6*, 433–439.
20. Patel, N.P.; Badheka, V.J.; Vora, J.J.; Upadhyay, G.H. Effect of oxide fluxes in activated TIG welding of stainless steel 316LN to low activation ferritic/martensitic steel (LAFM) dissimilar combination. *Trans. Indian Inst. Met.* **2019**, *72*, 2753–2761. [CrossRef]
21. Mishra, D.; Rajanikanth, K.; Shunmugasundaram, M.; Praveen Kumar, A.; Maneiah, D. Dissimilar resistance spot welding of mild steel and stainless steel metal sheets for optimum weld nugget size. *Mater. Today Proc.* **2021**, *46*, 913–924. [CrossRef]
22. Sharma, P.; Dwivedi, D.K. A-TIG welding of dissimilar P92 steel and 304H austenitic stainless steel: Mechanisms, microstructure and mechanical properties. *J. Manuf. Process.* **2019**, *44*, 166–178. [CrossRef]
23. Landowski, M.; Swierczynska, A.; Rogalski, G.; Fydrych, D. Autogenous fiber laser welding of 316L austenitic and 2304 lean duplex stainless steels. *Materials* **2020**, *13*, 2930. [CrossRef] [PubMed]
24. Khan, M.; Dewan, M.W.; Sarkar, M.Z. Effects of welding technique, filler metal and post-weld heat treatment on stainless steel and mild steel dissimilar welding joint. *J. Manuf. Process.* **2021**, *64*, 1307–1321. [CrossRef]
25. Chen, H.-C.; Ng, F.L.; Du, Z. Hybrid laser-TIG welding of dissimilar ferrous steels: 10 mm thick low carbon steel to 304 austenitic stainless steel. *J. Manuf. Process.* **2019**, *47*, 324–336. [CrossRef]
26. Pandey, C. Mechanical and metallurgical characterization of dissimilar P92/SS 304L welded joints under varying heat treatment regimes. *Metall. Mater. Trans. A Phys. Metall. Mater. Sci.* **2020**, *51*, 2126–2142. [CrossRef]
27. Maurya, A.K.; Kumar, N.; Chhibber, R.; Pandey, C. Study on microstructure mechanical integrity of the dissimilar gas tungsten arc weld joint of SSDs 2057/x-70 steels for marine applications. *J. Mater. Sci.* **2023**, *58*, 11392–11423. [CrossRef]
28. Kumar, A.; Pandey, S.M.; Bhattacharyya, A.; Fydrych, A.; Sirohi, S.; Pandey, C. Selection of electrode material for Inconel 617/P92 steel SMAW Dissimilar welds. *J. Press. Vessel Technol.* **2023**, *145*, 051503. [CrossRef]
29. Atul, B.; Kumar, A.; Jain, V.; Gupta, D. Enhancement of activated tungsten inert gas (A-TIG) welding using multi-component TiO_2-SiO_2-Al_2O_3 hybrid flux. *Measurement* **2019**, *148*, 106912.
30. Sanjay, G.; Nayee, A.; Vishvesh, J. Effect of oxide-based fluxes on mechanical and metallurgical properties of Dissimilar Activating Flux Assisted-Tungsten Inert Gas Welds. *J. Manuf. Process.* **2014**, *16*, 137–143.
31. Chaudhari, V.; Bodkhe, V.; Deokate, S.; Mali, B.; Mahale, R. Parametric optimization of TIG welding on SS 304 and MS using Taguchi approach. *Int. Res. J. Eng. Technol.* **2019**, *6*, 880–885.
32. Ramadan, N.; Boghdadi, A. Parametric Optimization of TIG Welding Influence on Tensile Strength of Dissimilar Metals SS-304 And Low Carbon Steel by Using Taguchi Approach. *Am. J. Eng. Res.* **2020**, *9*, 7–14.
33. Nagaraju, S.; Vasantharaja, P.; Chandrasekhar, N.; Vasudevan, M.; Jayakumar, T. Optimization of Welding Process Parameters for 9cr-1mo Steel Using RSM and GA. *Mater. Manuf. Process.* **2016**, *31*, 319–327. [CrossRef]
34. Ragavendran, M.; Chandrasekhar, N.; Ravikumar, R.; Saxena, R.; Vasudevan, M.; Bhaduri, A.K. Optimization of Hybrid Laser—TIG Welding of 316LN Steel Using Response Surface Methodology (RSM). *Opt. Lasers Eng.* **2017**, *94*, 27–36. [CrossRef]
35. Vora, J.J.; Abhishek, K.; Srinivasan, S. Attaining Optimized A-TIG Welding Parameters for Carbon Steels by Advanced Parameter-less Optimization Techniques: With Experimental Validation. *J. Braz. Soc. Mech. Sci. Eng.* **2019**, *41*, 261. [CrossRef]

36. Chihaoui Hedhibi, A.; Touileb, K.; Ouis, A.; Djoudjou, R.; Ahmed, M.Z. Mechanical Properties and Microstructure of TIG and ATIG Welded 316L Austenitic Stainless Steel with Multicomponent Flux Optimization using Mixing Design method and Particle Swarm Optimization (PSO). *Materials* **2021**, *14*, 7139. [CrossRef] [PubMed]
37. Kshirsagar, R.; Jones, S.; Lawrence, J.; Tabor, J. Prediction of Bead Geometry Using a Two-Stage SVM–ANN Algorithm for Automated Tungsten Inert Gas (TIG) Welds. *J. Manuf. Mater. Process.* **2019**, *3*, 39. [CrossRef]
38. Ran, L.; Manshu, D.; Hongming, G. Prediction of Bead Geometry with Changing Welding Speed Using Artificial Neural Network. *Materials* **2021**, *14*, 1494.
39. Kulkani, A.; Dwivedi, K.; Vasudevan, M. Effect of oxide fluxes on activated TIG welding of AISI 316L austenitic stainless steel. *Mater. Today Proc.* **2019**, *18*, 4695–4702. [CrossRef]
40. Vasudevan, M. Effect of A-TIG Welding Process on the Weld Attributes of Type 304LN and 316LN Stainless Steels. *J. Mater. Eng. Perform.* **2017**, *26*, 1325–1336. [CrossRef]
41. Touileb, K.; Ouis, A.; Djoudjou, R.; Chihaoui Hedhibi, A.; Alrobei, H.; Albaijan, I.; Alzahrani, B.; El-Sayed, M.; Hany, S. Effects of ATIG Welding on Weld Shape, Mechanical Properties and Corrosion Resistance of 430 Ferritic Stainless Steel Alloy. *Metals* **2020**, *10*, 404. [CrossRef]
42. Lin, H.-L.; Wu, T.-M. Effects of Activating Flux on Weld Bead Geometry of Inconel 718 Alloy TIG Welds. *Mater. Manuf. Process.* **2012**, *27*, 1457–1461. [CrossRef]
43. Albaijan, I.; Chihaoui Hedhibi, A.; Touileb, K.; Djoudjou, R.; Ouis, A.; Alrobei, H. Effect of binary oxide flux on ATIG 2205 duplex stainless steel welds. *Adv. Mater. Sci. Eng.* **2020**, *2020*, 5842741. [CrossRef]
44. Touileb, K.; Chihaoui Hedhibi, A.; Djoudjou, R.; Ouis, A.; Bensalama, A.; Albaijan, I.; Hany, S.; Mohamed, M.Z. Mechanical Microstructure, and Corrosion Characterization of Dissimilar Austenitic 316L and Duplex 2205 Stainless-Steel ATIG Welded Joints. *Materials* **2022**, *15*, 2470. [CrossRef] [PubMed]
45. Venkatesan, G.; George, J.; Sowmyasri, M.; Muthupandi, V. Effect of Ternary Fluxes on Depth of Penetration in A-TIG Welding of AISI 409 Ferritic Stainless Steel. *Procedia Mater. Sci.* **2014**, *5*, 2402–2410. [CrossRef]
46. Ahmed, A.N.; Noor, C.W.M.; Allawi, M.F.; El-Shafie, A. RBF-NN-based model for prediction of weld bead geometry in Shielded Metal Arc Welding (SMAW). *Neural Comput. Appl.* **2018**, *29*, 889–899. [CrossRef]
47. Kumar, R.; Saurav, S.K. Modeling of TIG welding process by regression analysis and neural network technique. *Int. J. Mech. Eng. Technol.* **2015**, *6*, 10–27.
48. Kshirsagar, R.; Jones, S.; Lawrence, J.; Tabor, J. Optimization of TIG Welding Parameters Using a Hybrid Nelder Mead-Evolutionary Algorithms Method. *J. Manuf. Mater. Process.* **2020**, *4*, 10. [CrossRef]
49. Boubaker, S.; Kamel, S.; Kolsi, L.; Kahouli, O. Forecasting of One-Day-Ahead Global Horizontal Irradiation Using Block-Oriented Models Combined with a Swarm Intelligence Approach. *Nat. Resour. Res.* **2020**, *30*, 1–26. [CrossRef]
50. Touileb, K.; Djoudjou, R.; Chihaoui Hedhibi, A.; Ouis, A.; Benselama, A.; Albaijan, I.; Hany, S.A.; Abdus Samad, U. Comparative Microstructural, Mechanical and Corrosion Study between Dissimilar ATIG and Conventional TIG Weldments of 316L Stainless Steel and Mild Steel. *Metals* **2022**, *12*, 635. [CrossRef]
51. Engineering Standards Manual ISD 341-2, Section WFP 2-01—Welding Fabrication Procedure. Available online: https://engstandards.lanl.gov/esm/welding/vol1/GWS%201-05-Att.09-R1.pdf (accessed on 30 June 2023).
52. *BS EN ISO 15614-1-2017*; Specification and Qualification of Welding Procedures for Metallic Materials—Welding Procedure Test. ISO: Geneva, Switzerland, 2017.
53. Coetzee, T. Phase chemistry of Submerged Arc Welding (SAW) fluoride based slag. *J. Mater. Res. Technol.* **2020**, *9*, 9766–9776. [CrossRef]
54. Leconte, S.; Paillard, P.; Chapelle, P.; Henrion, G.; Saindrenan, J. Effects of flux containing fluorides on TIG welding process. *Sci. Technol. Weld. Join.* **2007**, *12*, 120–126. [CrossRef]
55. Dong, J.; Han, X.; Xu, D.; Gao, X. Arc behavior of fluoride effects in the A-TIG welding of Ti6Al4V. *Trans. China Weld. Inst.* **2017**, *38*, 6–10.
56. Wu, W.; Hu, S.; Shen, J. Microstructure, mechanical properties and corrosion behavior of laser welded dissimilar joints between ferritic stainless steel and carbon steel. *Mater. Des.* **2015**, *65*, 855–861. [CrossRef]
57. Osoba, L.O.; Ayoola, W.A.; Adegbuji, Q.A.; Ajibade, O.A. Influence of Heat Inputs on Weld Profiles and Mechanical Properties of Carbon and Stainless Steel. *Niger. J. Technol. Dev.* **2021**, *18*, 135–143. [CrossRef]

Disclaimer/Publisher's Note: The statements, opinions and data contained in all publications are solely those of the individual author(s) and contributor(s) and not of MDPI and/or the editor(s). MDPI and/or the editor(s) disclaim responsibility for any injury to people or property resulting from any ideas, methods, instructions or products referred to in the content.

Article

Microstructural Optimization of Sn-58Bi Low-Temperature Solder Fabricated by Intense Pulsed Light (IPL) Irradiation

Hyeri Go [1], Taejoon Noh [2], Seung-Boo Jung [2] and Yoonchul Sohn [1,*]

[1] Department of Welding & Joining Science Engineering, Chosun University, 309 Pilmoon-daero, Dong-gu, Gwangju 61452, Republic of Korea; heili1235@chosun.ac.kr

[2] School of Advanced Materials Science and Engineering, Sungkyunkwan University, 2066 Seobu-ro, Jangan-gu, Suwon 16419, Republic of Korea; hokujoon@skku.edu (T.N.); sbjung@skku.edu (S.-B.J.)

* Correspondence: yoonchul.son@chosun.ac.kr

Abstract: In this study, intense pulsed light (IPL) soldering was employed on Sn-58Bi solder pastes with two distinct particle sizes (T3: 25–45 µm and T9: 1–8 µm) to investigate the correlation between the solder microstructure and mechanical properties as a function of IPL irradiation times. During IPL soldering, a gradual transition from an immature to a refined to a coarsened microstructure was observed in the solder, impacting its mechanical strength (hardness), which initially exhibited a slight increase followed by a subsequent decrease. It is noted that hardness measurements taken during the immature stage may exhibit slight deviations from the Hall–Petch relationship. Experimental findings revealed that as the number of IPL irradiation sessions increased, solder particles progressively coalesced, forming a unified mass after 30 sessions. Subsequently, after 30–40 IPL sessions, notable voids were observed within the T3 solder, while fewer voids were detected at the T9-ENIG interface. Following IPL soldering, a thin layered structure of Ni_3Sn_4 intermetallic compound (IMC) was observed at the Sn-58Bi/ENIG interface. In contrast, reflow soldering resulted in the abundant formation of rod-shaped Ni_3Sn_4 IMCs not only at the reaction interface but also within the solder bulk, accompanied by the notable presence of a P-rich layer beneath the IMC.

Keywords: intense pulsed light; Sn-58Bi; solder; Hall–Petch relationship; microstructure

Citation: Go, H.; Noh, T.; Jung, S.-B.; Sohn, Y. Microstructural Optimization of Sn-58Bi Low-Temperature Solder Fabricated by Intense Pulsed Light (IPL) Irradiation. *Crystals* **2024**, *14*, 465. https://doi.org/10.3390/cryst14050465

Academic Editors: Reza Beygi, Mahmoud Moradi, Ali Khalfallah and Marek Sroka

Received: 31 March 2024
Revised: 23 April 2024
Accepted: 15 May 2024
Published: 16 May 2024

Copyright: © 2024 by the authors. Licensee MDPI, Basel, Switzerland. This article is an open access article distributed under the terms and conditions of the Creative Commons Attribution (CC BY) license (https://creativecommons.org/licenses/by/4.0/).

1. Introduction

Modern electronic packaging technology, which facilitates the interconnection of semiconductor chips and packages through solder bonding, stands as a pivotal component in ensuring device reliability [1–5]. While conventional reflow soldering techniques utilizing convection ovens have been extensively employed, there is a persistent drive to address substrate warpage stemming from disparate coefficients of thermal expansion among electronic package materials, alongside endeavors to curtail soldering expenses [6,7]. This impetus has fueled the continual evolution of technologies such as laser-assisted bonding (LAB) and intensive pulsed light (IPL) soldering.

LAB technology, exploiting a laser wavelength that exhibits heightened absorption characteristics in Si chips relative to polymer materials, offers the notable advantage of mitigating substrate heating and minimizing warpage [8–10]. Through decades of dedicated research, a repertoire of materials and methodologies applicable to LAB, including flux-free underfills, has been cultivated. Nonetheless, LAB equipment is encumbered by the requisite inclusion of a homogenizer to transform the laser point source into a planar source, and achieving large-area irradiation poses a formidable challenge. Conversely, IPL technology has traditionally found utility in aggregating and sintering minute nanoparticles via diffusion-mediated reactions. Recently, there has been a burgeoning interest in repurposing this technology for soldering processes. These advancements underscore the relentless pursuit within the electronic packaging industry of ways to bolster the reliability of soldering operations, address substrate warpage concerns, and optimize cost efficiency.

Only a limited number of documented instances involve the application of IPL soldering to Sn-based solder formulations within operational electronic packages [11–16]. Jung et al. [11] presented findings detailing the IPL soldering of Sn-58Bi, illustrating a mere 2.5% duration requirement relative to conventional reflow techniques, alongside a 40% enhancement in mechanical strength. Ha et al. [12] conducted an evaluation of the reliability impact on Sn–3.0Ag–0.5Cu ball grid array (BGA) packages with Electroless Nickel Electroless Palladium Immersion Gold (ENEPIG) surface treatment, employing the IPL soldering methodology. Experimental observations under optimized IPL parameters revealed an approximate 6.7-fold increase in the number of failure cycles compared to reflow soldering. Furthermore, Min et al. [13] provided comprehensive experimental results contrasting IPL soldering against conventional reflow soldering. Their analysis indicated significantly lower power consumption during the IPL radiation soldering process (17.95 kWh) compared to the traditional convection reflow process (29.50 kWh). Moreover, IPL-soldered joints exhibited a substantial improvement in drop impact reliability (with a recorded number of drops to failure of 277) compared to reflow-soldered joints (with a recorded number of drops to failure of 103). These findings underscore the considerable potential of IPL soldering in reducing process duration, controlling intermetallic compound (IMC) thickness to enhance drop impact reliability, and optimizing power consumption.

In addition to the aforementioned advantages, IPL soldering exhibits distinctive attributes. For instance, in reflow soldering, post-process grain growth in solders is prominently observed. Conversely, the IPL process affords facile manipulation of the degree to which metal particles constituting the solder paste are liquefied, through adjustments in process parameters such as frequency, pulse width, and intensity. This inherent flexibility facilitates the generation of varied microstructural configurations, spanning from refined microstructures with diminutive grain sizes to those characterized by larger grain dimensions. In this study, meticulous scrutiny of the microstructural evolution of Sn-58Bi solder paste was undertaken, spanning from its nascent stages to later stages of reaction. The investigation encompassed an analysis of the correlation between grain size and resultant mechanical properties. It is posited that such investigations furnish valuable insights pertinent to future applications of IPL in soldering processes.

2. Materials and Methods

Two distinct sizes of Sn-58Bi solder paste were procured from BBEIN: type 3 (BBI-LESP04, 25–45 µm) and type 9 (BBI-NCLFSP048, 1–8 µm). A printed circuit board (PCB) featuring Electro-less Nickel Immersion Gold (ENIG)-treated 400 µm metal pads was fabricated for substrate application, possessing dimensions of 77×132 mm^2 with a 6×12 array for ball alignment. Solder paste deposition onto the PCB was facilitated using a metal mask with a thickness of 0.2 mm. For IPL soldering, the charge voltage and peak current were maintained at 1500 V and 2000 A, respectively. The IPL irradiation area spanned 20×30 cm^2. Process variables for IPL, including frequency and pulse width, were set at 3 Hz and 2.0 ms, respectively, with IPL irradiation sessions ranging from 10 to 70 during soldering operations. Figure 1 provides a graphical representation of the IPL process variable configurations for enhanced comprehension. Furthermore, the experimental parameters utilized in this investigation are summarized in Table 1. Depending on the experimental conditions, the applied energy increased proportionally with IPL irradiation sessions, ranging from 100 to 700 J/cm^2, while maintaining a consistent power consumption of 18 kWh.

For comparative analysis, a reflow-soldered joint was fabricated utilizing a reflow machine (model: BT301N) manufactured by Autotronik-SMT GmbH, Amberg, Germany. The peak temperature during reflow was set at 180 °C, and the specific reflow profile employed in the experimental setup is delineated in Figure 2. Microstructural and compositional characterization of the solder joints was conducted utilizing field emission scanning electron microscopy (FE-SEM, model: JEOL JSM-7900) coupled with energy dispersive spectroscopy (EDS, model: JEOL JXA-8500F). The thickness of the IMC in the SEM micrographs was quantified by employing image analysis software, whereby the thickness of the IMC layer was defined as the total area occupied by the phase divided by its length. Average thickness values were derived from measurements acquired from six distinct regions within each reaction specimen. Electron Backscattered Diffraction (EBSD) analysis was performed utilizing FE-SEM (model: HITACHI S-5000) equipped with EDAX Velocity super. The hardness of the solder joints post-IPL irradiation was assessed utilizing MMT-X7A equipment from MATSUZAWA. The hardness assessment involved the averaging of measurements taken at six distinct locations for each solder, with a measurement interval of 100 μm, using a load of 5 gf and a dwell time of 10 s.

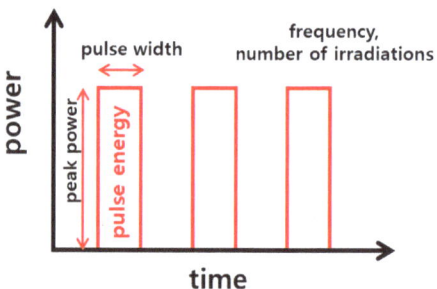

Figure 1. Process variables for IPL soldering.

Table 1. IPL soldering conditions and sample identification.

Sample ID	Frequency	Pulse Width	Number of IPL Irradiation	Total Energy	Power Consumption	Time
	(Hz)	(ms)	(n)	(J/cm²)	(kWh)	(sec)
IPL 10	3	2	10	100	18	3.3
IPL 20	3	2	20	200	18	6.6
IPL 30	3	2	30	300	18	9.9
IPL 40	3	2	40	400	18	13.2
IPL 50	3	2	50	500	18	16.5
IPL 60	3	2	60	600	18	19.8
IPL 70	3	2	70	700	18	23.1
reflow						600

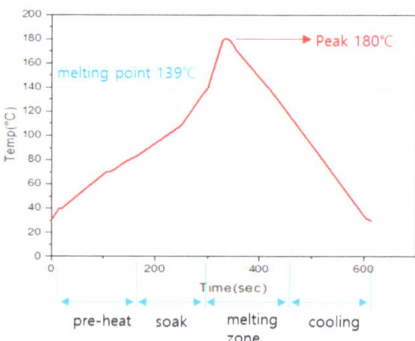

Figure 2. Reflow profile for Sn-58Bi soldering.

3. Results and Discussion

3.1. Microstructural Evolution during IPL Soldering

When IPL soldering was conducted under the prescribed conditions of a 3 Hz frequency and a 2 ms pulse width, the evolution of macrostructural features in Sn-58Bi solder paste was observed as a function of the number of IPL irradiation sessions, as illustrated in Figure 3. Following 10 IPL irradiation sessions, it is evident that a substantial portion of both type 3 (T3) and type 9 (T9) solder paste particles remained unmelted, retaining their initial particle size distribution. Notably, in the case of T9, partial melting and recombination of particles can be observed predominantly in the upper region, resulting in an enlargement of particle dimensions from the initial 1–8 µm range to several tens of micrometers.

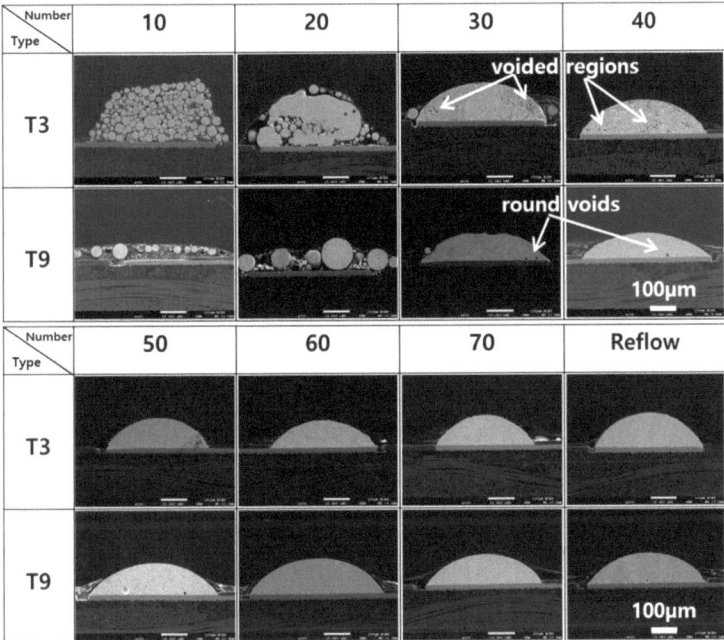

Figure 3. Macroscopic morphological changes in Sn-58Bi solder pastes with increasing sessions of IPL irradiation (3 Hz, 2 ms).

After 20 IPL irradiation sessions, notable particle bonding occurred, particularly evident in the T3 solder paste despite its larger initial particle size. Significant solder lump formation, primarily concentrated in the upper portion, can be observed, with partial lump formation occurring in regions interfacing with the ENIG substrate. Notably, remnants of smaller particles persist in the mid-section. Conversely, in T9, the agglomeration phenomenon initiated in the uppermost region propagates downward, albeit with limited advancement observable at the lower extremity.

Following 30 IPL irradiation sessions, complete fusion of solder particles from both T3 and T9 culminated in the formation of a singular, substantial solder bump. Subsequent to 30–40 IPL irradiation sessions, a considerable prevalence of voids can be observed within the T3 solder bump, whereas T9 exhibits only sparse, round voids primarily localized at the lower extremity of the solder bump. The occurrence of voids during IPL soldering can be attributed to the size and aggregation process of the initial solder particles in the solder paste, as well as the wetting characteristics on ENIG, the substrate surface treatment. In the T3 sample, substantial aggregates form as a unified entity, with voids emerging in between, whereas in the T9 sample, initial formation comprises small and medium-sized aggregates that subsequently coalesce into a single large aggregate. Notably, for the T9 sample, the initial solder particle size is smaller, and there is improved spreading on ENIG, resulting in a minimal occurrence of voids. Beyond 50 IPL irradiation sessions, both T3 and

T9 configurations manifest a densified microstructure with minimal void formation, with discernible convergence in structural characteristics.

Under the conditions of a 3 Hz frequency and a 2.0 ms pulse width, the alterations in the microstructural characteristics of Sn-58Bi solder paste as a function of the number of IPL irradiation sessions were as depicted in Figure 4. Initial inspection of the Sn-58Bi solder particles after 10–20 IPL sessions revealed a notably finer microstructure in type 9 (T9) compared to type 3 (T3). Notably, the bright, protruding regions within the solder particles corresponded to Bi, while the darker, recessed regions represented Sn, forming an alternately arranged lamellar structure. The lamellar width was in the order of several micrometers in T3, whereas it diminished to submicron dimensions in T9.

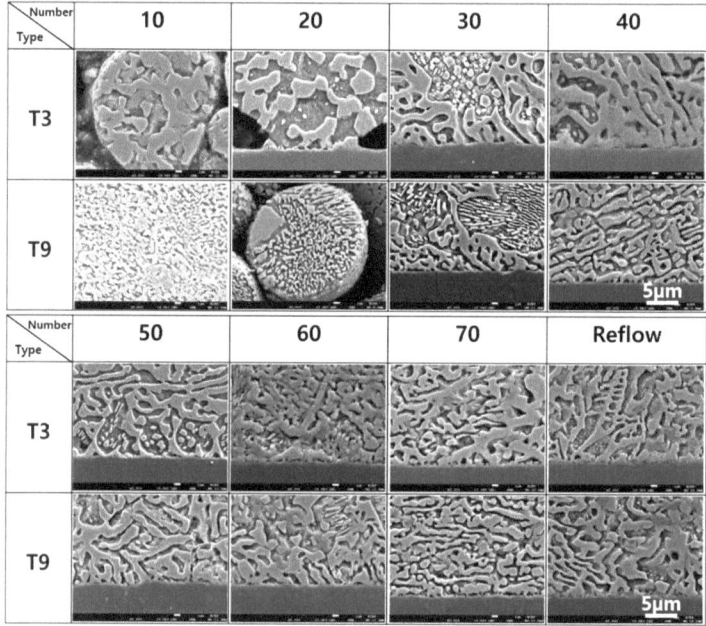

Figure 4. Changes in microstructure of Sn-58Bi solders with increasing sessions of IPL irradiation (3 Hz, 2 ms).

As IPL reached 30–40 sessions, the emergence of interparticle bonding and the coarsening of lamellar structures within the particles became increasingly apparent. Particularly in the case of T9, the microstructure following 30 IPL iterations exhibited a coexistence of regions characterized by submicron-sized lamellae and coarsened regions featuring lamellae several micrometers wide. Subsequent IPL irradiation sessions led to a predominant transformation of the microstructure, where the solder area predominantly comprised coarsened regions. Beyond 50 IPL sessions, discerning differences in microstructural features between T3 and T9 became challenging.

It is established that the reaction between Sn-58Bi solder and ENIG results in the formation of Ni_3Sn_4 IMCs at the interface [17–24]. Initially, when the number of IPL irradiation sessions is limited, the thickness of the formed IMC remains minimal. However, as the number of irradiation sessions increases to 70, Ni_3Sn_4 IMCs with thicknesses comparable to those observed in conventional reflow soldering can be discerned at the interface. Notably, the degree of IMC formation appears independent of the solder particle size, indicating a consistent reaction mechanism across different particle sizes.

3.2. Correlation between Microstructure and Mechanical Properties of Sn-58Bi Solder

The morphological features of the Sn-58Bi/ENIG reaction interfaces following 70 rounds of IPL irradiation and conventional reflow processing are depicted in Figure 5. Post 70 IPL rounds, a thin layered structure of Ni_3Sn_4 IMCs can be observed at the interface, alongside the presence of a minor quantity of feather-shaped IMCs within the Sn-rich region within the solder matrix. In contrast, after the reflow process, rod-shaped Ni_3Sn_4 IMCs are evident at the interface, accompanied by a notable abundance of feather-shaped IMCs distributed throughout the solder volume. Notably, a distinct P-rich layer atop the Ni-P layer is prominently visible in the reflow specimen, presenting as a thin, dark band [25–28]. Conversely, this P-rich layer is conspicuously absent from the IPL specimen. These observations suggest a heightened level of chemical reactivity occurring at the Sn-58Bi/ENIG interface during the conventional reflow process.

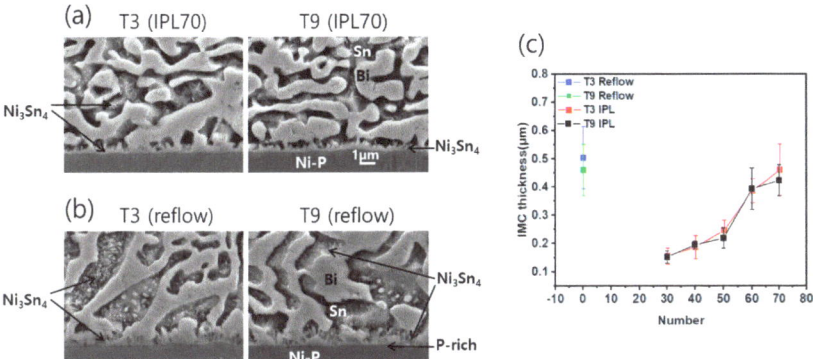

Figure 5. Close observation of the Ni_3Sn_4 IMCs formed at the Sn-58Bi/ENIG interface after 70 IPL sessions (**a**) and after the reflow process (**b**). (**c**) Measured thickness of the Ni_3Sn_4 IMCs.

During the IPL process, the intermittent application of instantaneous light followed by rapid cooling cycles leads to inadequate interdiffusion between the solder and ENIG. In contrast, the reflow process maintains the solder in a molten state for an extended duration, facilitating sufficient interdiffusion between the solder and ENIG. Consequently, once Ni atoms are introduced into the solder, they can diffuse over an extended period and react with Sn to form additional Ni_3Sn_4 IMCs within the solder matrix.

Initially, the thickness of Ni_3Sn_4 IMCs formed at the interface during the IPL process is minimal, and the growth rate is sluggish. However, after 50 IPL irradiation sessions, the IMC growth rate accelerates significantly compared to the initial stages of the reaction, as shown in Figure 5c. Nonetheless, even after 70 IPL irradiation sessions, the overall thickness of the IMCs remains inferior to that observed in the reflow process.

Following IPL and reflow soldering, EBSD analysis was conducted to assess the microstructural characteristics and grain orientation within the solder specimens. Phase maps and inverse pole figures (IPFs) for each solder variant were generated, as depicted in Figure 6. In the phase map, the red region signifies Sn while the green region corresponds to Bi. Post-IPL and reflow soldering, discernible variations in grain orientation were observed within the solder specimens.

Upon IPL soldering for 40 cycles, the orientation of Sn grains was situated within the intermediary region between the (001) and (110) planes in both T3 and T9 specimens. With an increase in the number of IPL irradiation sessions to 70, T9 specimens maintained a similar orientation, while T3 specimens exhibited an orientation spanning from (001) to (100). Conversely, after reflow soldering, a dominance of orientation within the intermediate region of the (100) and (110) planes was observed in both T3 and T9 specimens, with the left-hand side of T9 grains exhibiting a direct (110) orientation in addition.

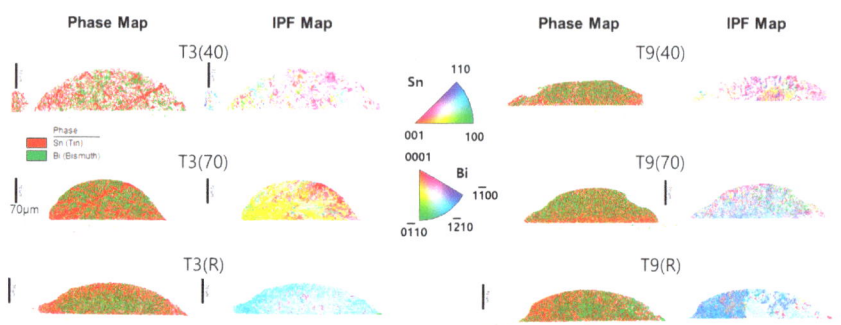

Figure 6. Results of EBSD analysis for the specimens fabricated with IPL soldering (after 40 and 70 sessions) and reflow soldering.

Regarding Bi grains, the $(1\bar{2}10)$ orientation emerged as the predominant orientation across all specimens, irrespective of the soldering method employed. Additionally, the presence of the (0001) orientation was noted alongside the $(1\bar{2}10)$ orientation exclusively within the grains of 70-IPL-session specimens. This observed shift in solder grain orientation contingent upon the soldering method or degree of irradiation presents an intriguing discovery warranting further investigation.

Figure 7 illustrates the variations in average Sn grain size and hardness across each solder variant, as determined through EBSD analysis and hardness testing. The average grain size was quantified based on the enumeration and dimensions of grains within a $180 \times 550\ \mu m^2$ area of each specimen. Notably, the grain size of Sn, constituting the solder, exhibited a nadir at 40 IPL sessions, subsequently increasing with additional IPL irradiation sessions. Conversely, the hardness value displayed a maximum at 40 IPL sessions, declining with escalating IPL irradiation counts.

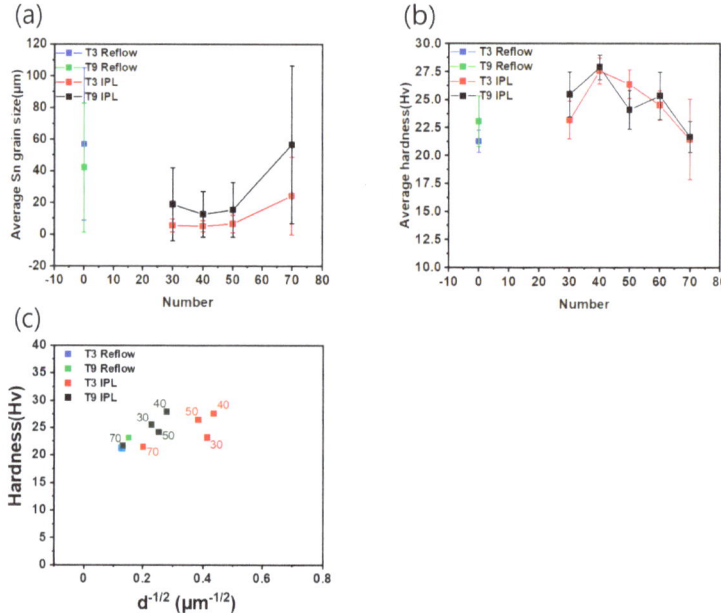

Figure 7. (**a**) Average Sn grain size and (**b**) average hardness measured for the solders fabricated by IPL or reflow soldering; (**c**) Hall–Petch relationship taken from the measured grain size and hardness.

It is widely acknowledged that as the microstructure coarsens, mechanical properties such as hardness and strength tend to degrade [29–33]. Specifically, hardness often conforms to the well-established Hall–Petch relationship, which correlates hardness with the inverse square root of grain size [29,30].

$$H = H_0 + kd^{-1/2} \tag{1}$$

where H is hardness and d the average grain size. H_0 and k are constants. Figure 7c displays a graph illustrating the relationship between the inverse square root of average grain size ($d^{-1/2}$) and hardness (H), as determined by substituting the measured grain size and hardness values from each solder variant produced via IPL and reflow soldering into the Hall–Petch relationship. Overall, it is evident that, with the exception of the IPL T3 specimen irradiated 30 times, the graph demonstrates a predominantly linear correlation between $d^{-1/2}$ and H, thus adhering to the Hall–Petch relationship.

In the realm of solder compositions, studies have highlighted that mechanical strength tends to improve with the refinement of the internal structure. For instance, Li et al. [32] investigated composite solders comprising Sn-58%Bi alloyed with 0, 0.03, 0.05, and 0.1 wt% Cu_6Sn_5 nanoparticles. They observed that the addition of Cu_6Sn_5 nanoparticles, particularly those sized at 10 nm, led to microstructural refinement of the Sn-58%Bi solder, with the solder containing 0.05 wt% Cu_6Sn_5 nanoparticles exhibiting superior performance. This enhancement in solder strength was attributed to the refinement of matrix phases, aligning with the principles outlined in the Hall–Petch equation. Similarly, Zhang et al. [33] demonstrated that Ga additions in Sn-XGa alloys (where X = 0.5, 1.0, 2.0 wt%) could moderately refine β-Sn grains while rendering the alloys more prone to recrystallization during cross-sectioning and polishing processes. Notably, Sn-1.0Ga exhibited significantly elevated yields and tensile strengths compared to pure Sn, attributed to the robust solid-solution strengthening effect of Ga and the refinement of β-Sn grains.

In this study, an intriguing observation was made where, despite an increase in the number of IPL irradiation sessions from 30 to 40 during the initial stages of the IPL process, the hardness of Sn-58Bi solders exhibited an increase, contrary to the expectations based on the Hall–Petch relationship. This anomaly is believed to be associated with the densification of the solders. As depicted in Figure 3, the solder initially possessed a coarse microstructure with internal voids at 30 IPL sessions, gradually transitioning into a denser structure with diminishing voids as the number of IPL sessions increased. This transformation in the microstructure of the solder likely contributed to the enhancement of its mechanical properties.

Min et al. [13] explored the feasibility of IPL soldering for BGA package assembly under varying IPL soldering conditions. During the soldering process, Ag and Cu components within the Sn 3.0Ag 0.5Cu solder underwent transformation into Ag_3Sn and Cu_6Sn_5 IMCs. At lower IPL energies (or fewer irradiation sessions), partially unreacted Ag was detected within the solders, resulting in lower shear strength and hardness values. However, with an increase in IPL energy (or irradiation count), complete transformation of Ag to Ag_3Sn occurred, leading to improved strength and hardness. From these findings, it can be inferred that during the initial stages of the IPL soldering process, all solder particles melt and coalesce into a unified lump, yet the resulting solder lump may still exhibit an immature microstructure. The specimens subjected to 30 IPL sessions in this study likely resembled such cases, displaying an immature structure characterized by internal voids and consequently exhibiting lower hardness values. The deviation of the 30-IPL specimen from the linear relationship depicted in Figure 7c can thus be attributed to this phenomenon.

4. Conclusions

In this study, Sn-58Bi solder with two different particle sizes of 1–8 μm (T9) and 25–45 μm (T3) was used and the number of IPL irradiation sessions was increased under the frequency and pulse width conditions of 3 Hz and 2.0 ms. Through an examination of alterations in solder microstructure and hardness, the following conclusions were drawn.

(1) As the number of IPL irradiation sessions increases, the initial solder particles gradually agglomerate and aggregate into a single lump after 30 irradiation sessions. After 30–40 IPL sessions, a large number of voids was observed overall inside the T3 solder, and a small number of round voids was discovered at the interface with ENIG in T9.

(2) After IPL soldering, a thin layered structure of Ni_3Sn_4 IMC forms at the Sn-58Bi/ENIG interface, and the P-rich layer is not prominently visible. In contrast, after reflow soldering, rod-shaped Ni_3Sn_4 IMCs are abundantly formed not only at the reaction interface but also within the solder bulk, accompanied by the notable presence of a P-rich layer beneath the IMC.

(3) After IPL soldering, the orientation of Sn grains is situated within the intermediary region between the (001) and (110) planes. On the other hand, after reflow soldering, a dominance of orientation within the intermediate region of the (100) and (110) planes can be observed. Regarding Bi grains, the ($1\bar{2}10$) orientation emerges as the predominant orientation across all specimens, irrespective of the soldering method employed.

(4) During IPL soldering, as the number of irradiation sessions increases gradually, a progression from an immature to a fine to a coarsened microstructure occurs in the solder. Consequently, the mechanical strength (hardness) of the solder exhibits a tendency to initially increase slightly, before decreasing. The hardness measured in the immature stage may deviate slightly from the range predicted by the Hall–Petch relationship.

(5) Based on the experimental findings, for IPL at a frequency of 3 Hz and a pulse width of 2 ms, optimal outcomes were achieved within the IPL exposure range of 40–50 sessions, characterized by a diminutive grain size, elevated hardness, and minimal IMC thickness.

Author Contributions: Conceptualization, methodology, writing—original draft, and funding acquisition, Y.S.; investigation, methodology, and funding acquisition, H.G.; investigation, data curation, and visualization, T.N.; methodology, writing—review editing, and formal analysis, S.-B.J. and Y.S. All authors have read and agreed to the published version of the manuscript.

Funding: This study was supported by a research fund from Chosun University, 2022.

Data Availability Statement: The original contributions presented in the study are included in the article, further inquiries can be directed to the corresponding author/s.

Conflicts of Interest: The authors declare no conflicts of interest.

References

1. Feldmann, K.; Franke, J.; Schüßler, F. Development of micro assembly processes for further miniaturization in electronics production. *CIRP Ann.* **2010**, *59*, 1–4. [CrossRef]
2. Sitaraman, S.; Sukumaran, V.; Pulugurtha, M.R.; Wu, Z.; Suzuki, Y.; Kim, Y.; Sundaram, V.; Kim, J.; Tummala, R.R. Miniaturized bandpass filters as ultrathin 3-D IPDs and embedded thinfilms in 3-D glass modules. *IEEE Trans. Compon. Packag. Manuf. Technol.* **2017**, *7*, 1410–1418. [CrossRef]
3. Kim, M.S.; Pulugurtha, M.R.; Kim, Y.; Park, G.; Cho, K.; Smet, V.; Sundaram, V.; Kim, J.; Tummala, R. Miniaturized and high-performance RF packages with ultra-thin glass substrates. *Microelectron. J.* **2018**, *77*, 66–72. [CrossRef]
4. Zhang, P.; Xue, S.; Wang, J. New challenges of miniaturization of electronic devices: Electromigration and thermomigration in lead-free solder joints. *Mater. Des.* **2020**, *192*, 108726. [CrossRef]
5. Xiong, M.Y.; Zhang, L. Interface reaction and intermetallic compound growth behavior of Sn-Ag-Cu lead-free solder joints on different substrates in electronic packaging. *J. Mater. Sci.* **2019**, *54*, 1741–1768. [CrossRef]
6. Lau, C.S.; Khor, C.Y.; Soares, D.; Teixeira, J.C.; Abdullah, M.Z. Thermo-mechanical challenges of reflowed lead-free solder joints in surface mount components: A review. *Solder. Surf. Mt. Technol.* **2016**, *39*, 88. [CrossRef]
7. Feng, J.; Xu, D.E.; Tian, Y.; Mayer, M. SAC305 solder reflow: Identification of melting and solidification using in-process resistance monitoring. *IEEE Trans. Compon. Packag. Manuf. Technol.* **2019**, *9*, 1623–1631. [CrossRef]
8. Jung, Y.; Ryu, D.; Gim, M.; Kim, C.; Song, Y.; Kim, J.; Yoon, J.; Lee, C. Development of Next Generation Flip Chip Interconnection Technology Using Homogenized Laser-Assisted Bonding. In Proceedings of the IEEE 66th Electronic Components and Technology Conference (ECTC), Las Vegas, NV, USA, 31 May–3 June 2016; pp. 88–94.
9. Tian, W.; Ma, Z.; Cao, X.; Lin, J.; Cui, Y.; Huang, X. Application of metal interconnection process with micro-LED display by laser-assisted bonding technology. *J. Mater. Sci. Mater. Electron.* **2023**, *34*, 2253. [CrossRef]

10. Kim, S.; Yu, D.; Kim, Y.; Son, J.; Byun, D.; Bang, J. Interfacial Properties of Sn-Cu-xCr Alloy using Laser-Assisted Bonding. *J. Weld. Join.* **2023**, *41*, 291–298. [CrossRef]
11. Jung, K.H.; Min, K.D.; Lee, C.J.; Jeong, H.; Kim, J.H.; Jung, S.B. Ultrafast Photonic Soldering with Sn–58Bi Using Intense Pulsed Light Energy. *Adv. Eng. Mater.* **2020**, *22*, 2000179. [CrossRef]
12. Ha, E.; Min, K.D.; Lee, S.; Hwang, J.S.; Kang, T.; Jung, S.B. Intense Pulsed Light Soldering of Sn–3.0Ag–0.5Cu Ball Grid Array Component on Au/Pd(P)/Ni(P) Surface-Finished Printed Circuit Board and Its Drop Impact Reliability. *Adv. Eng. Mater.* **2023**, *25*, 10. [CrossRef]
13. Min, K.D.; Ha, E.; Lee, S.; Hwang, J.S.; Kang, T.; Joo, J.; Jung, S.B. Proposal of intense pulsed light soldering process for improving the drop impact reliability of Sn–3.0Ag–0.5Cu ball grid array package. *J. Manuf. Process.* **2023**, *98*, 19–28. [CrossRef]
14. Jang, J.H.; Lee, C.J.; Hwang, B.U.; Min, K.D.; Kim, J.H.; Jung, S.B. Microstructural evolution and mechanical properties of SAC305 with the intense pulsed light soldering process under high-temperature storage test. In Proceedings of the IEEE 71st Electronic Components and Technology Conference (ECTC), Virtual, 1 June–4 July 2021; pp. 2314–2319.
15. Ho, M.J.; Li, Y.; Liu, L.; Li, Y.; Gao, C. Study of an Ultra-fast Photonic Soldering Technology without Thermal Damage in Display Module Package. *SID Symp. Dig. Tech. Pap.* **2021**, *52*, 115–118. [CrossRef]
16. Jang, Y.R.; Jeong, R.; Kim, H.S.; Park, S.S. Fabrication of solderable intense pulsed light sintered hybrid copper for flexible conductive electrodes. *Sci. Rep.* **2021**, *11*, 14551. [CrossRef] [PubMed]
17. Myung, W.R.; Kim, Y.; Jung, S.B. Evaluation of the bondability of the epoxy-enhanced Sn-58Bi solder with ENIG and ENEPIG surface finishes. *J. Electron. Mater.* **2015**, *44*, 4637–4645. [CrossRef]
18. Choi, H.; Kim, C.L.; Sohn, Y. Diffusion Barrier Properties of the Intermetallic Compound Layers Formed in the Pt Nanoparticles Alloyed Sn-58Bi Solder Joints Reacted with ENIG and ENEPIG Surface Finishes. *Materials* **2022**, *15*, 8419. [CrossRef] [PubMed]
19. Lee, S.M.; Yoon, J.W.; Jung, S.B. Electromigration effect on Sn-58% Bi solder joints with various substrate metallizations under current stress. *J. Mater. Sci. Mater. Electron.* **2016**, *27*, 1105–1112. [CrossRef]
20. Chen, C.; Wang, C.; Sun, H.; Yin, H.; Gao, X.; Xue, J.; Ni, D.; Bian, K.; Gu, Q. Interfacial Microstructure and Mechanical Reliability of Sn-58Bi/ENEPIG Solder Joints. *Processes* **2022**, *10*, 295. [CrossRef]
21. Kim, J.; Park, D.Y.; Ahn, B.; Bang, J.; Kim, M.S.; Park, H.S.; Sohn, Y.; Ko, Y.H. Effective (Pd, Ni)Sn$_4$ diffusion barrier to suppress brittle fracture at Sn-58Bi-xAg solder joint with Ni(P)/Pd (P)/Au metallization pad. *Microelectron. Reliab.* **2022**, *129*, 114472. [CrossRef]
22. Hao, Q.; Tan, X.F.; McDonald, S.D.; Sweatman, K.; Akaiwa, T.; Nogita, K. Investigating the Effects of Rapid Precipitation of Bi in Sn on the Shear Strength of BGA Sn-Bi Alloys. *J. Electron. Mater.* **2024**, *53*, 1223–1238. [CrossRef]
23. Zhou, S.; He, S.; Nishikawa, H. Effect of Zn addition on interfacial reactions and mechanical properties between eutectic Sn58Bi solder and ENIG substrate. *J. Nanosci. Nanotechnol.* **2020**, *20*, 106–112. [CrossRef] [PubMed]
24. Fan, Y.; Wu, Y.; Dale, T.F.; Lakshminarayana, S.A.P.; Greene, C.V.; Badwe, N.U.; Aspandiar, R.F.; Blendell, J.E.; Subbarayan, G.; Handwerker, C.A. Influence of Pad Surface Finish on the Microstructure Evolution and Intermetallic Compound Growth in Homogeneous Sn-Bi and Sn-Bi-Ag Solder Interconnects. *J. Electron. Mater.* **2021**, *50*, 6615–6628. [CrossRef]
25. Elsener, B.; Crobu, M.; Scorciapino, M.A.; Rossi, A. Electroless deposited Ni–P alloys: Corrosion resistance mechanism. *J. Appl. Electrochem.* **2008**, *38*, 1053–1060. [CrossRef]
26. Liu, P.L.; Xu, Z.; Shang, J.K. Thermal stability of electroless-nickel/solder interface: Part A. Interfacial chemistry and microstructure. *Metall. Mater. Trans. A* **2000**, *31*, 2857–2866. [CrossRef]
27. Lin, J.D.; Chou, C.T. The influence of phosphorus content on the microstructure and specific capacitance of etched electroless Ni-P coatings. *Surf. Coat. Technol.* **2019**, *368*, 126–137. [CrossRef]
28. Alam, M.O.; Chan, Y.C.; Hung, K.C. Reliability study of the electroless Ni–P layer against solder alloy. *Microelectron. Reliab.* **2002**, *42*, 1065–1073. [CrossRef]
29. Fecht, H.J.; Hellstern, E.; Fu, Z.; Johnson, W.L. Nanocrystalline metals prepared by high-energy ball milling. *Metall. Trans. A* **1990**, *21*, 2333. [CrossRef]
30. Zheng, Z.; Chiang, P.C.; Huang, Y.T.; Wang, W.T.; Li, P.C.; Tsai, Y.H.; Chen, C.M.; Feng, S.P. Study of grain size effect of Cu metallization on interfacial microstructures of solder joints. *Microelectron. Reliab.* **2019**, *99*, 44–51. [CrossRef]
31. Kang, H.; Rajendran, S.H.; Jung, J.P. Low Melting Temperature Sn-Bi Solder: Effect of Alloying and Nanoparticle Addition on the Microstructural, Thermal, Interfacial Bonding, and Mechanical Characteristics. *Metals* **2021**, *11*, 364. [CrossRef]
32. Li, J.F.; Agyakwa, P.A.; Johnson, C.M. A numerical method to determine interdiffusion coefficients of Cu$_6$Sn$_5$ and Cu$_3$Sn intermetallic compounds. *Intermetallics* **2013**, *40*, 50–59. [CrossRef]
33. Zhang, H.; Ma, Z.; Yang, S.; Fan, M.; Cheng, X. Microstructure and mechanical properties of Sn-xGa alloys and solder joints. *J. Mater. Res. Technol.* **2023**, *26*, 3830–3839. [CrossRef]

Disclaimer/Publisher's Note: The statements, opinions and data contained in all publications are solely those of the individual author(s) and contributor(s) and not of MDPI and/or the editor(s). MDPI and/or the editor(s) disclaim responsibility for any injury to people or property resulting from any ideas, methods, instructions or products referred to in the content.

Article

Experimental and Computational Study of Microhardness Evolution in the HAZ for Al–Cu–Li Alloys

Stavroula Maritsa [1], Stavros Deligiannis [2], Petros E. Tsakiridis [2] and Anna D. Zervaki [1,*]

[1] Shipbuilding Technology Laboratory, School of Naval Architecture and Marine Engineering, National Technical University of Athens, Zografou, 157 80 Athens, Greece; maritsastavroula@mail.ntua.gr
[2] Laboratory of Physical Metallurgy, School of Mining and Metallurgical Engineering, National Technical University of Athens, Zografou, 157 80 Athens, Greece; stavrosdel@metal.ntua.gr (S.D.); ptsakiri@central.ntua.gr (P.E.T.)
* Correspondence: annazervaki@mail.ntua.gr; Tel.: +30-210-772-3691

Abstract: The Laser Beam Welding (LBW) of aluminum alloys has attracted significant interest from industrial sectors, including the shipbuilding, automotive and aeronautics industries, as it expects to contribute to significant cost reduction associated with the production of high-quality welds. To comprehend the behavior of welded structures in regard to their damage tolerance, the application of fracture mechanics serves as the instrumental tool. However, the methods employed overlook the changes in the microstructure within the Heat-Affected Zone (HAZ), which leads to the degradation of the mechanical properties of the material. The purpose of this study is to simulate microhardness evolution in the HAZ of AA2198-T351 LBW. The material represents the latest generation of Al-Cu-Li alloys, which exhibit improved mechanical properties, enhanced damage tolerance behavior, lower density and better corrosion and fatigue crack growth resistance than conventional Al-Cu alloys. In this work, the microhardness profile of LBW AA2198 was measured, and subsequently, through isothermal heat treatments on samples, the microhardness values of the HAZ were replicated. The conditions of the heat treatments (T, t) were selected in line with the thermal cycles that each area of the HAZ experienced during welding. ThermoCalc and DICTRA were employed in order to identify the strengthening precipitates and their evolution (dissolution and coarsening) during the weld thermal cycle. The microstructure of the heat-treated samples was studied employing LOM and TEM, and the strengthening precipitates and their characteristics (volume fraction and size) were defined and correlated to the calculations and the experimental conditions employed during welding. The main conclusion of this study is that it is feasible to imitate the microstructure evolution within the HAZ through the implementation of isothermal heat treatments. This implies that it is possible to fabricate samples for fatigue crack growth tests, enabling the experimental examination of the damage tolerance behavior in welded structures.

Keywords: laser beam welding; Al–Cu–Li alloys; heat affected zone; microstructure; microhardness

Citation: Maritsa, S.; Deligiannis, S.; Tsakiridis, P.E.; Zervaki, A.D. Experimental and Computational Study of Microhardness Evolution in the HAZ for Al–Cu–Li Alloys. *Crystals* **2024**, *14*, 246. https://doi.org/10.3390/cryst14030246

Academic Editors: Reza Beygi, Mahmoud Moradi and Ali Khalfallah

Received: 21 February 2024
Revised: 26 February 2024
Accepted: 27 February 2024
Published: 1 March 2024

Copyright: © 2024 by the authors. Licensee MDPI, Basel, Switzerland. This article is an open access article distributed under the terms and conditions of the Creative Commons Attribution (CC BY) license (https://creativecommons.org/licenses/by/4.0/).

1. Introduction

Al–Cu–Li alloys have attracted strong interest from the industry due to their increased tensile strength, elastic modulus, fatigue resistance, ductility and lower density [1–3]. Each 1 wt.% Li added to aluminum leads to a 3% decrease in density and a 6% increase in the elastic modulus. By adding 2 wt.% Li, the density of the alloy decreases by 10%, and the elastic modulus increases by 25–35% [2,4]. By replacing conventional high-strength Al alloys with Al–Li alloys, the structure's weight is reduced by 10–20%, and the elastic behavior is increased by 25–35% [4]. Therefore, these alloys are ideal materials for the aerospace and aeronautical industry, since the economic benefits are significant, such as increased payload and better fuel efficiency [5]. Furthermore, the potential Al–Li alloys exhibit in marine industry is significant [6,7].

In the third generation of Al–Li alloys, copper (Cu) content is increased (2–4 wt.%), while the amount of Li is reduced (0.75–1.8 wt.%). These alloys exhibit higher hardness, corrosion resistance, good crack propagation resistance and improved weldability [4,5,8–10]. Other micro-alloying elements, such as Mg, Mn, Zn, Ag, Zr and Fe, are added to improve mechanical strength. For example, Mg and Ag additions favor the formation of the T_1 phase, and the addition of Mg can stimulate the nucleation of GP zones [11]. The third generation of Al–Cu–Li alloys has been widely used as structural materials in the aerospace industry and has attracted significant research interest. These alloys are amenable to artificial aging heat treatment, leading to final products with increased strength and hardness. In addition, natural aging can be applied, during which the precipitates that are formed lead to improved ductility compared to artificial aging [4,12].

AA2198 is a popular widely used Al–Cu–Li alloy of this generation, exhibiting a good combination of hardness, corrosion resistance and ductility [2]. In the United States, alloy 2198-T8 (3.6–6.1 mm thick plate) has been successfully utilized for the construction of the Falcon 9 rocket to manufacture the primary and secondary fuel tanks and fairing components [4]. Furthermore, AA2198 belongs to the typical Al–Cu–Li grades that are widely used for the construction of aircrafts, such as A330/340/350/380 in Europe, Boeing 747/777/787 in America and Comac's C919 in China [13].

The hardness of AA2198 can be increased by natural or artificial aging due to the formation of strengthening precipitates, as overviewed in the Al-rich corner of the ternary Al–Cu–Li phase diagram given in Figure 1 [14]. A complex precipitation sequence has been observed during the aging process of Al–Cu–Li alloy systems. The precipitation sequence impacts the interaction mechanism between dislocations and precipitates during plastic deformation, consequently influencing the mechanical properties of the aged alloys [7,15]. Several precipitates have been observed through aging, including GP zones, θ'(Al_2Cu), β'(Al_3Zr), T_1(Al_2CuLi), δ'(Al_3Li) and S'(Al_2CuMg) [2,4,15,16]. The Li content has been found to be crucial in the precipitation of these alloys. High Li content (>2%) leads to a primary strengthening role attributed to δ' precipitates, with some contribution from S', while lower Li content (<2%) results in T_1 being the primary strengthening precipitates, alongside minor θ' [15,16]. Regarding the precipitation sequence of AA2198, where the Li content is low, a study [16] has shown that after natural aging, the microstructure is primarily composed of Cu-rich clusters, although they do not fit the description of single-layer Cu-rich GPI zones. When the temperature increases, these clusters become unstable and appear to dissolve entirely. The further increase of the temperature results in the artificial aging of AA2198 (155 °C/16 h in [16]) where the formation of T_1 initiates and becomes dominant, contributing to the alloy's strength. Additionally, other less prevalent precipitates are present, based on the Al–Cu sequence (GPI, GPII and θ') [16]. The alloy's composition, especially the Cu/Li ratio is a dominant factor affecting the sequence. Furthermore, a higher Cu/Li ratio seems to produce a more rapid nucleation of T_1 in low-Li Al–Cu–Li alloys [16].

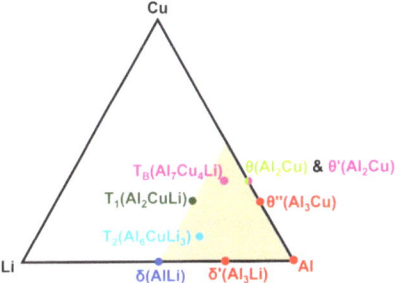

Figure 1. Al–Cu–Li ternary phase diagram near the Al-rich region. The different colors of the phases indicate their lattice (red: FCC, pink: cubic CaF_2, navy blue: BCC, green: BCT, olive green: hexagonal and sky blue: icosahedral). Reprinted/adapted with permission from Ref. [14]. 2024, Elsevier.

The introduction of welding in Al–Cu–Li alloy structures offers significant positive effects, such as lower weight and improved performance, while replacing conventional joining techniques [17,18]. Regarding the AA2198 alloy, two welding methods are usually utilized: Friction Stir Welding (FSW) [17,19–27] and laser welding methods [18,28–31]. Studies have confirmed the positive effect of micro-jet cooling for the welding of Al alloys [32,33]. Laser Beam Welding (LBW) is a popular method in the aerospace industry, as it offers high productivity, manufacturing flexibility and efficiency compared to other welding methods. Using the LBW method has proven that the total construction cost can be reduced by up to 40%, and a weight reduction of up to 28% can be achieved compared to conventional joining techniques [28]. In addition, LBW results in a narrower Heat-Affected Zone (HAZ) and a faster welding speed compared to FSW [29].

During the LBW of AA2198, a significant decrease in mechanical properties is observed near the Weld Zone (WZ) [17,34]. This is attributed to the fact that AA2198 is a heat-treatable alloy, and the mechanical properties depend mainly on the size and distribution of strengthening precipitates. The microstructural changes due to the welding thermal cycles, such as the coarsening or dissolution and re-precipitation of certain phases or precipitates, significantly affect the mechanical properties of an AA2198 weld [18]. Even after optimizing welding parameters, strength loss in alloy welds is considered inevitable [17]. During the welding of AA2198, the phases T1, δ', S' and θ' are completely dissolved in the Al matrix, and after cooling, they re-precipitate [17,24]. The heterogeneous distributions of these phases in the HAZ depend on the local thermal cycles that the HAZ is subjected to [18].

The study of Fatigue Crack Growth (FCG) behavior in welds is of great importance, as it changes along the length of the weld seam due to the different thermal cycles to which the material is subjected. Studies have shown that the FCG rate is higher in the HAZ [35,36]. Investigating the mechanical and fatigue properties of the HAZ of a laser weld is a demanding task, due to the narrowness and heterogeneity of the region [37]. The main aim of this study is to simulate the reduction of the microhardness in the HAZ of an Al–Cu–Li 2198–T351 alloy by implementing several isothermal heat treatment schemes. This implies that it could be possible to fabricate samples that possess similar mechanical properties with different regions of the HAZ [38]. This, in turn, leads to the conclusion that these specimens could be subjected to FCG tests instead of the welds themselves [35,37,39].

2. Materials and Methods

2.1. Material and Experimental Procedure

The material to be studied is the AA2198–T351 Al–Cu–Li alloy in the form of a 3.8 mm thick sheet. The chemical composition of the alloy is shown in Table 1.

Table 1. Composition of AA2198 alloy (wt. %) in this study.

Si	Fe	Cu	Mn	Mg	Cr	Zn	Zr	Li	Ag	Al
0.08	0.10	3.50	0.5	0.80	0.05	0.35	0.18	1.10	0.50	Rem.

In this study, the Laser Beam Welding (LBW) method was utilized for the bead-on-plate welding of the alloy, with shielding gas 50% Ar (17.5 L/mm) and 50% He (17.5 L/mm) [40]. The welding conditions are shown in Table 2.

Table 2. AA2198 laser beam welding conditions.

Power (W)	Speed (m/min)	Heat Input Rate (J/mm)	Arc Efficiency
3441	2	103	0.47

During welding of the AA2198 alloy, various metallurgical phenomena occur in the HAZ, leading to its division into two sub-zones. The first will be referred to as HAZ_1, where the maximum temperature exceeds the dissolution temperature of the main strengthening precipitates, causing them to dissolve completely and then re-precipitate. This phenomenon is equivalent to aging. The second one is HAZ_2, where the temperature is lower, and therefore, coarsening of the strengthening precipitates occurs, which is equivalent to over-aging. In fact, there is no exact boundary between these two regions; between them, a third sub-zone exists where both dissolution and coarsening occur simultaneously [41]. To simplify the problem in the present work, it will be assumed that the distinct boundary of the two sub-zones lies at the dissolution temperature of the main strengthening precipitate. To simulate the microhardness profile in the HAZ through isothermal heat treatments, appropriate aging and over-aging heat treatment schemes must be selected, which will be performed on samples sectioned from the parent material. Therefore, the microhardness values of HAZ_1 are replicated through solution heat treatments followed by natural aging, and those of HAZ_2 are replicated through over-aging heat treatments. This procedure is illustrated schematically in Figure 2. The conditions of the heat treatments, shown in Table 3, were selected in line with the Rosenthal solution model and DICTRA results, which are described in the next paragraph. Different temperatures were chosen for the solution heat treatment (450, 500 and 550 °C) to study whether the temperature affects the precipitation trend.

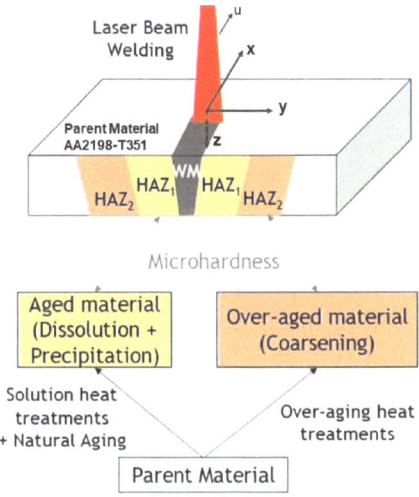

Figure 2. Schematic illustration of the experimental procedure in this study.

Table 3. Conditions (T,t) of the heat treatments carried out, followed by water quenching.

Solution Heat Treatment + Natural Aging	Over-Aging
450 °C/5, 10, 20, 40, 60 min	200 °C/8, 16, 24, 32, 48, 80, 153 h
500 °C/10, 20, 40, 60 min	250 °C/2, 4, 8, 16, 24, 32 h
550 °C/5, 10, 20, 40, 60 min	300 °C/8, 16 h

The heat-treated samples were metallographically prepared by embedding, grinding and polishing using 3 and 1 μm alumina paste. Keller's reagent was employed for etching the material. The microhardness measurements on the weld seam and the heat-treated samples were performed using the 402MVD Wolpert Wilson microhardness tester (Wilson Instruments, Norwood, MA, USA), with a load set at 0.2 kgf ($HV_{0.2}$) applied for 10 s. The macrostructure of the weld-seam was studied by utilizing a Leica Wilz M3Z stereoscope. The investigation of the precipitates formed during natural aging and over-aging was performed with a JEOL JEM-2100 LaB6 (JEOL Ltd., Tokyo, Japan) Transmission Electron

Microscope (TEM), operating at 200 kV. TEM specimens, with a 3 mm diameter, were prepared in the form of 30 μm-thick disk-type plates via mechanical polishing, followed by ion-polishing with a precision Ar-ion polishing system (PIPS) (Gatan model 691). Elemental analyses were carried out using an Oxford X-Max 100 Silicon Drift Energy Dispersive X-ray spectrometer (EDS) (Oxford Instruments, Abingdon, Oxfordshire, UK), connected to the TEM. Data were acquired in areas ranging from 2 to 5 nm in STEM mode.

2.2. Computational Procedure

A computational part preceded the experimental process in order to select the conditions of the heat treatments and predict the evolution of the main strengthening precipitates.

Thermocalc software (version 2022a, TCAL7 database) and Diffusion Module (DICTRA) (version 2022a, TCAL7/MOBAL5 databases) [42] were employed for the selection of the appropriate heat treatment time and the determination of the behavior of the main strengthening precipitates during the isothermal heat treatments.

Furthermore, the conditions of the heat treatments were chosen according to the thermal cycles the material was subjected to during welding. For this purpose, the Rosenthal solution (Equations (1)–(5)) [43] was implemented using suitable computing software. A coordinate (w, y, z) system is used (Figure 2), which moves at the same speed as the welding arc (u):

$$w = x - ut \tag{1}$$

$$T - T_o = \frac{Q}{2\pi k} e^{-\frac{u}{2\alpha}w} \left[\frac{e^{-\frac{u}{2\alpha}R}}{R} + \sum_{n=1}^{\infty} \left(\frac{e^{-\frac{u}{2\alpha}R_n}}{R_n} + \frac{e^{-\frac{u}{2\alpha}R'_n}}{R'_n} \right) \right] \tag{2}$$

$$R = \sqrt{w^2 + y^2 + z^2} \tag{3}$$

$$R_n = \sqrt{w^2 + y^2 + (2nH - z)^2} \tag{4}$$

$$R'_n = \sqrt{w^2 + y^2 + (2nH + z)^2} \tag{5}$$

The symbols used for the parameters are explained in Table 4.

Table 4. Model parameters and their numerical values.

Model Parameter (Unit)	Symbol	Numerical Value
Temperature (°C)	T	Calculated
Initial temperature (°C)	T_o	25
Arc thermal power (arc efficiency × power) (W)	Q	1617
Thermal conductivity (W/mm°C)	k	0.125
Arc speed (mm/s)	u	33.3
Thermal diffusivity (mm^2/s)	α	51.63
Thickness of Al sheet (mm)	H	3.8

Temperature field developed during LBW was calculated using Equation (2), entering the values depicted in Table 4. The results were validated via experimental observations regarding the boundary between the WM and the HAZ and are discussed in Section 3.2.

3. Results

3.1. Strengthening Precipitates and Their Evolution during the Welding Thermal Cycle

The boundary between HAZ$_1$ and HAZ$_2$ is set at the temperature at which the main strengthening precipitate(s) dissolve completely. For the AA2198, these are the T1, θ' and δ' precipitates. According to ThermoCalc calculations, the T1 and θ' precipitates dissolve at 493 °C, so this temperature is chosen (Figure 3). Therefore, in Figure 3 the temperature boundaries of the two sub-zones are shown on the calculated phase diagram. The upper boundary of HAZ1 is set to the Liquidus Temperature (651 °C), whereas the lower boundary of HAZ$_2$ corresponds to the artificial aging temperature (175 °C).

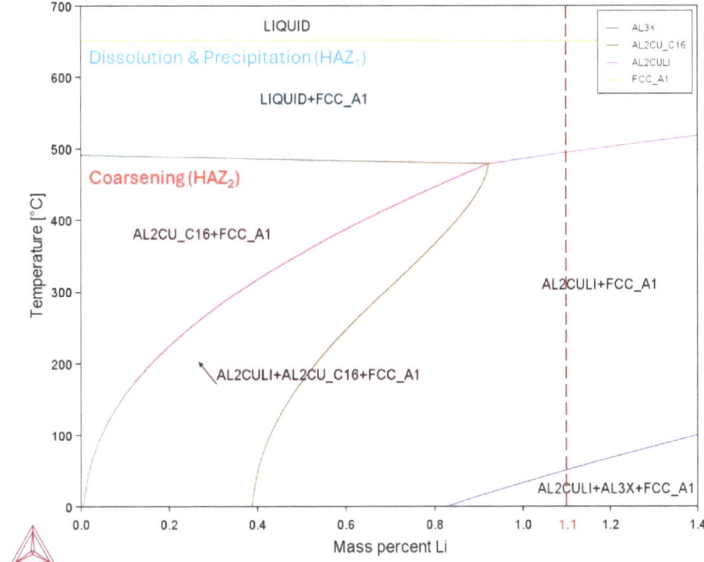

Figure 3. Calculated phase diagram for AA2198 alloy, along with the temperature boundaries of the two subzones HAZ$_1$ and HAZ$_2$.

The conditions of the isothermal heat treatments to replicate the microhardness reduction of these two zones were selected according to the Diffusion Module calculations. In HAZ$_2$ where coarsening takes place, temperatures ranging from 180 °C to 300 °C were tested with a step of 25 °C. As shown in Figure 4, as the temperature increases, the maximum diameter of θ′ increases. According to the calculations, the maximum diameter increases up to 63% at 300 °C in comparison with the initial mean diameter of θ′.

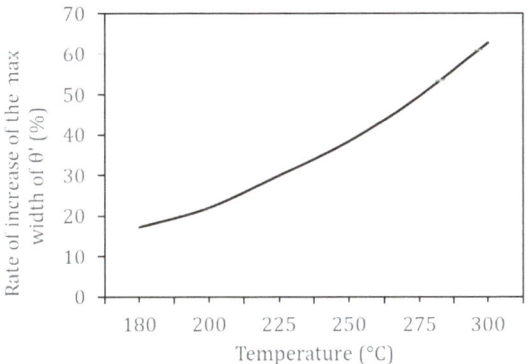

Figure 4. Rate of increase of the maximum width (nm) of the precipitate θ′ in relation to the initial width for various temperatures.

3.2. Temperature Field Developed during Welding

The calculated temperature field that develops in the sheet during LBW is shown in Figure 5a.

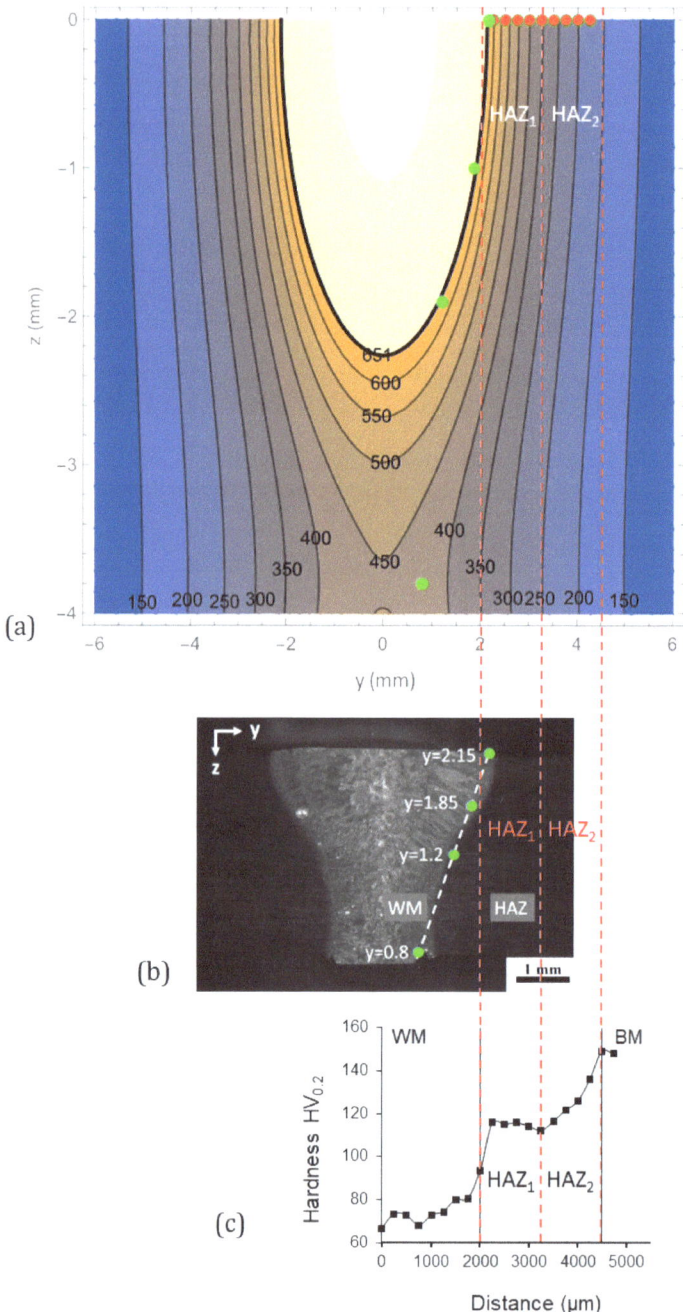

Figure 5. (a) Calculated temperature field in relation to the depth (z) and width (y) of the sheet, at x = 0. The green marks depict the experimental boundary between the WM and HAZ on the micrograph of the weld seam, while the red marks show the locations of the microhardness tests; (b) OM micrograph of the weld seam. Dashed line represents the WM/HAZ boundary where the max temperature during welding is equal to the liquidus, while the green marks on the boundary are the ones plotted in (a); (c) microhardness profile of the weld seam. The red dashed lines represent the comparison between the theoretical and experimental boundaries of HAZ_1 and HAZ_2.

In Figure 5a, the green dots depict the experimental points that lie on the actual boundary between the WM and the HAZ (Figure 5b), where the maximum temperature of the material during welding is equal to the liquidus (T_{liq} = 651 °C). The microhardness test results of the weld seam are presented in Figure 5c. The initial hardness of the material is 150 $HV_{0.2}$, while the microhardness decreases while moving from the center of the weld. Two sub-zones seem to be formed in the HAZ: the first at 2250–3250 μm where the microhardness value remains constant at approximately 115 $HV_{0.2}$ and the second at 3500–4250 μm, where the microhardness ranges between 116–136 $HV_{0.2}$.

As shown in Figure 6, the points on the experimental T_{liq} are very close to the points of the calculated T_{liq}, except for the last point (y = 0.8, z = 3.8). This is attributed to the fact that the Rosenthal solution does not take into account parameters such as the keyhole method that has been used. With this method, the arc reaches a greater depth in the material, which, however, cannot be predicted through the computational model. However, this does not negate the fact that the calculations correspond to the experimental data. The percentages in Figure 6 indicate the error divergence of the theoretical model in relation to the experimental data.

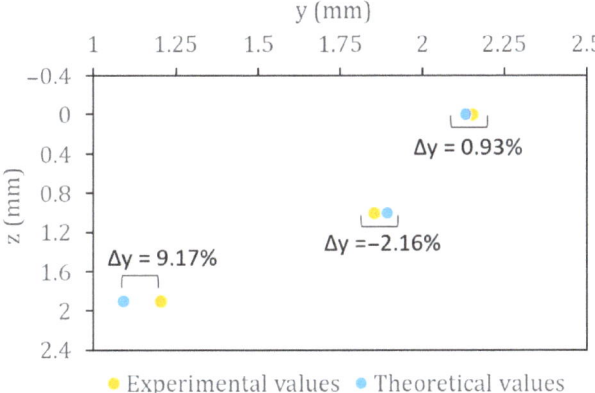

Figure 6. Comparison of the points on the experimental and theoretical WM/HAZ where the maximum temperature reached is equal to the T_{liq} = 651 °C. This indicates that the theoretical model is close to the real values, as the Δy% is relatively low.

Time is another parameter to consider in the calculations. Therefore, the thermal cycles to which different points of the HAZ are subjected are shown in Figure 7. These points are depicted in red in Figure 5a and are located at the surface of the sheet (z – 0 mm). At points from 2.25 to 3 mm, the temperature exceeds 493 °C during welding, causing the strengthening phases to dissolve completely and re-precipitate. At 3.25 mm, the temperature reaches up to 460 °C, resulting in a significant dissolution of the strengthening precipitates. Therefore, points 2.25 to 3.25 mm belong to HAZ_1 sub-zone. At points 3.5–4.25 mm, the temperature does not exceed the temperature of complete dissolution; therefore, these points belong to the HAZ_2 sub-zone where coarsening occurs.

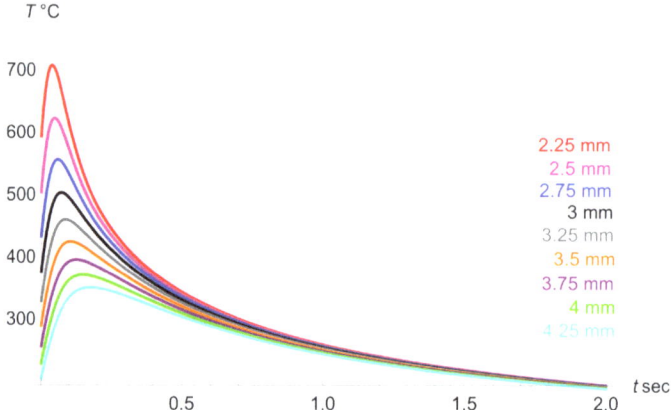

Figure 7. Calculated thermal cycles induced at various points on the HAZ. These points are depicted in Figure 5a in red. The legend indicates the y(mm) coordinate of each point.

3.3. Microhardness Tests Results

The microhardness tests results of the naturally aged samples are presented in Figure 8. The tests were carried out immediately after the heat treatment (0 h) and at intervals of 1 h up to 18 days afterward in order to investigate the effect of natural aging on the microhardness of the alloy.

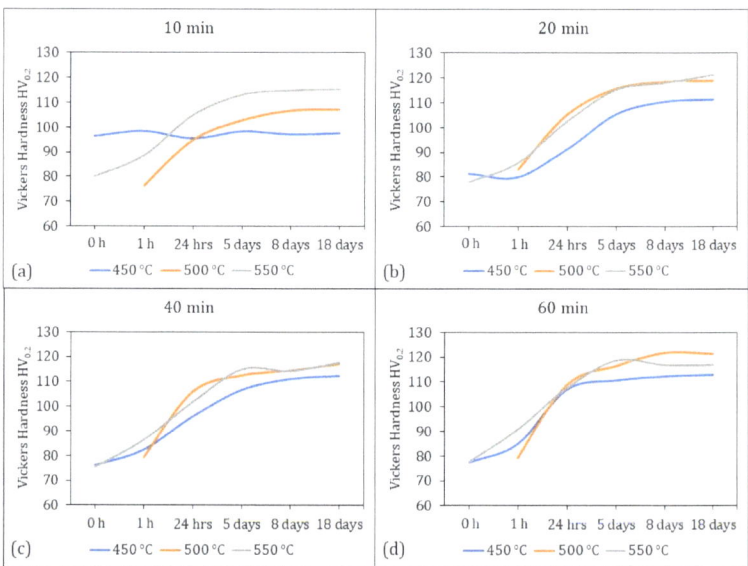

Figure 8. Microhardness test results for the solution heat-treated samples for hold times (**a**) 10 min, (**b**) 20 min, (**c**) 40 min and (**d**) 60 min. Afterward, the samples were naturally aged for various time intervals between 1 h and 18 days. Time 0 h indicates the tests performed immediately after the dissolution.

After the solution heat treatment, the microhardness of the alloy decreases from 150 $HV_{0.2}$ to about 80 $HV_{0.2}$ in almost all cases. The microhardness increases over time due to natural aging, reaching a mean value of 120 $HV_{0.2}$. Microhardness appears to stabilize after approximately 8 days.

As shown in Figure 9, the microhardness of the over-aged samples drops to 132 $HV_{0.2}$ after 8 h at 200 °C, while it stabilizes after approximately 48 h. For the samples heated to 250 °C, the microhardness decreases to 128 $HV_{0.2}$ after 2 h and stabilizes around 85 $HV_{0.2}$ after 24 h. For the samples heated to 300 °C, the decrease in microhardness is approximately 75 $HV_{0.2}$ after 8 h and stabilizes around 65 $HV_{0.2}$.

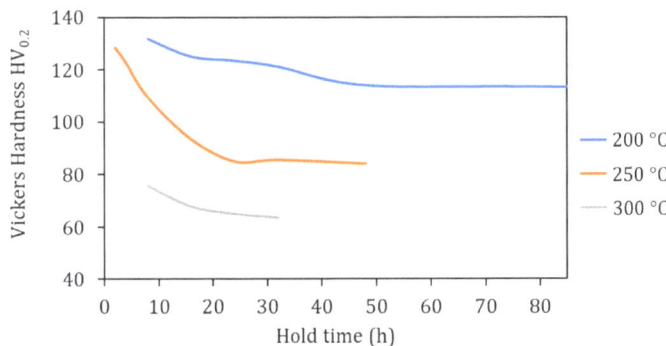

Figure 9. Microhardness test results for the overaged heat-treated samples.

4. Discussion

4.1. Simulation of the HAZ Microhardness Profile with Isothermal Heat Treatments

One of the objectives of this study is to replicate the microhardness profile of the HAZ with appropriate isothermal heat treatment schemes. In Figure 10, the HAZ microhardness profile is compared with the nearest microhardness values obtained through the heat treatments.

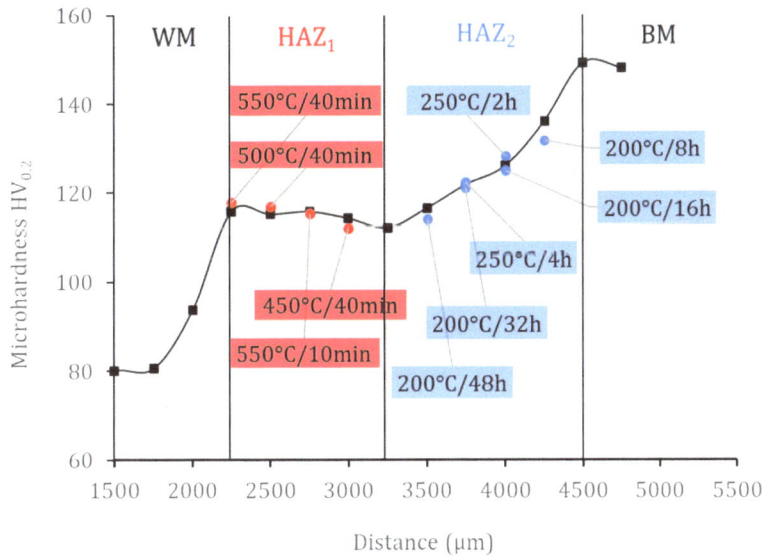

Figure 10. The microhardness profile of the weld seam compared to microhardness values resulting from isothermal heat treatments (red and blue points).

At the 2.25–3 mm region of the HAZ microhardness profile, the temperature during welding exceeded 493 °C (Figure 7); therefore, dissolution and precipitation occurred. The microhardness of this zone (112–116 HV$_{0.2}$) can be simulated with solution heat treatments followed by natural aging, as shown in Figure 10, where the microhardness values of the isothermal treatments are correlated to the microhardness profile. At the 3.5–4.25 mm region, the maximum temperature did not exceed the temperature of complete dissolution (Figure 7); therefore, coarsening occurred. As shown in Figure 10, the microhardness profile of HAZ$_2$, ranging between 116–136 HV$_{0.2}$, can be sufficiently simulated by over-aging heat treatments. The heat treatments that closely correspond in value to the microhardness profile are 200 °C/48 h, 200 °C/32 h, 250 °C/4 h, 250 °C/2 h, 200 °C/16 h and 200 °C/8 h.

4.2. Correlation between Microhardness and Microstructure Changes

The reduction in the microhardness that occurs after dissolution/re-precipitation and coarsening are attributed to microstructure changes that involve the T$_1$ and θ' precipitates. The TEM analysis of the as-received, naturally aged (after solution heat-treat 500 °C/20 min) and overaged (200 °C/48 h) material are presented in Figure 11.

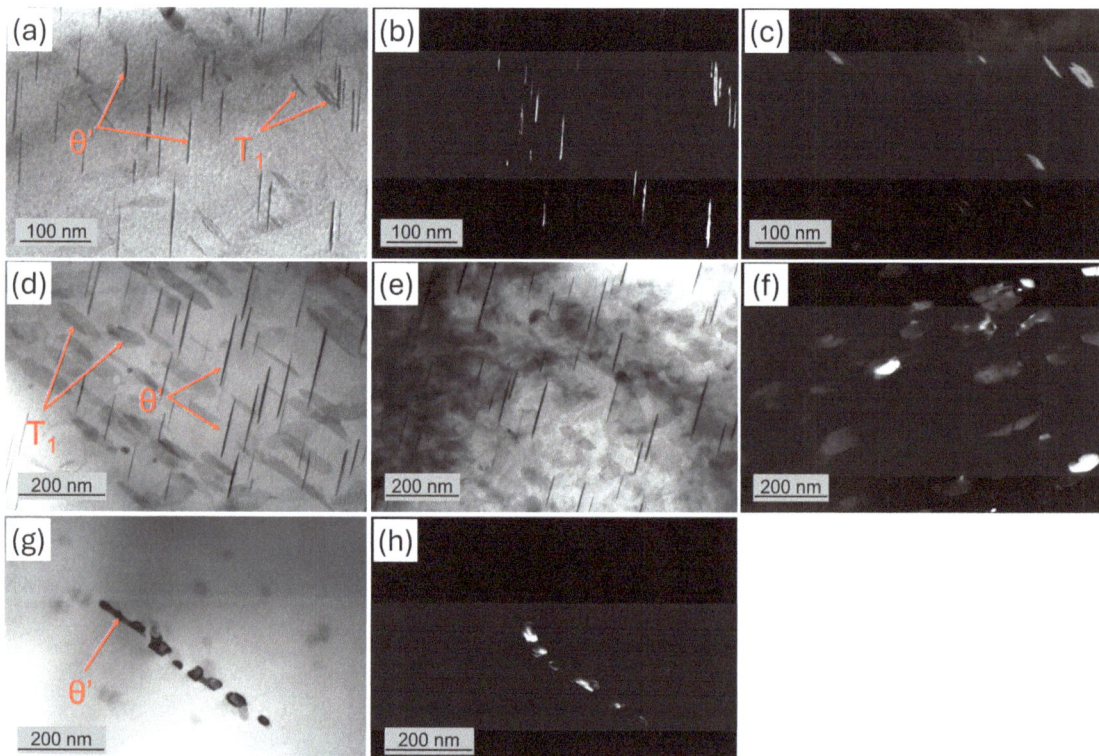

Figure 11. TEM BF and DF micrographs of (**a–c**) the as-received sample where θ' and T1 phases are evident, (**d–f**) the solution heat treated at 500 °C for 20 min and naturally aged sample, (**g,h**) the overaged sample at 200 °C for 48 h.

As shown in Figure 11a–c, in the as-received state, two types of needle-shaped intermetallics crystallize with different orientations within the same grain. According to the EDS analyses, one type is θ', whose length varies between 20–200 nm and whose mean width is 2.5 ± 0.6 nm. The second intermetallic is T$_1$, which shows a greater width (5.85 ± 2.3 nm) and a shorter length than θ'. After dissolution and natural aging (Figure 11d–f), two types of intermetallics are formed in the microstructure as well: θ' and T$_1$. After re-precipitation, the mean width of θ' is 4.86 ± 1.63 nm, which is slightly increased compared to the as-received state. On the contrary, T_1 precipitates with a significantly increased mean width of 21.61 ± 9.25 nm, compared to T351 temper state. This is attributed to the fact that the formation of T_1 phase is dependent on the presence of dislocations, which is why pre-deformation favors the nucleation and precipitation of T_1 [44]. In the as-received state, the specimen, being in temper T351, has undergone cold deformation, while the solution heat-treated (500 °C/20 min) specimen has undergone dissolution and re-precipitation through natural aging without any pre-deformation. Previous studies have shown that the mean thickness of T_1 decreases with increasing pre-deformation [44]. Therefore, due to the absence of pre-deformation, T_1 precipitates with a significantly greater width, which leads to a reduction in microhardness. After over-aging, θ' (Figure 11g,h) seems to prevail in the microstructure. The considerable difference in its mean width after overaging is evident, as from 2.5 nm in the as-received state, after heat treatment, the width increased to an average value of 25.5 ± 9 nm. This is to be expected, as over-aging leads to the coarsening of the strengthening phases, negatively affecting the microhardness. In Table 5, the mean width and volume fraction of the two strengthening phases are presented. Volume fractions were calculated through a two-dimensional method.

Table 5. Comparison of the mean width and volume fraction (f) of the θ' and T_1 phases in three as-received, naturally aged and overaged conditions in combination with the microhardness values.

Phase	As-Received		HT 500 °C/20 min + Natural Aging		HT 200 °C/48 h	
	Width (nm)	f (%)	Width (nm)	f (%)	Width (nm)	f (%)
θ' (Al$_2$Cu)	2.5 ± 0.6	3.42 ± 1.03	4.86 ± 1.63	2.59 ± 0.97	25.5 ± 9	3.42 ± 1.03
T1 (Al$_2$CuLi)	5.85 ± 2.3	1.17 ± 0.29	21.61 ± 9.25	8.46 ± 2.3	-	-
HV$_{0.2}$	150		119		114.2	

5. Conclusions

In the present work, the reduction of the HAZ microhardness in an AA2198-T351 LBW was simulated through isothermal heat treatments. The HAZ was divided into two subzones: HAZ$_1$ and HAZ$_2$. Within the first zone, the temperature field developing during welding exceeds the dissolution temperature of the main strengthening precipitates, so the material is subjected to aging. In HAZ$_2$, the temperature is lower than that limit, which leads to the coarsening of the strengthening phases (over-aging). In both cases, the microhardness decreases. To decide on the appropriate conditions (T, t) for the heat treatments, the thermal cycles at different points of the HAZ were calculated through the application of the Rosenthal solution, and ThermoCalc software (version 2022a) and Diffusion Module (DICTRA) (version 2022a) were utilized. Microhardness tests were performed on the naturally aged and over-aged samples, and the values were matched with the HAZ microhardness profile. Our conclusions are as follows:

1. The calculated temperature field in the weld seam matches the microhardness profile; therefore, the theoretical and experimental data are in agreement.
2. After solution heat-treatments and natural aging, the microhardness of the material decreases from 150 to approximately 120 HV$_{0.2}$ due to the re-precipitation of T_1 with significantly increased width.
3. After over-aging heat treatments, the width of θ' phase is almost ten times higher compared to the as-received sample, which is responsible for the reduction in microhardness. Microhardness decreases at a faster rate as the temperature increases. At 200 °C, it decreased to the value of 112 HV$_{0.2}$ for 250 °C to 85 HV$_{0.2}$ and for 300 °C to 64 HV$_{0.2}$.
4. With appropriate isothermal heat treatments, the microhardness profile of the HAZ during LBW can be replicated accurately, implying that it is possible to fabricate samples for the experimental study of damage tolerance behavior in the HAZ.

Author Contributions: S.M. methodology, experimental investigation, validation, software, original draft preparation. S.D. methodology, investigation, review and editing. P.E.T. methodology, investigation, Validation, review and editing. A.D.Z. Conceptualization, methodology, supervision, Validation, review and editing. All authors have read and agreed to the published version of the manuscript.

Funding: This research received no external funding.

Data Availability Statement: The raw and processed data required to reproduce these findings cannot be shared at this time due to technical and time limitations. The data can be shared through direct contact with the corresponding author.

Acknowledgments: Acknowledgements to the Director of the Laboratory of Materials of the Department of Mechanical Engineering UTH G.N. Haidemenopoulos for providing the material and his support.

Conflicts of Interest: The authors declare no conflicts of interest.

References

1. Duan, S.-W.; Matsuda, K.; Wang, T.; Zou, Y. Microstructures and Mechanical Properties of a Cast Al–Cu–Li Alloy during Heat Treatment Procedure. *Rare Met.* **2021**, *40*, 1897–1906. [CrossRef]
2. Lv, K.; Zhu, C.; Zheng, J.; Wang, X.; Chen, B. Precipitation of T_1 Phase in 2198 Al–Li Alloy Studied by Atomic-Resolution HAADF-STEM. *J. Mater. Res.* **2019**, *34*, 3535–3544. [CrossRef]
3. Eswara Prasad, N.; Gokhale, A.A.; Wanhill, R.J.H. Aluminium–Lithium Alloys. In *Aerospace Materials and Material Technologies*; Prasad, N.E., Wanhill, R.J.H., Eds.; Springer: Singapore, 2017; pp. 53–72, ISBN 9789811021336.
4. Zhang, S.; Zeng, W.; Yang, W.; Shi, C.; Wang, H. Ageing Response of a Al–Cu–Li 2198 Alloy. *Mater. Des.* **2014**, *63*, 368–374. [CrossRef]
5. Hajjioui, E.A.; Bouchaâla, K.; Faqir, M.; Essadiqi, E. A Review of Manufacturing Processes, Mechanical Properties and Precipitations for Aluminum Lithium Alloys Used in Aeronautic Applications. *Heliyon* **2023**, *9*, e12565. [CrossRef]
6. Wu, H.; Wu, G.; Xiang, L.; Tao, J.; Zheng, Z.; Sun, J.; Li, W.; Huang, C.; Lan, X. Impact of Corrosion on the Degradation of the Mechanical Properties of 2195 and 2297 Al Alloys in the Marine Environment. *Metals* **2022**, *12*, 1371. [CrossRef]
7. Kramer, L.S.; Langan, T.J.; Pickens, J.R.; Last, H. Development of Al-Mg-Li Alloys for Marine Applications. *J. Mater. Sci.* **1994**, *29*, 5826–5832. [CrossRef]
8. Tsivoulas, D.; Prangnell, P.B. Comparison of the Effect of Individual and Combined Zr and Mn Additions on the Fracture Behavior of Al-Cu-Li Alloy AA2198 Rolled Sheet. *Met. Mater. Trans. A* **2014**, *45*, 1338–1351. [CrossRef]
9. Zou, Y.; Chen, X.; Chen, B. Corrosion Behavior of 2198 Al–Cu–Li Alloy in Different Aging Stages in 3.5 wt% NaCl Aqueous Solution. *J. Mater. Res.* **2018**, *33*, 1011–1022. [CrossRef]
10. Starke, E.A. Historical Development and Present Status of Aluminum–Lithium Alloys. In *Aluminum-Lithium Alloys*; Elsevier: Amsterdam, The Netherlands, 2014; pp. 3–26, ISBN 9780124016989.
11. Hirosawa, S.; Sato, T.; Kamio, A.; Flower, H.M. Classification of the Role of Microalloying Elements in Phase Decomposition of Al Based Alloys. *Acta Mater.* **2000**, *48*, 1797–1806. [CrossRef]
12. Jambor, M.; Nový, F.; Bokůvka, O.; Trško, L. The Natural Aging Behavior of the AA 2055 Al-Cu-Li Alloy. *Transp. Res. Procedia* **2019**, *40*, 42–45. [CrossRef]
13. Li, S.; Yue, X.; Li, Q.; Peng, H.; Dong, B.; Liu, T.; Yang, H.; Fan, J.; Shu, S.; Qiu, F.; et al. Development and Applications of Aluminum Alloys for Aerospace Industry. *J. Mater. Res. Technol.* **2023**, *27*, 944–983. [CrossRef]
14. Liu, S.; Wróbel, J.S.; LLorca, J. First-Principles Analysis of the Al-Rich Corner of Al-Li-Cu Phase Diagram. *Acta Mater.* **2022**, *236*, 118129. [CrossRef]
15. Li, Y.; Shi, Z.; Lin, J. Experimental Investigation and Modelling of Yield Strength and Work Hardening Behaviour of Artificially Aged Al-Cu-Li Alloy. *Mater. Des.* **2019**, *183*, 108121. [CrossRef]
16. Decreus, B.; Deschamps, A.; De Geuser, F.; Donnadieu, P.; Sigli, C.; Weyland, M. The Influence of Cu/Li Ratio on Precipitation in Al–Cu–Li–x Alloys. *Acta Mater.* **2013**, *61*, 2207–2218. [CrossRef]
17. Tao, Y.; Zhang, Z.; Xue, P.; Ni, D.R.; Xiao, B.L.; Ma, Z.Y. Effect of Post Weld Artificial Aging and Water Cooling on Microstructure and Mechanical Properties of Friction Stir Welded 2198-T8 Al-Li Joints. *J. Mater. Sci. Technol.* **2022**, *123*, 92–112. [CrossRef]
18. Zhao, T.; Sato, Y.S.; Xiao, R.; Huang, T.; Zhang, J. Hardness Distribution and Aging Response Associated with Precipitation Behavior in a Laser Pressure Welded Al–Li Alloy 2198. *Mater. Sci. Eng. A* **2021**, *808*, 140946. [CrossRef]
19. Gao, C.; Zhu, Z.; Han, J.; Li, H. Correlation of Microstructure and Mechanical Properties in Friction Stir Welded 2198-T8 Al–Li Alloy. *Mater. Sci. Eng. A* **2015**, *639*, 489–499. [CrossRef]
20. Cavaliere, P.; Cabibbo, M.; Panella, F.; Squillace, A. 2198 Al–Li Plates Joined by Friction Stir Welding: Mechanical and Microstructural Behavior. *Mater. Des.* **2009**, *30*, 3622–3631. [CrossRef]
21. Gao, C.; Ma, Y.; Tang, L.; Wang, P.; Zhang, X. Microstructural Evolution and Mechanical Behavior of Friction Spot Welded 2198-T8 Al-Li Alloy during Aging Treatment. *Mater. Des.* **2017**, *115*, 224–230. [CrossRef]

22. Zeng, Q.; Zeng, S.; Wang, D. Stress-Corrosion Behavior and Characteristics of the Friction Stir Welding of an AA2198-T34 Alloy. *Int. J. Miner. Met. Mater.* **2020**, *27*, 774–782. [CrossRef]
23. Texier, D.; Zedan, Y.; Amoros, T.; Feulvarch, E.; Stinville, J.C.; Bocher, P. Near-Surface Mechanical Heterogeneities in a Dissimilar Aluminum Alloys Friction Stir Welded Joint. *Mater. Des.* **2016**, *108*, 217–229. [CrossRef]
24. Rao, J.; Payton, E.J.; Somsen, C.; Neuking, K.; Eggeler, G.; Kostka, A.; Dos Santos, J.F. Where Does the Lithium Go?—A Study of the Precipitates in the Stir Zone of a Friction Stir Weld in a Li-containing 2xxx Series Al Alloy. *Adv. Eng. Mater.* **2010**, *12*, 298–303. [CrossRef]
25. Bitondo, C.; Prisco, U.; Squillace, A.; Giorleo, G.; Buonadonna, P.; Dionoro, G.; Campanile, G. Friction Stir Welding of AA2198-T3 Butt Joints for Aeronautical Applications. *Int. J. Mater. Form.* **2010**, *3*, 1079–1082. [CrossRef]
26. Li, W.; Jiang, R.; Zhang, Z.; Ma, Y. Effect of Rotation Speed to Welding Speed Ratio on Microstructure and Mechanical Behavior of Friction Stir Welded Aluminum–Lithium Alloy Joints. *Adv. Eng. Mater.* **2013**, *15*, 1051–1058. [CrossRef]
27. Ma, Y.E.; Xia, Z.C.; Jiang, R.R.; Li, W. Effect of Welding Parameters on Mechanical and Fatigue Properties of Friction Stir Welded 2198 T8 Aluminum–Lithium Alloy Joints. *Eng. Fract. Mech.* **2013**, *114*, 1–11. [CrossRef]
28. Examilioti, T.N.; Kashaev, N.; Enz, J.; Klusemann, B.; Alexopoulos, N.D. On the Influence of Laser Beam Welding Parameters for Autogenous AA2198 Welded Joints. *Int. J. Adv. Manuf. Technol.* **2020**, *110*, 2079–2092. [CrossRef]
29. Zhao, T.; Sato, Y.S.; Xiao, R.; Huang, T.; Zhang, J. Laser Pressure Welding of Al-Li Alloy 2198: Effect of Welding Parameters on Fusion Zone Characteristics Associated with Mechanical Properties. *High Temp. Mater. Process.* **2020**, *39*, 146–156. [CrossRef]
30. Yin, S.; Lü, J.; Xiao, R. Corrosion Behavior of Laser Beam Welded Joint of 2198 Aluminium-Lithium Alloy with Filler Wire. *Chin. J. Laser* **2016**, *43*, 0403007. [CrossRef]
31. Lin, K.; Yang, W.; Huang, T.; Xiao, R. Laser Welding of 2198-T851 Al-Li Alloy with Filler Wire. *Guang Xue Xue Bao* **2015**, *35*, s216001. [CrossRef]
32. Węgrzyn, T.; Piwnik, J.; Łazarz, B.; Wieszała, R.; Hadryś, D. Parameters of Welding with Micro-Jet Cooling. *Arch. Mater. Sci. Eng.* **2012**, *54*, 86–92.
33. Szczucka-Lasota, B.; Węgrzyn, T.; Krzysztoforski, M. Aluminum Alloys Welding with Micro-Jet Cooling in Busducts. *Weld. Tech. Rev.* **2018**, *91*, 24–30. [CrossRef]
34. David, S.A.; Babu, S.S.; Vitek, J.M. Welding. In *Encyclopedia of Materials: Science and Technology*; Elsevier: Amsterdam, The Netherlands, 2003; pp. 1–9, ISBN 9780080431529.
35. Song, W.; Man, Z.; Xu, J.; Wang, X.; Liu, C.; Zhou, G.; Berto, F. Fatigue Crack Growth Behavior of Different Zones in an Overmatched Welded Joint Made with D32 Marine Structural Steel. *Metals* **2023**, *13*, 535. [CrossRef]
36. Song, Y.; Chai, M.; Han, Z. Experimental Investigation of Fatigue Crack Growth Behavior of the 2.25Cr1Mo0.25V Steel Welded Joint Used in Hydrogenation Reactors. *Materials* **2021**, *14*, 1159. [CrossRef] [PubMed]
37. Smaili, F.; Lojen, G.; Vuherer, T. Fatigue Crack Initiation and Propagation of Different Heat Affected Zones in the Presence of a Microdefect. *Int. J. Fatigue* **2019**, *128*, 105191. [CrossRef]
38. Tzamtzis, A. *Fatigue Crack Growth Prediction under Mode I Loading in Friction Stir Aluminum Alloy Weld*; University of Thessaly, School of Engineering, Department of Mechanical Engineering: Volos, Greece, 2015.
39. Kim, S.; Kang, D.; Kim, T.-W.; Lee, J.; Lee, C. Fatigue Crack Growth Behavior of the Simulated HAZ of 800MPa Grade High-Performance Steel. *Mater. Sci. Eng. A* **2011**, *528*, 2331–2338. [CrossRef]
40. Vriami, D. Experimental Study of Welding of New Aircraft Aluminum Alloys with High Power Beams (LBW, EBW). Diploma Thesis, Department of Mechanical Engineering, University of Thessaly, Volos, Greece, 2010. (In Greek)
41. Zervaki, A.D.; Haidemenopoulos, G.N. Computational Kinetics Simulation of the Dissolution and Coarsening in the HAZ during Laser Welding of 6061-T6 Al-Alloy. *Weld. Res.* **2007**, *86*, 211–221.
42. Andersson, J.-O.; Helander, T.; Höglund, L.; Shi, P.; Sundman, B. Thermo-Calc & DICTRA, Computational Tools for Materials Science. *Calphad* **2002**, *26*, 273–312. [CrossRef]
43. Masubuchi, K. *Analysis of Welded Structures*; Elsevier: Amsterdam, The Netherlands, 1980; pp. 60–71, ISBN 9780080227146.
44. Yang, Y.; He, G.; Liu, Y.; Li, K.; Wu, W.; Huang, C. Quantitative Contribution of T1 Phase to the Strength of Al-Cu-Li Alloys. *J. Mater. Sci.* **2021**, *56*, 18368–18390. [CrossRef]

Disclaimer/Publisher's Note: The statements, opinions and data contained in all publications are solely those of the individual author(s) and contributor(s) and not of MDPI and/or the editor(s). MDPI and/or the editor(s) disclaim responsibility for any injury to people or property resulting from any ideas, methods, instructions or products referred to in the content.

Article

Heat-Affected Zone Microstructural Study via Coupled Numerical/Physical Simulation in Welded Superduplex Stainless Steels

Leonardo Oliveira Passos da Silva [1], Tiago Nunes Lima [2], Francisco Magalhães dos Santos Júnior [1], Bruna Callegari [2], Luís Fernando Folle [2] and Rodrigo Santiago Coelho [1,2,*]

[1] Programa de Pós-Graduação GETEC/MPDS/MCTI, Centro Universitário SENAI CIMATEC, Av. Orlando Gomes 1845, Salvador 41650-010, Brazil; leonardo.passos@fieb.org.br (L.O.P.d.S.); francisco.santos@senaicni.com.br (F.M.d.S.J.)

[2] SENAI CIMATEC, Instituto SENAI de Inovação em Conformação e União de Materiais (CIMATEC ISI C&UM), Av. Orlando Gomes 1845, Salvador 41650-010, Brazil; tiago.nunes@fieb.org.br (T.N.L.); bruna.callegari@fieb.org.br (B.C.)

* Correspondence: rodrigo.coelho@fieb.org.br

Citation: da Silva, L.O.P.; Lima, T.N.; Júnior, F.M.d.S.; Callegari, B.; Folle, L.F.; Coelho, R.S. Heat-Affected Zone Microstructural Study via Coupled Numerical/Physical Simulation in Welded Superduplex Stainless Steels. *Crystals* **2024**, *14*, 204. https://doi.org/10.3390/cryst14030204

Academic Editors: Reza Beygi, Mahmoud Moradi and Ali Khalfallah

Received: 26 January 2024
Revised: 9 February 2024
Accepted: 12 February 2024
Published: 21 February 2024

Copyright: © 2024 by the authors. Licensee MDPI, Basel, Switzerland. This article is an open access article distributed under the terms and conditions of the Creative Commons Attribution (CC BY) license (https://creativecommons.org/licenses/by/4.0/).

Abstract: Superduplex stainless steels (SDSS) are known for their combination of good mechanical properties and excellent corrosion resistance, enabled by the microstructural balance between austenite and ferrite and an amount of alloying elements. Their application in welded components is, however, limited by the possibility of the precipitation of intermetallic phases and microstructural misbalance, which might hinder their properties, especially in the heat-affected zone (HAZ). This work introduces a methodology that relies simultaneously on physical and numerical simulations to study the HAZ in a UNS S32750 SDSS. Dimensions of the fusion zone and thermal cycles were calibrated for a numerical model using preliminary welding trials. Numerically simulated cycles for each heat input (HI) were physically reproduced in a Gleeble® simulator, and the heat-treated samples were characterized and compared with real specimens welded using the same parameters. Thermal curves resulting from the numerical simulations were successfully replicated by the Gleeble®, indicating adequate application of the desired HI. The hardness and microstructural results from simulated and welded specimens were also found to be quite similar. Therefore, the proposed methodology showed itself adequate not only for the study of duplex stainless steels, but also of materials with similar thermal and mechanical properties, including the extrapolation of welding parameters.

Keywords: welding; numerical simulation; physical simulation; superduplex stainless steel

1. Introduction

Many industrial sectors demand advanced materials with excellent mechanical and corrosion properties. One example of such materials are the duplex stainless steels [1]. Frequently, components wherein these materials are used involve a welding step in their manufacturing process. Therefore, the weldability of duplex stainless steels has been the focus of extensive research [2].

Stainless steels are widely used in a variety of applications due to their corrosion resistance achieved by a passive Cr_2O_3 film, which prevents future oxidation, ensuring resistance to corrosion in these alloys. The family of stainless steels includes martensitic, ferritic, austenitic, precipitation-hardened, and duplex steels, classified according to their microstructural phases and chemical composition. Duplex stainless steels emerged from industry's increasing need for materials that combine corrosion resistance and mechanical strength. These alloys consist of roughly equal fractions of ferrite (δ) and austenite (γ), combining the favorable properties of austenitic and ferritic stainless steels. Among duplex steels, superduplex stainless steels stand out for their higher content of alloying elements and a pitting resistance equivalent number (PREN) greater than 40, developed to enhance the properties of conventional duplex steels. The application of these materials provides a significant reduction in the thickness of equipment such as pressure vessels and heat exchangers, and a consequent weight reduction [3–5].

The heavy amount of alloying elements in duplex stainless steels enables high levels of mechanical and corrosive properties. On the other hand, these elements also facilitate the appearance of intermetallic phases in certain time/temperature conditions, such as those typically faced during welding processes [6]. In these steels, secondary phases can precipitate in temperatures ranging between 300 °C and 1000 °C, preferentially at grain boundaries, leading to embrittlement and negatively affecting their corrosion resistance [7]. One of the main deleterious phases is sigma (σ), which can reduce both the impact and corrosion resistance of duplex stainless steels, leading to as much as a 90% decrease in toughness when its quantity lies close to 4% [8–10].

Due to its likely deleterious effects, it is absolutely important to understand the effect of the welding thermal cycle on the fabrication and operation of industrial equipment, with the need to control both composition and microstructure in the weld region [7]. Properties of the fusion zone (FZ) are influenced by the choice of filler and shielding gases, which can contain some alloying elements to compensate for possible losses that occur during the process [3]. In its turn, properties of the heat-affected zone (HAZ) are affected by the thermal cycle alone, which is why it is crucial to understand its behavior upon exposure to heat [11]. In some materials, such as duplex stainless steels, this region is extremely narrow and difficult to isolate, imposing limitations on its characterization, especially in the case of multi-pass welding [2,3,12].

Simulation plays a key role in several manufacturing stages, e.g., casting, welding and heat treatment [13]. Regarding welding, numerical and physical simulations allow the prediction of the material's response when exposed to thermal cycles, thus enabling an enlargement of the HAZ for subsequent analysis [14–18]. Computational simulations are useful to predict thermal cycles and microstructural evolution, which converts into the time reduction and economy of resources necessary to evaluate the effect of parameter variation on the behavior of the welded joint [19]. Likewise, physical simulation reproduces thermal cycles in larger-scale specimens for subsequent laboratory testing to study in detail the region that is subjected to a specific thermal cycle [20]. However, studies that combine physical and numerical simulations to analyze and predict HAZ microstructure and behavior are scarce, mainly when it comes to multi-pass welding as compared with conventional HAZ studies using simple welding procedures [19,21]. The lack of information about thermomechanical properties during heating and cooling in welding, mainly in multi-pass variations, constitutes a challenge for the development and calibration of numerical models [19].

Rikken et al. [22] used numerical and thermal simulations to evaluate the development of residual stresses in welded joints of S355G10+M steel. For the numerical model, material properties were collected using dilatometry and high-temperature tensile testing, whereas thermal properties were calculated using JMatPro® 4.0 software. Thermal cycles were acquired upon bead-on plate gas metal arc welding (GMAW) runs for model validation. The study correlated mechanical properties, microstructure, and residual stresses, and agreement was observed between simulated and experimental results, although no assessment of the effect of different heat inputs and cooling rates was carried out.

In the work by Hosseini and Karlsson [23], nitrogen loss and microstructural evolution in the high temperature heat-affected zone (HTHAZ) of an AISD SAF 2507 steel were investigated by means of numerical simulation tools validated by physical simulation and autogenous bead-on-plate gas tungsten arc welding (GTAW) weldments. Such an approach allowed for the correlation between real welding and physical simulation results from data obtained by numerical simulation, comparing both real and simulated microstructures. Nonetheless, welding data were not used to calibrate the numerical simulation-derived thermal cycle for physical simulation. Moreover, no variation of heat input was conducted.

Deepu and Phanikumar [13] used an ICME-based methodology (ICME—Integrated Computational Materials Engineering) to study the microstructure and property evolution in the HAZ of a DP980 steel. In order to validate the finite element simulation, a preliminary bead-on-plate GMAW welding procedure was performed, using an ER70S-6 electrode, for subsequent comparison between experimental and numerically simulated thermal cycles. Again, results suggested good correlation between results, but no investigation of heat input variation was conducted for methodology validation.

The so-called microstructure-predicting methodology (MPM) proposed by Dornelas et al. [19] used numerical, thermodynamic, and physical simulation to predict the influence of the t8/5 time (cooling time between 800 and 500 °C) on the microstructure and mechanical behavior of the coarse-grained heat-affected zone (CGHAZ) of multi-pass Cr–Mo low-alloy steel weldments produced by GTAW and SAW. Physical properties of the material necessary for model calibration were obtained using JMatPro® software. The simulation was validated by comparison between real and simulated thermal cycles using t8/5 = 15, 30, 80, and 210 s obtained using Sysweld software and reproduced in a Gleeble® simulator equipped with a dilatometer. Sysweld's thermal profile, the comparison between predicted and obtained microhardness, and the microstructure prediction model were used to validate the methodology, which showed itself to be an adequate tool for microstructure and hardness prediction in the CGHAZ zone, despite not correlating the thermal cycles with the corresponding heat input.

Nonetheless, welding parameters are predominantly defined through trial and error, making the process costly in terms of time, material, and expense. The use of physical simulations proves to be an effective strategy, aiming to expand the region of the HAZ for analysis and achieve more precise control over the cooling rates to which the samples are exposed. These simulations, including computational ones, can predict welding thermal cycles and microstructural changes, resulting in a significant reduction in the time and resources needed to assess the effects of parameter modifications on the behavior of welded joints. This approach not only allows for a more in-depth analysis of the HAZ but also contributes to a more informed and efficient decision-making process during welding operations.

Facing the exposed context and with the aim of filling the gaps observed in previous studies [13,19,22,23], the objective of this study is to propose a complete methodology that combines numerical and physical simulations to study the behavior of the HAZ of a multi-pass weld in UNS S32750 superduplex stainless steel, with posterior validation in specimens welded using the same parameters. This methodology involves the calibration of the numerical model, properly validated by bead-on-plate welding trials, using material properties obtained using JMatPro® 9.0 software. The proven model was then used to derive thermal cycle curves for different heat inputs to be subsequently simulated in a Gleeble® system. Microstructure and hardness of both simulated and multi-pass welded specimens were assessed and compared.

2. Materials and Methods

2.1. General Procedure

A flowchart of the developed study is shown in Figure 1. It includes numerical simulation, properly calibrated by previous multi-pass welding and using data properties obtained using JMatPro®, simulation of the thermal cycle in a Gleeble® system, and GTAW/GMAW processes for comparison with simulated specimens.

The welding procedure described in Sections 2.3.1 and 2.3.2 was performed with the purpose of calibrating the numerical simulation (Section 2.3.3) which, in its turn, was used to convert the target heat inputs (Section 2.4.1) into a thermal cycle to be physically simulated (Section 2.4.2). Test specimens produced by physical simulation were subsequently compared with actual welds (Section 2.4.3) in terms of microstructure and hardness (Section 2.5).

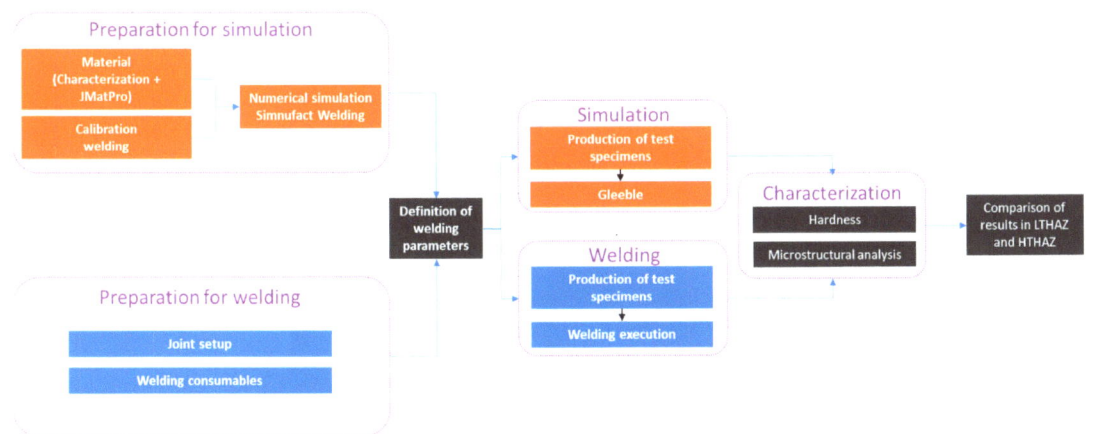

Figure 1. Flowchart showing the steps comprised in the proposed methodology.

2.2. Materials

Superduplex stainless steel (SDSS) UNS S32750 plates with a thickness of 10 mm were used for the experiments. The chemical composition of as-received material and the ER2594 [24] filler used in welding experiments is shown in Table 1. The initial austenite/ferrite ratio of the base steel was 50/50% ± 2%. The table also includes its PREN (pitting resistance equivalent number), which was higher than 40, as expected for super duplex steels [25].

Table 1. Chemical composition (% weight; Fe: balance) of the UNS S32750 superduplex steel and of the ER2594 filler.

Material	C	Cr	Ni	Mo	Mn	Si	N	P	S	Cu	W	PREN [1]
SDSS	0.015	24.72	6.88	3.80	0.74	0.42	0.27	0.029	0.01	-	-	41.58
ER2594	0.003	24.0–27.0	8.0–10.5	2.5–4.5	2.5	0.03	0.20–0.30	0.03	0.02	1.5	1.0	35.45–46.65

[1] PREN = Cr + 3.3 Mo + 16 N.

2.3. Calibration Welding

The calibration welding procedure of the steel was carried out to provide the thermal response of the material, information regarding fusion zone dimensions, and to acquire the thermal cycle (temperature vs. time curve) corresponding to the employed heat input. This information was used to calibrate the numerical simulation model. Thermal cycles were recorded by thermocouples, and the heat input ("HI") of each pass was calculated using monitored tension, current, and welding speed according to Equation (1) as follows:

$$HI = \frac{60 \times U \times I}{1000 \times V} \left[\frac{kJ}{mm}\right], \qquad (1)$$

where "V" is the welding speed [mm/min], "U" is the tension [V], and "I" is the current [A] [17]. Welds with and without weaving were produced by manual GTAW and mechanized GMAW techniques.

2.3.1. Calibration Welding with Weaving

Weave bead welding was carried out with five passes by manual GTAW (root and reinforcement) and mechanized GMAW (three final passes), using an ER2594 filler with 2.4 mm diameter. Joint dimensions, length of passes, and thermocouple positions for temperature monitoring are shown in Figure 2. After the two first GTAW passes, a transverse cut (perpendicular to the welding direction) was made for analysis of the fusion zone (FZ), and the same procedure was repeated after the final three GMAW passes.

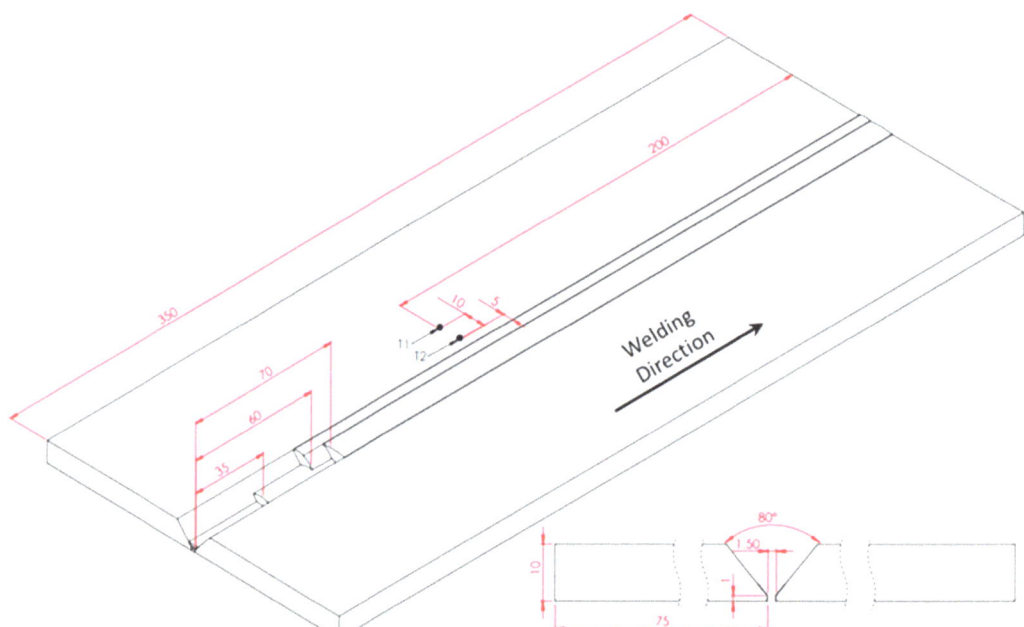

Figure 2. Schematic drawing, in mm, of the designed joint for weave bead welding. T1 and T2 indicate the positions of the thermocouples used.

For the manual GTAW root pass, an Origo Arc 3000i welding source with direct current was used. Argon with 99.99% purity served as purge and shielding gas, and an ER2594 stick was utilized as filler. For the mechanized GMAW welding, the same filler, but in wire form, and a Miller Electric CST 280 source were used. Ar 98% and 2% N were employed as purge and shielding gases, respectively, and a Lincoln Electric robotic manipulator was utilized for torch control. Open arc duration, voltage, current, and gas flow were obtained via a Portable Welding Process Monitoring System—SAP V4, and temperature data were acquired via two thermocouples and a thermographic camera. The average welding speed was indirectly measured by the division of the total joint length by the total open arc duration.

After welding, the cross section of the joint was sectioned and polished for microstructural analysis and for the measurement of each welding bead. Cutting sections were aligned with the thermocouples' positions for an accurate correlation with the thermal profile. Information about thermocouple positioning and welding pool dimensions are extremely relevant parameters for the calibration of a numerical simulation [9]. Moreover, obtained thermal cycles were useful to generate the computational model from which data were extrapolated for different heat inputs.

2.3.2. Calibration Welding without Weaving

For a better understanding of the effect of heat input on the fusion zone dimensions, weldments without weaving were also produced. Bead-on-plate autogenous GTAW and ER2594-filler GMAW were performed with equal monitoring of voltage, current, average speed, and heat input. Information drawn from this procedure served to calibrate the heat source for numerical simulation. For a more assertive calibration, five and nine beads were deposited for GMAW and GTAW processes, respectively, with a variation of welding parameters, resulting in different heat inputs.

2.3.3. Numerical Simulation

GMAW and GTAW welding of the superduplex steel were simulated with the finite element method (FEM) implemented in the commercial software Simufact Welding® 2023, using input data obtained from the calibration welding. Local mesh refinement was applied to the weld area, as shown in Figure 3. The mesh used was hexahedral with sizes of 0.5 mm in the most refined part (front), 2 mm in the coarsest part (front), and 5 mm in depth. Material properties were obtained using JmatPro® 9.0 software, assuming that the material is homogeneous. Heat source parameters were calibrated based on the experimentally observed weld pool dimensions. Table 2 shows the values collected from the experiments, and the efficiency was also obtained by calibrating the simulation with the experimental observation of deposited weld pools. Table 3 shows the parameters used in the simulation.

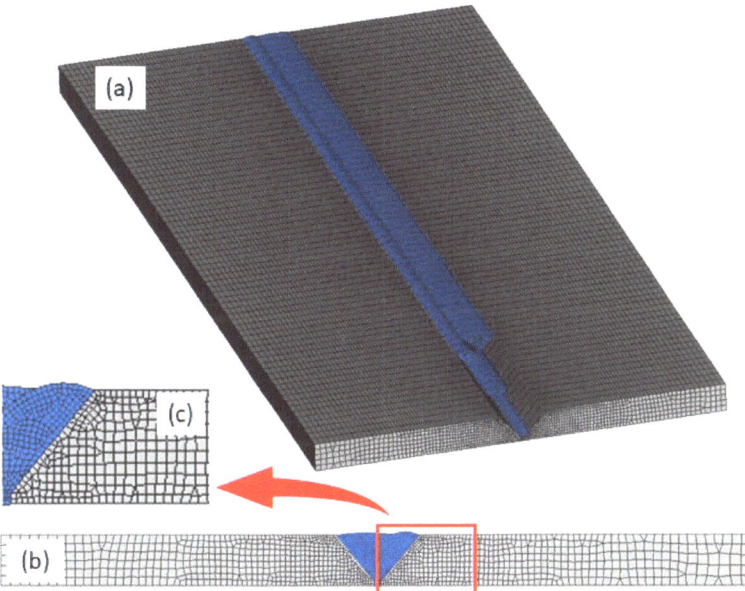

Figure 3. Mesh used in the simulation. 3D view (**a**), frontal view (**b**) and HAZ zoomed section showing mesh details (**c**).

Table 2. Data collected from experimental tests.

Parameter	PASS 1	PASS 2	PASS 3	PASS 4	PASS 5
Start (s)	0	703	2626	4298.6	5073
End (s)	360	1030	2940	4625	5172
Interpass waiting time (s)	0	343	1596	1358.6	448
Welding time (s)	360	327	314	326.4	99
Bead size (mm)	350	350	315	290	280
Speed (mm/s)	2.70	2.77	6.09	6.09	5.42
Voltage (V)	12.5	12.5	32	32	32
Current (A)	154	146	114	112	103
Efficiency (η)	0.4	0.55	0.45	0.5	0.6

Table 3. Parameters used for FEM simulation of welding in Simufact Welding® 2023 software.

Parameter	GTAW	GMAW	Unit
Front length (af)	2	4.5	mm
Rear length (ar)	8	14	mm
Depth (d)	4	7	mm
Width (b)	1	5	mm
Gaussian parameter (M)	3	3	-
Convective heat transfer coefficient (h)	20	20	W/m^2k
Contact heat transfer coefficient (α)	1000	1000	W/m^2k
Emissivity (ε)	0.6	0.6	-

Numerical simulation calibrated with results obtained from the welding procedures described above was carried out to reproduce heat input effects in the chosen points on the lower face of the plates, where heating and reheating effects are usually more pronounced. The aim of numerical simulation was to obtain thermal cycles for the heat inputs of interest to later be input into the Gleeble® system for physical simulation. Chosen HAZ regions were high-temperature HAZ (HTHAZ) and low-temperature HAZ (LTHAZ), as defined by their respective peak temperatures of 1350 °C and 1000 °C [9]. The fusion zone's profile of each pass was modeled with SolidWorks® 2020 software, imported into the Apex® 2023 software for mesh creation and later imported again into Simufact Welding® for the simulation itself.

To calibrate the heat source, an ellipsoid Goldak model was selected [13,26]. For the calculation, information such as depth, width, and front and rear length of the heat source, data resulting from the calibration welding, and both pass and interpass times were required [13]. Additional parameters, e.g., room temperature (25 °C) and thermophysical properties of the material—density, thermal conductivity, and specific heat—obtained with JMatPro® 9.0 software, were also necessary for the simulation [13,27,28]. Simulation configuration also considered boundary conditions such as joint geometry and restraints in six degrees of freedom, including the use of hooks to avoid displacement [13]. From the calibrated numerical model, three heat input conditions were extrapolated, and the thermal cycles were extracted to be reproduced in the physical simulation, as detailed in Section 2.4.2.

2.4. Production of Test Specimens

2.4.1. Parameter Selection

Practical knowledge regarding superduplex stainless steel welding defines that primary heat input values should comprise between 0.7 and 1.2 kJ/mm for 7–20 mm plate thickness; the heat input of the second pass should lie around 75–85% of the first pass's input; and the interpass temperature should not exceed 100 °C. Three heat input values were chosen as targets for this study: 70%, 110%, and 165% of the upper limit (1.20 kJ/mm), to validate the methodology for a wide heat input range. For the second pass, values of 75% with respect to the first pass were defined. Preliminary studies showed that lower heat input values resulted in insufficient fusion of the base material. The summary of conditions is shown in Table 4.

Table 4. Heat input and interpass temperature target conditions used in this study.

Condition	1st Pass (GTAW) [kJ/mm]	2nd Pass (GTAW) [kJ/mm]	Subsequent Passes (GMAW) [kJ/mm]	Interpass Temperature [°C]
CD1	0.84	0.63	0.84	100
CD2	1.32	0.99	1.32	100
CD3	1.98	1.49	1.98	100

2.4.2. Physical Simulation

A Gleeble 540® welding simulator was used for physical simulation of HTHAZ and LTHAZ microstructures associated with the three heat input conditions, using specimens with dimensions of 95 × 25 × 10 mm. Thermal cycles obtained via numerical simulation were input into Gleeble's script, QuikSim® 2 software. A free span of 28 mm between copper grips was defined. A K-type thermocouple was spot-welded onto the specimen's surface in its mid-length position for temperature control and acquisition. All subsequent characterization was performed on the cross-sections of specimens aligned with the thermocouple's placement.

2.4.3. Real Welding

The aim of this step was to reproduce real joints under the same conditions of the physical simulation (Table 4). However, in practice, the measured parameters indicated in Table 5 were not consistently achieved due to variations during execution, as it involved manual welding. The same joint configuration as the one shown in Figure 2 was used, with the first two passes made by GTAW and the remaining ones by GMAW. The process was monitored by an ALX III device developed by The Validation Center (TVC) with an acquisition of voltage, current, speed, gas flow, and open arc duration. A Lincoln Electric Speedtec 505 SP welding source was used, and the purge/shielding gas flow was kept between 10 and 15 L/min. GTAW was executed with a mechanized torch and manual consumable feeding. Heat input values were indirectly calculated from the data acquired by the ALX III device. For the lower heat input condition, CD1, it was necessary to perform seven passes to completely fill the welding joint.

Table 5. Measured welding parameters for the three heat input conditions studied.

Condition	Pass	Speed [mm/min]	Voltage [V]	Current [A]	HI [kJ/mm]
CD1	1	79.96 ± 7.01	10.29 ± 0.52	114.00 ± 0.01	0.88
	2	105.62 ± 9.23	10.20 ± 0.47	114.00 ± 0.01	0.66
	3–7	360–370	23.89 ± 0.10	160–175	0.66–0.68
CD2	1	79.96 ± 7.01	10.29 ± 0.52	114.00 ± 0.01	0.88
	2	105.62 ± 9.23	10.20 ± 0.47	114.00 ± 0.01	0.66
	3–7	360–370	23.89 ± 0.10	160–175	0.66–0.68
CD3	1	54.45 ± 4.89	12.23 ± 0.34	148.00 ± 0.01	1.99
	2	108.43 ± 5.24	13.17 ± 0.43	219.00 ± 0.01	1.60
	3–5	200–230	29.25 ± 0.10	240–260	1.83–2.15

2.5. Microstructural Characterization

For microstructural analysis, including ferrite quantification and grain size measurement, specimens were cut; embedded in resin; ground with progressively finer sandpaper grits (from #180 to #1500); and polished with glycol-based 3 μm and 1 μm diamond suspensions using an alcohol-based lubricant. After polishing, an electrolytic etching at 3 V for around 7 s was conducted using a modified Murakami etchant (40% KOH solution) [29].

For the measurement of the fusion zone dimensions, low-magnification (20×) images were acquired with a Wild M3C Heerbrugg Switzerland Type-S optical microscope. Measurements were carried out in ImageJ 1.53k software. Ferrite and austenite quantification was performed according to the ASTM E 562 standard [30]. Images of the microstructure were acquired at a magnification of 500× using a Zeiss AxioScope optical microscope. A 14 × 14 grid with 196 points was used for a higher area coverage to improve measurement precision. Phase quantification was carried out for welded and physically simulated specimens.

Physically simulated and welded specimens also underwent Vickers microhardness measurements with a 0.2 kgf load, (sufficiently low to allow for adequate indentations in the narrow HAZ regions of welded specimens [3,12]), in a Shimadzu HMV2TE tester.

Twenty-four measurements were taken in physically simulated specimens and four in each one of the two studied HAZ sub-regions, being two on each side of the weld. HTHAZ indentations were done closer to the FZ, whereas LTHAZ ones were done in the region with the most intense microstructural variation, in terms of grain morphology. Hardness testing was carried out according to the ASTM A370 standard [31] after electrolytic etching, so that different regions of the weld could be identified. At the end, microstructure and hardness were compared to validate the proposed methodology.

3. Results and Discussion

3.1. Welding Calibration and Numerical Simulation

The comparison between thermal curves and FZ profile obtained during weave bead welding and by numerical simulation is shown in Figure 4. It can be observed that the modeled results lie relatively close to the real ones, with the temperature peak difference ranging from a minimum of 2% for thermocouple 1 in pass 5 (calibration welding, 321.6 °C, and Simufact Welding®, 314 °C) to a maximum of 31% for thermocouple 2 in pass 1 (calibration welding, 196.5 °C, and Simufact Welding®, 135 °C).

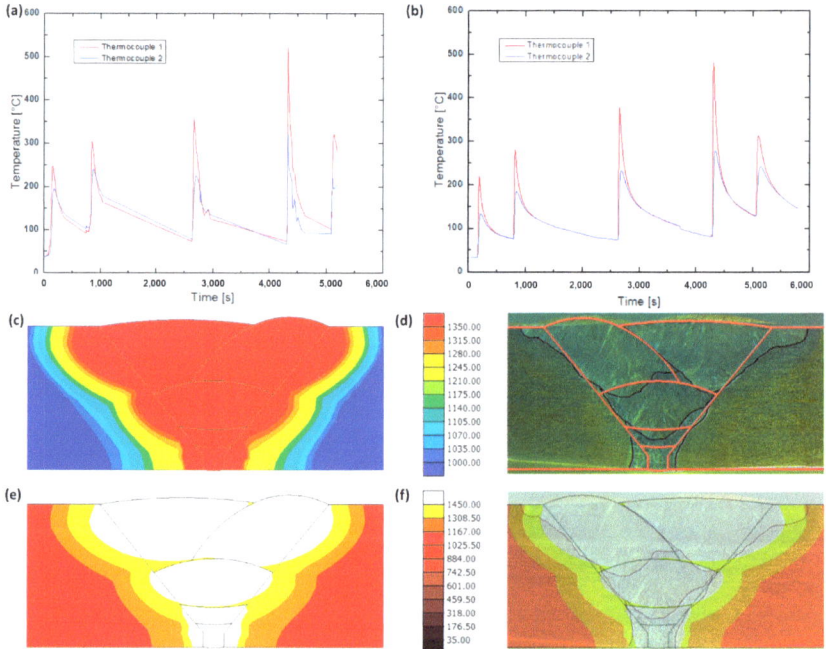

Figure 4. Thermal cycle measured in calibration welding (**a**); thermal cycle obtained in Simufact Welding simulation (**b**); temperature maps (**c**,**e**); pass geometry in (**d**,**f**).

3.2. Numerical Simulation and Physical Simulation

The execution of Gleeble® simulations required a series of trials for the validation of curves, involving free span and script adjustments, so that physically simulated curves could approximate as much as possible to the modeled ones. The reduction of free span between the copper grips allowed higher cooling rates, but increased the thermal gradient along the specimen's length, reducing the useful region for subsequent characterization [9]. Another issue is correlated with the temperature overshoot effect caused by the thermal inertia of the system when facing relatively rapid heating and/or cooling, as a result of a delay in the electronic response from the simulator and the physical response from the material [32], leading to peak temperatures higher than the programmed ones.

After parameter and program optimization, thermal cycles were carried out for the heat inputs of interest. Resulting curves were compared with numerically modeled ones to assess the representativeness of the applied cycles. As can be seen in Figure 5, all physically simulated curves coincide well with the ones obtained by numerical simulation, with a maximum peak temperature difference of 6% between simulations conducted using Gleeble® physical simulation and numerical simulation via Simufact Welding®, in the LTHAZ of the third pass of CD2 (Gleeble® 481.6 °C and Simufact Welding® 453.3 °C) [9].

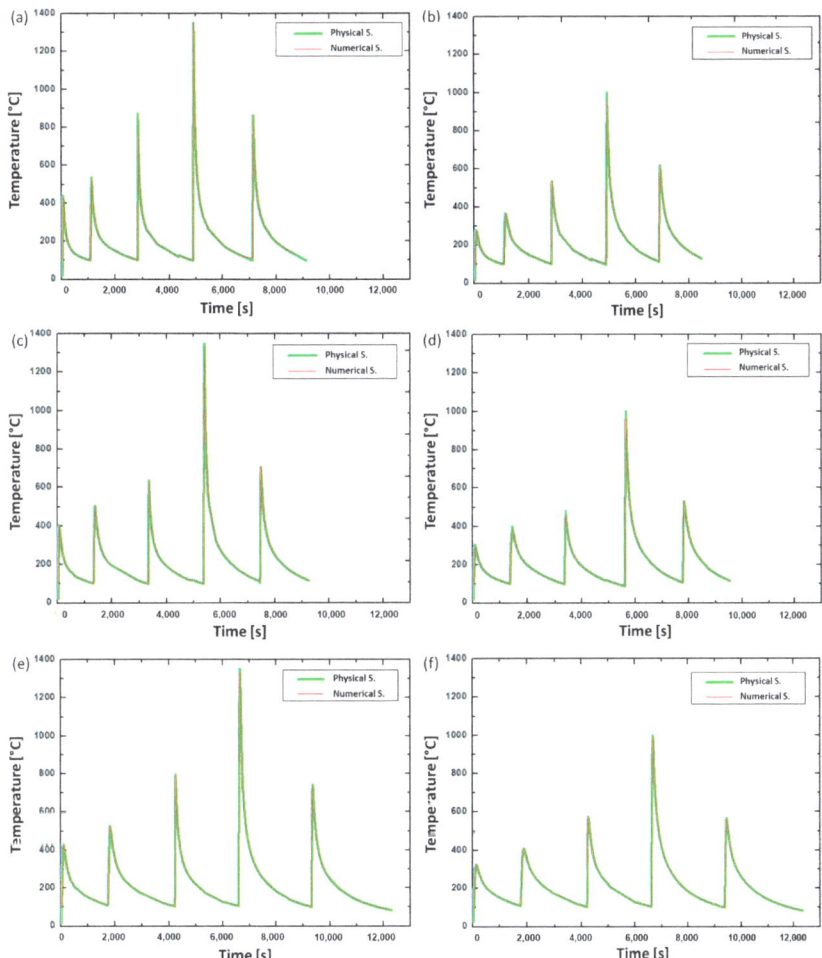

Figure 5. Physical (green line) and numerical (red line) simulation curves of HTHAZ (**a**,**c**,**e**) and LTHAZ (**b**,**d**,**f**) regions for condition CD1 (**a**,**b**), CD2 (**c**,**d**), and CD3 (**e**,**f**).

The microstructure of LTHAZ is like that of the as-received material, Figure 6, with elongated grains arising from the rolling process. HTHAZ, on the other hand, presents a ferritic matrix with allotriomorphic grain boundary austenite (GBA), intragranular austenite (IGA), and Widmanstätten austenite (WA) [2,9,11,33], as seen in Figure 7. In the images, darker grains are ferrite and lighter ones are austenite. The microstructures observed correspond to other findings reported in the literature, where lower heat inputs, resulting in higher cooling rates, lead to higher volume fractions of ferrite. In contrast, high heat inputs, with consequently lower cooling rates, promote prolonged exposure to the temperature range in which the precipitation of intermetallic phases may take place, in addition to the presence of forms of austenite, such as GBA, IGA, and WA [11,34–37].

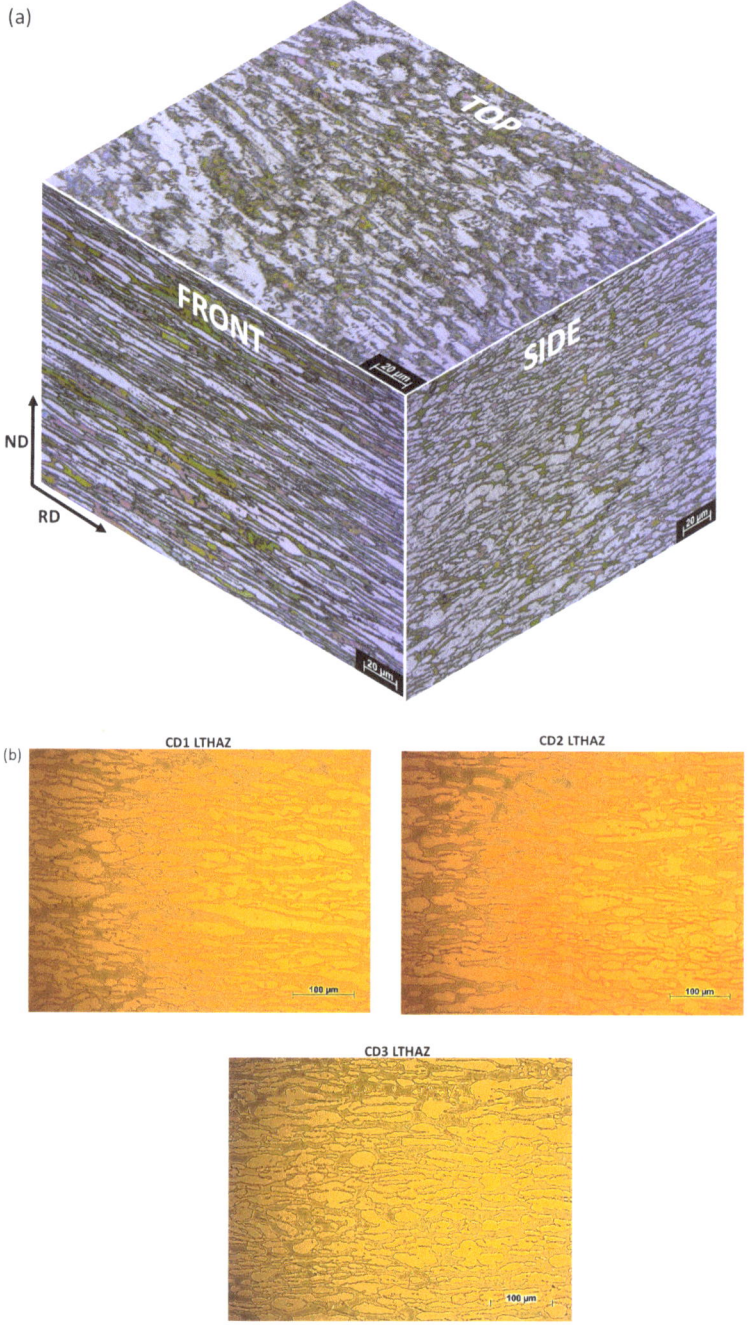

Figure 6. Metallography in three dimensions of AISI UNS S32750 material, 10 mm as received. 200× magnification. Electrolytic etching with 40% KOH, 3V for about 7 s. RD—rolling direction; ND—normal direction (**a**), microstructure of LTHAZ CD1, CD2, and CD3 from physical samples (**b**).

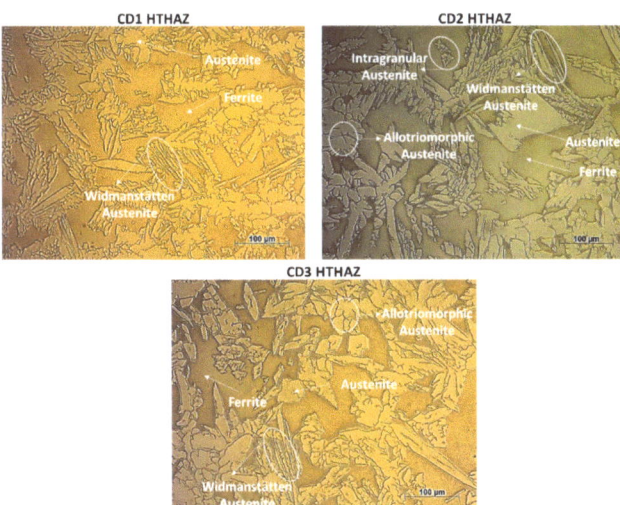

Figure 7. Different microstructures identified in samples subjected to the thermal cycle of physical simulation using the Gleeble®.

3.3. Physical Simulation and Real Welding

The final goal of the proposed methodology was to compare the microstructural assessment (microscopy and microhardness) results of the physically simulated and welded specimens. Physical simulation enables the augmentation of the useful area for analysis, leading to a better understanding of the specific morphology related to the thermal cycle of each sub-region, which show themselves relatively narrow in a real weld HAZ [38,39].

Microhardness values measured in Gleeble® specimens were close to those found in the literature [3]. Welded specimens, however, presented somewhat lower hardness levels, which might be ascribed to the fact that sub-region boundaries are difficult to define, meaning that indentations might have been partially placed in different regions, Figure 8. The graphs in Figure 9 indicate that samples from Gleeble® and welded samples exhibit values with a standard deviation within the range of error, and through analysis of variance (ANOVA), it was identified that there is no significant variation between Gleeble® and welded samples in terms of percentage of ferrite and microhardness. Both simulated and welded microstructures present similarities in microconstituent morphology.

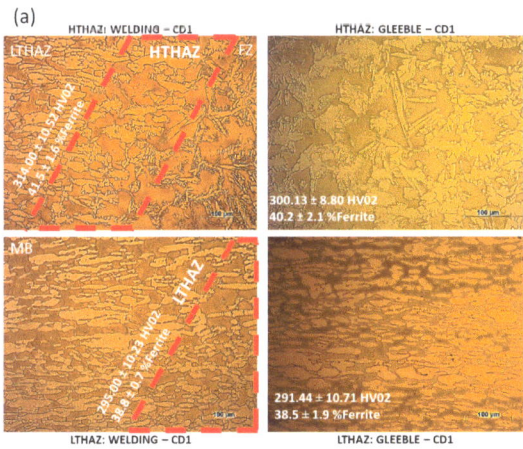

Figure 8. *Cont.*

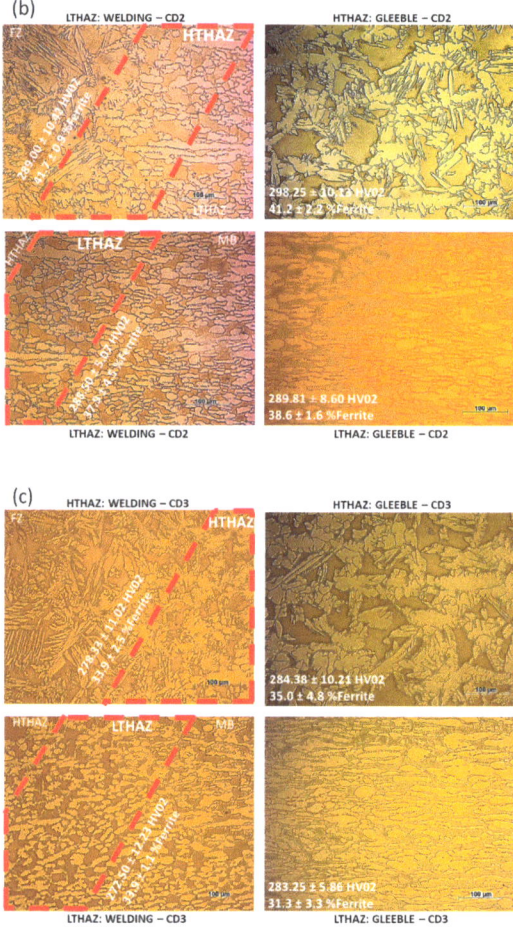

Figure 8. Microstructure comparison between HAZ sub-regions in welded specimens (regions highlighted with a dashed red line), and in Gleeble® specimens in conditions CD1 (**a**), CD2 (**b**), and CD3 (**c**).

The ferrite content and microhardness values may slightly differ between the results obtained from physical simulation and actual welding, within the margin of error, as seen in Table 6. This variance occurs because the physical simulation is able to replicate the heat-affected zone (HAZ) and expand this area compared to the HAZ of a real welded joint. In the actual weld, it is difficult to differentiate between the high-temperature HAZ and low-temperature HAZ, due to their small dimensions [3,40]. The values of both the HTHAZ and the LTHAZ are quite close, with small relevant variations between the sub-regions. In this context, it is important to consider these results in conjunction with the morphological analysis of the phases present. Additionally, it is necessary to emphasize the difficulty of identifying a specific region of HAZ in the welded joint, due to its reduced dimensions and the microstructural gradient between adjacent sub-regions.

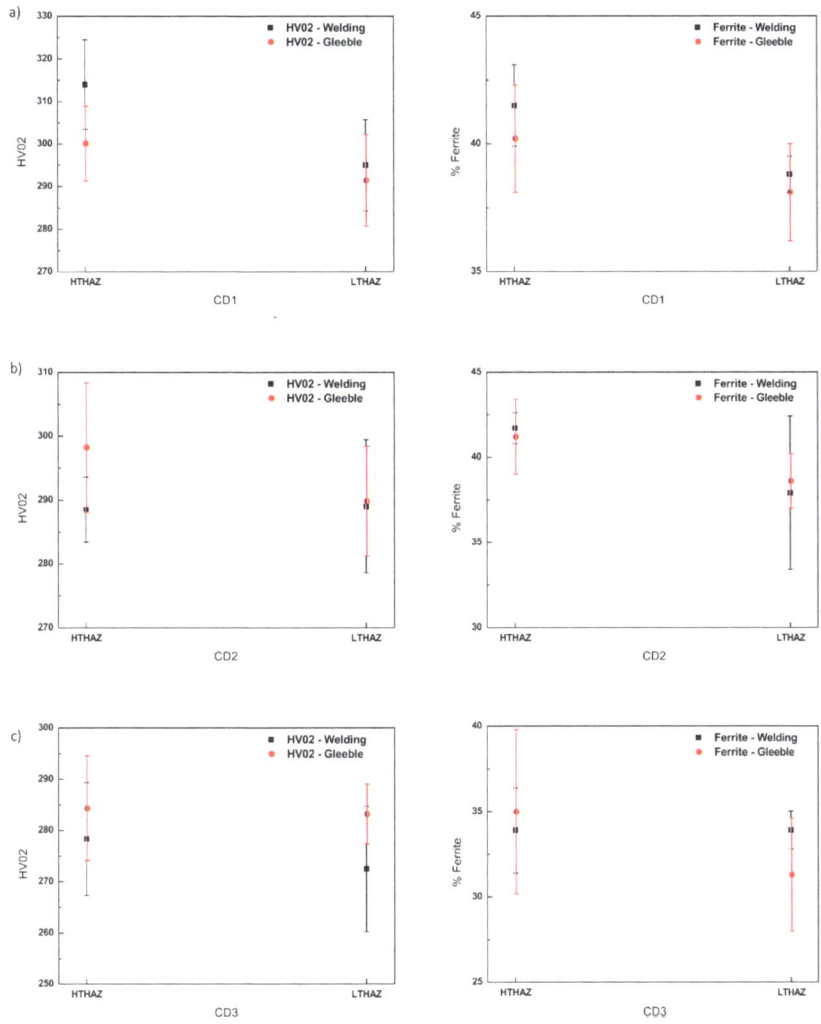

Figure 9. Comparison between HAZ sub-regions in welded specimens (black) and in Gleeble® (red) specimens, both in term of percentage ferrite and microhardness, in conditions CD1 (**a**), CD2 (**b**), and CD3 (**c**).

Table 6. Ferrite content (%) and microhardness (HV02) comparison of Gleeble® and welded samples.

Region	CD1		CD2		CD3	
	Gleeble (%)	Welded (%)	Gleeble (%)	Gleeble (%)	Welded (%)	Gleeble (%)
LTHAZ	38.5 ± 1.9	38.8 ± 0.7	38.6 ± 1.6	37.9 ± 4.5	31.3 ± 3.3	33.9 ± 1.1
HTHAZ	40.2 ± 2.1	41.5 ± 1.6	41.2 ± 2.2	41.7 ± 0.9	35 ± 4.8	33.9 ± 2.5
	Gleeble (HV02)	Welded (HV02)	Gleeble (HV02)	Welded (HV02)	Gleeble (HV02)	Welded (HV02)
LTHAZ	291.4 ± 10.7	295 ± 10.2	289.8 ± 8.6	288.5 ± 5.0	283.2 ± 5.8	272.5 ± 12.2
HTHAZ	300.1 ± 8.8	314.0 ± 10.5	298.2 ± 10.1	289 ± 10.4	284.4 ± 10.2	278.3 ± 11.02

Given that the heat input conditions were selected based on "good" conditions deemed by standards and the literature, the absence of intermetallic phases was expected, as confirmed by the exposure time at peak temperature (less than one thousand (1000) seconds, in all conditions and regions of the HAZ), which is insufficient for nucleation and formation of these intermetallic phases. Typically, the precipitation of secondary phases occurs more rapidly in ferrite or at austenite/ferrite phase boundaries than inside austenite grains. In all heat inputs, whether low or high, corresponding to fast or slow cooling rates, and across all the steel types analyzed, no intermetallic phases, such as sigma phase, among others, were observed. The nondetection of intermetallic phases does not necessarily imply nonformation; however, it can be inferred that the quantity formed may be insignificant, justifying their nondetection in the microstructural analysis conducted in this study, akin to the findings of Fonseca et al., who investigated the same material under more extreme heat inputs of 0.5 kJ/mm and 3.2 kJ/mm [37,40].

4. Conclusions

In this study, a new methodology was proposed and assessed to predict the microstructure of a welding HAZ in superduplex stainless steels, based on coupled numerical and physical simulations. The focus on the HAZ is justified by its high complexity and the possibility of the formation of different microstructures that can be harmful to the mechanical performance of materials.

One of the major challenges in studying the HAZ of duplex and superduplex stainless steels is identifying their subregions due to their small dimensions. Each subregion exhibits distinct characteristics and possibilities for the formation of different intermetallic phases, and expanding this HAZ for a more accurate study is crucial to identify and understand their characteristics. This study introduces a novel methodology for investigating the HAZ of superduplex stainless steel through simulations. It began with a preliminary welding process to calibrate a numerical simulation, from which three variations of heat input were extrapolated, chosen according to standards and the literature. These heat input values were then converted into thermal cycles for physical simulation using the Gleeble® system, and adapted into welding parameters for the final welding process. Subsequently, samples from both the physical simulation and the real welding were analyzed and compared in relation to ferrite content and microhardness.

The use of numerical simulation previously calibrated by data acquired from a preliminary welding procedure, made it possible to derive thermal cycles for each heat input studied in HTHAZ and LTHAZ. The computationally generated curve simulated different heating and cooling rates in each subregion during welding of UNS S32750 superduplex stainless steel, and served as input for physical simulation. Numerically and physically simulated curves were sufficiently close to each other, suggesting that Gleeble® can successfully reproduce conditions studied and simulated by Simufact Welding®. Therefore, the methodology has proven itself adequate to predict relevant characteristics of the welding process, such as thermal cycles and HAZ microstructure in its different regions. Results showed it is possible to accurately reproduce microstructural features under different heat input conditions. Values of microhardness and ferrite content were shown to be relatively close, as well as microconstituent morphologies.

This methodology can be applied to the development of better-suited welding procedures for specific materials, and also for welding studies of new special alloys. The preliminary trial-and-error welding stage for parameter optimization can be adequately replaced with simulation tools, which offers the advantages of defining parameters more assertively for further onsite applications and allowing better control of thermal cycles, which is difficult to attain during real welding procedures. From model validation, other welding parameters, different from those chosen in this study, can be selected for the same material and, alternatively, the methodology can be applied to the welding of other metals and alloys.

Due to the requirement for robust calibration and expensive equipment, the methodology has a higher initial cost as compared with traditional studies that do not involve simulation tools. Nonetheless, if the procedure is extrapolated to various welding conditions and materials, its costs can be easily compensated in due course. Coupled numerical and physical simulations allow for an enhanced control of parameters and an enlargement of the analyzed region, with high repeatability without imparting relevant additional costs.

Author Contributions: Conceptualization, T.N.L. and R.S.C.; methodology, B.C. and T.N.L.; formal analysis, B.C., L.O.P.d.S., F.M.d.S.J. and L.F.F.; investigation, L.O.P.d.S. and F.M.d.S.J.; resources, L.O.P.d.S. and F.M.d.S.J.; data curation, L.O.P.d.S., F.M.d.S.J. and L.F.F.; writing—original draft preparation, L.O.P.d.S. and L.F.F.; writing—review and editing, B.C., T.N.L. and R.S.C.; visualization, B.C.; supervision, R.S.C.; project administration, T.N.L.; funding acquisition, R.S.C. All authors have read and agreed to the published version of the manuscript.

Funding: This research was funded by the ANP/PETROBRAS program (SAP 4600580712).

Data Availability Statement: The raw data supporting the conclusions of this article will be made available by the authors on request.

Acknowledgments: The authors would like to acknowledge the post-graduate program GETEC at SENAI CIMATEC and ANP/PETROBRAS for their financial support to this project.

Conflicts of Interest: The authors declare no conflicts of interest.

References

1. Zhao, G.; Zhang, R.; Li, J.; Li, H.; Ma, L.; Li, Y. Study on Microstructure and Tensile Damage Evolution of 2205 Duplex Steel at Different Solution Temperatures. *Mater. Today Commun.* **2022**, *33*, 104472. [CrossRef]
2. Calderon-Uriszar-aldaca, I.; Briz, E.; Garcia, H.; Matanza, A. The Weldability of Duplex Stainless-Steel in Structural Components to Withstand Corrosive Marine Environments. *Metals* **2020**, *10*, 1475. [CrossRef]
3. Arun, D.; Devendranath Ramkumar, K.; Vimala, R. Multi-Pass Arc Welding Techniques of 12 mm Thick Super-Duplex Stainless Steel. *J. Mater. Process Technol.* **2019**, *271*, 126–143. [CrossRef]
4. Tavares, S.S.M.; Silva, V.G.; Pardal, J.M.; Corte, J.S. Investigation of Stress Corrosion Cracks in a UNS S32750 Superduplex Stainless Steel. *Eng. Fail. Anal.* **2013**, *35*, 88–94. [CrossRef]
5. Pardal, J.M.; Tavares, S.S.M.; Fonseca, M.D.P.C.; Souza, J.A.D.; Vieira, L.M.; Abreu, H.F.G.D. Deleterious Phases Precipitation on Superduplex Stainless Steel UNS S32750: Characterization by Light Optical and Scanning Electron Microscopy. *Mater. Res.* **2010**, *13*, 401–407. [CrossRef]
6. Freitas, G.C.L.D.; da Fonseca, G.S.; Moreira, L.P.; Leite, D.N.F. Phase Transformations of the Duplex Stainless Steel UNS S31803 under Non-Isothermal Conditions. *J. Mater. Res. Technol.* **2021**, *11*, 1847–1851. [CrossRef]
7. de Farias Azevedo, C.R.; Boschetti Pereira, H.; Wolynec, S.; Padilha, A.F. An Overview of the Recurrent Failures of Duplex Stainless Steels. *Eng. Fail. Anal.* **2019**, *97*, 161–188. [CrossRef]
8. Fonseca, C.S.; da Silva, S.N.; Gomes, H.G.; Silva, D.W.; Júnior, J.B.G.; Pinheiro, I.P. Evaluation of the Sigma Phase Precipitation Susceptibility in Duplex Stainless Steel SAF2205 Welding. *Rev. Mater.* **2019**, *24*, e12470. [CrossRef]
9. Acuna, A.; Ramirez, A.J. Sigma Phase Formation Kinetics in Hyper Duplex Stainless Steel Welding Filler Metal. *Mater. Charact.* **2023**, *200*, 112832. [CrossRef]
10. Singh, J.; Shahi, A.S. Impact Toughness, Fatigue Crack Growth and Corrosion Behavior of Thermally Aged UNS S32205 Duplex Stainless Steel. *Trans. Indian Inst. Met.* **2019**, *72*, 1497–1502. [CrossRef]
11. Chaudhari, A.N.; Dixit, K.; Bhatia, G.S.; Singh, B.; Singhal, P.; Saxena, K.K. Welding Behaviour of Duplex Stainless Steel AISI 2205: A Review. *Mater. Today Proc.* **2019**, *18*, 2731–2737. [CrossRef]
12. Zhang, Z.; Zhang, H.; Hu, J.; Qi, X.; Bian, Y.; Shen, A.; Xu, P.; Zhao, Y. Microstructure Evolution and Mechanical Properties of Briefly Heat-Treated SAF 2507 Super Duplex Stainless Steel Welds. *Constr. Build. Mater.* **2018**, *168*, 338–345. [CrossRef]
13. Deepu, M.J.; Phanikumar, G. ICME Framework for Simulation of Microstructure and Property Evolution During Gas Metal Arc Welding in DP980 Steel. *Integr. Mater. Manuf. Innov.* **2020**, *9*, 228–239. [CrossRef]
14. Hosseini, V.A.; Cederberg, E.; Hurtig, K.; Karlsson, L. A Physical Simulation Technique for Cleaner and More Sustainable Research in Additive Manufacturing. *J. Clean. Prod.* **2021**, *285*, 124910. [CrossRef]
15. Gáspár, M.; Sisodia, R.P.S.; Dobosy, A. Physical Simulation-Based Characterization of HAZ Properties in Steels. Part 2. Dual-Phase Steels. *Strength. Mater.* **2019**, *51*, 805–815. [CrossRef]
16. Genchev, G.; Doynov, N.; Ossenbrink, R.; Michailov, V.; Bokuchava, G.; Petrov, P. Residual Stresses Formation in Multi-Pass Weldment: A Numerical and Experimental Study. *J. Constr. Steel Res.* **2017**, *138*, 633–641. [CrossRef]
17. Varbai, B.; Pickle, T.; Májlinger, K. Effect of Heat Input and Role of Nitrogen on the Phase Evolution of 2205 Duplex Stainless Steel Weldment. *Int. J. Press. Vessel. Pip.* **2019**, *176*, 103952. [CrossRef]
18. Sakata, M.; Kadoi, K.; Inoue, H. Age-Hardening Behaviors of the Weld Metals of 22% Cr and 25% Cr Duplex Stainless Steels at 400 °C. *J. Nucl. Mater.* **2023**, *581*, 154438. [CrossRef]
19. Dornelas, P.H.G.; Filho, J.d.C.P.; Farias, F.W.C.; e Oliveira, V.H.P.M.; Moraes, D.d.O.; Júnior, P.Z. FEM-Thermodynamic Simulation Methodology to Predict the Influence of T8/5 on the Coarse Grain Heat-Affected Zone of a Cr-Mo Low-Alloy Steel Pipe. *J. Manuf. Process* **2020**, *60*, 520–529. [CrossRef]

20. Palmieri, M.E.; Lorusso, V.D.; Tricarico, L. Investigation of Material Properties of Tailored Press Hardening Parts Using Numerical and Physical Simulation. *Procedia Manuf.* **2020**, *50*, 104–109. [CrossRef]
21. Hosseini, V.A.; Hurtig, K.; Karlsson, L. Effect of Multipass TIG Welding on the Corrosion Resistance and Microstructure of a Super Duplex Stainless Steel. *Mater. Corros.* **2017**, *68*, 405–415. [CrossRef]
22. Rikken, M.; Pijpers, R.; Slot, H.; Maljaars, J. A Combined Experimental and Numerical Examination of Welding Residual Stresses. *J. Mater. Process Technol.* **2018**, *261*, 98–106. [CrossRef]
23. Hosseini, V.A.; Karlsson, L. Physical and Kinetic Simulation of Nitrogen Loss in High Temperature Heat Affected Zone of Duplex Stainless Steels. *Materialia* **2019**, *6*, 100325. [CrossRef]
24. American Welding Society. *AWS A5.9/A5.9:2017 Welding Consumables–Wire Electrodes, Strip Electrodes, Wires, and Rods for Arc Welding of Stainless and Heat Resisting Steels–Classification*, 9th ed.; The American Society of Mechanical Engineers: New York, NY, USA, 2017; ISBN 978-0-7918-7079-2.
25. Chail, G.; Kangas, P. Super and Hyper Duplex Stainless Steels: Structures, Properties and Applications. *Procedia Struct. Integr.* **2016**, *2*, 1755–1762. [CrossRef]
26. Goldak, J.; Chakravarti, A.; Bibby, M. A New Finite Element Model for Welding Heat Sources. *Metall. Trans. B* **1984**, *15*, 299–305. [CrossRef]
27. Hemmesi, K.; Mallet, P.; Farajian, M. Numerical Evaluation of Surface Welding Residual Stress Behavior under Multiaxial Mechanical Loading and Experimental Validations. *Int. J. Mech. Sci.* **2020**, *168*, 105127. [CrossRef]
28. Wang, Z.; Wang, K.; Liu, Y.; Zhu, B.; Zhang, Y.; Li, S. Multi-Scale Simulation for Hot Stamping Quenching & Partitioning Process of High-Strength Steel. *J. Mater. Process Technol.* **2019**, *269*, 150–162. [CrossRef]
29. Putz, A.; Althuber, M.; Zelić, A.; Westin, E.M.; Willidal, T.; Enzinger, N. Methods for the Measurement of Ferrite Content in Multipass Duplex Stainless Steel Welds. *Weld. World* **2019**, *63*, 1075–1086. [CrossRef]
30. *ASTM E562*; Standard Test Method for Determining Volume Fraction by Systematic Manual Point Count. ASTM International: West Conshohocken, PA, USA, 2000.
31. *ASTM A370*; Standard Test Methods and Definitions for Mechanical Testing of Steel Products. ASTM International: West Conshohocken, PA, USA, 2019.
32. Vieira, D.; Andrade, E.; Jorge, V.L.; Araújo, D.B.; Scotti, A. Determinação de "Overshoot" Em Simulação Física de ZAC de Soldagem. In Proceedings of the XLIV CONSOLDA—Congresso Nacional de Soldagem, Uberlândia, MG, Brazil, 10–13 September 2018; Volume 12.
33. Nuñez de la Rosa, Y.E.; Calabokis, O.P.; Pena Uris, G.M.; Borges, P.C. Pitting and Crevice Corrosion Behavior of the Duplex Stainless Steel UNS S32205 Welded by Using the GTAW Process. *Mater. Res.* **2022**, *25*, e20220179. [CrossRef]
34. Mohammed, G.; Ishak, M.; Aqida, S.; Abdulhadi, H. Effects of Heat Input on Microstructure, Corrosion and Mechanical Characteristics of Welded Austenitic and Duplex Stainless Steels: A Review. *Metals* **2017**, *7*, 39. [CrossRef]
35. Biezmaa, M.V.; Berlangab, C.; Argandonac, G. Relationship between Microstructure and Fracture Types in a UNS S32205 Duplex Stainless Steel. *Mater. Res.* **2013**, *16*, 965–969. [CrossRef]
36. Chen, L.; Tan, H.; Wang, Z.; Li, J.; Jiang, Y. Influence of Cooling Rate on Microstructure Evolution and Pitting Corrosion Resistance in the Simulated Heat-Affected Zone of 2304 Duplex Stainless Steels. *Corros. Sci.* **2012**, *58*, 168–174. [CrossRef]
37. Da Fonseca, G.S.; Barbosa, L.O.R.; Ferreira, E.A.; Xavier, C.R.; De Castro, J.A. Microstructural, Mechanical, and Electrochemical Analysis of Duplex and Superduplex Stainless Steels Welded with the Autogenous TIG Process Using Different Heat Input. *Metals* **2017**, *7*, 538. [CrossRef]
38. Olaya-Luengas, L.; Morejón, J.A.P.; Rezende, M.C.; Bott, I.d.S. Análise Microestrutural Da Zta Simulada Na Gleeble Do Aço Duplex Saf 2205. In Proceedings of the 71th ABM Annual Congress, Rio de Janeiro, RJ, Brazil, 27–29 September 2016; pp. 956–965. [CrossRef]
39. Verma, J.; Taiwade, R.V. Effect of Welding Processes and Conditions on the Microstructure, Mechanical Properties and Corrosion Resistance of Duplex Stainless Steel Weldments—A Review. *J. Manuf. Process* **2017**, *25*, 134–152. [CrossRef]
40. Hosseini, V.A.; Valiente Bermejo, M.A.; Gårdstam, J.; Hurtig, K.; Karlsson, L. Influence of Multiple Thermal Cycles on Microstructure of Heat-Affected Zone in TIG-Welded Super Duplex Stainless Steel. *Weld. World* **2016**, *60*, 233–245. [CrossRef]

Disclaimer/Publisher's Note: The statements, opinions and data contained in all publications are solely those of the individual author(s) and contributor(s) and not of MDPI and/or the editor(s). MDPI and/or the editor(s) disclaim responsibility for any injury to people or property resulting from any ideas, methods, instructions or products referred to in the content.

Article

Structure–Property Correlation between Friction-Welded Work Hardenable Al-4.9Mg Alloy Joints

Aditya M. Mahajan [1], K. Vamsi Krishna [1], M. J. Quamar [1], Ateekh Ur Rehman [2], Bharath Bandi [3] and N. Kishore Babu [1,*]

[1] Department of Metallurgical and Materials Engineering, National Institute of Technology Warangal, Warangal 506004, India; ammm20118@student.nitw.ac.in (A.M.M.); vamshik1@student.nitw.ac.in (K.V.K.); mj22mmr1r03@student.nitw.ac.in (M.J.Q.)

[2] Department of Industrial Engineering, College of Engineering, King Saud University, Riyadh 11451, Saudi Arabia; arehman@ksu.edu.sa

[3] Warwick Manufacturing Group (WMG), University of Warwick, Coventry CV4 7AL, UK; bharath.bandi@warwick.ac.uk

* Correspondence: kishorebabu@nitw.ac.in

Abstract: Friction welding of aluminum alloys holds immense potential for replacing riveted joints in the structural sections of the aeronautical and automotive sectors. This research aims to investigate the effects on the microstructural and mechanical properties when AA5083 H116 joints are subjected to rotary friction welding. To evaluate the quality of the welds, optical and scanning electron microanalysis techniques were utilized, revealing the formation of sound welds without porosity. The microstructural examination revealed distinct weld zones within the weldment, including the dynamically recrystallized zone (DRZ), thermo-mechanically affected zone (TMAZ), heat-affected zone (HAZ), and base metal (BM). During the friction-welding process, grain refinement occurred, leading to the development of fine equiaxed grains in the DRZ/weld zone. Tensile testing revealed that the weldment exhibited higher strength (YS: 301 ± 6 MPa; UTS: 425 ± 7 MPa) in the BM region compared to the base metal (YS: 207 ± 5 MPa; UTS: 385 ± 9 MPa). However, the weldment demonstrated slightly lower elongation (%El: 13 ± 2) compared to the base metal (%El: 15 ± 3). The decrease in ductility observed in the weldment can be attributed to the presence of distinct weld zones within the welded sample. Also, the tensile graph of the BM showed serrations throughout the curve, which is a characteristic phenomenon known as the Portevin–Le Chatelier effect (serrated yielding) in Al-Mg alloys. This effect occurs due to the influence of dynamic strain aging on the material's macroscopic plastic deformation. Fractography analysis showcased a wide range of dimple sizes, indicating a ductile fracture mode in the weldment. These findings contribute to understanding the microstructural and mechanical behavior of AA5083 H116 joints subjected to rotary friction welding.

Keywords: rotary friction welding; aluminum alloys; AA5083 H116; microstructural analysis; mechanical properties

1. Introduction

Aluminum is a versatile metal bearing excellent properties, such as low density, high thermal and electrical conductivity, along with high strength-to-weight ratios. These properties make it an ideal choice for the aerospace, automotive, and shipping industries. Nonetheless, these properties are enhanced or altered by the effect of various alloying elements based on the specified application [1]. The 5xxx series is one such alloy, with magnesium as the primary alloying element. These alloys are considered to be among the strongest of all non-heat-treatable aluminum alloys, and this strength is attained through cold working [2]. The temper designation "H116" in alloys belonging to the 5xxx series signifies that the material has undergone a distinctive blend of cold working and thermal treatment processes. This treatment imparts exceptional resistance to corrosion caused by water and high-humidity conditions. Furthermore, it minimizes the potential impact of stress-corrosion sensitization resulting from exposure to high temperatures [3]. In alloys like AA5059, AA5086, and AA5083, these properties impart excellent weldability, corrosion resistance, and a high strength-to-weight ratio, making them ideal for building shipbuilding and vehicle bodies [4]. Due to these excellent properties and wide range of applications, it becomes essential to join these metals.

Conventionally, these alloys are joined using different fusion-welding techniques, such as metal inert gas (MIG) welding, tungsten inert gas (TIG) welding, etc. However, these joining techniques often result in the formation of various welding defects, such as porosity, distortion, and hot cracking [5]. Samiuddin et al. [6] studied the microstructural and mechanical properties of TIG welded Al-5083 butt joints. They observed that porosity and solidification cracking were the most common defects, resulting in an 18.26% loss of base metal strength after welding. Porosity is caused by the entrapment of hydrogen gas during the solidification process, as hydrogen is highly soluble in molten aluminium, leading to the formation of bubble-like nucleation [7]. Another frequent defect observed in the joining of these alloys is the degradation of the heat-affected zone (HAZ), which is more common in commercial alloys due to the higher weight percentage of alloying elements [8]. As mentioned, cold working is the major strengthening mechanism in 5xxx series alloys and is negatively affected by fusion welding. Cetinel et al. [9] investigated the weldment of AA5083 and AA6061 joined through TIG welding. They observed a significant loss of cold work after welding, resulting in the degradation of strength and hardness. Some reports suggest that the evaporation of volatile elements at high temperatures and the formation of columnar grains during solidification are the main causes of strength loss and weld metal degradation during TIG welding [10,11]. Furthermore, welding defects like solidification cracking in these alloys are highly influenced by the weld metal composition [12]. The composition of the welding metal is influenced by both the filler metal used and the extent of dilution that occurs during welding at elevated temperatures. Therefore, it is essential to select the correct filler metal as it can adulterate the homogeneity of the base metal.

Solid-state welding is one of the more advantageous joining methods for aluminum alloys, as it avoids melting and the defects caused during these processes, such as porosity and solidification cracking. Rotary friction welding (RFW) is a versatile solid-state welding technique currently being used as a promising method for welding aluminum alloys. The principal operation of the rotary friction-welding technique can be divided into three major steps: (i) Softening of the material caused by the frictional heat generated due to the movement between the fraying surfaces, (ii) intermixing of the material or plastic deformation, and (iii) forging of the hot metal [13]. These welding techniques overcome flaws such as the loss of volatile alloying elements and distortion, thereby providing a high-quality joint [14]. Pouraliakbar et al. [15] studied the dynamic and softening phenomena of the Al-6Mg alloy during the two-step deformation through the hot compression test. They reported that the softening of the material can be influenced by strain rate. Lienert et al. [16] successfully joined SiC-reinforced 8009 aluminum by inertia friction welding with a minor loss in tensile and hardness properties. Kimura et al. [17] conducted a study to investigate how different friction welding conditions impact the mechanical properties of friction-welded joints in the AA5052 aluminum alloy. The optimum friction speed was found to be 27.5 s^{-1}, and an excellent joint with approximately 93% joint efficiency was observed. This phenomenon was attributed to the mild softening occurring in the peripheral region of the weld, as well as the distinct anisotropic properties observed between the longitudinal and radial directions of the base metal. Elumalai et al. [18] optimized the welding parameters for AA7068 weldment joined by rotary friction welding, and they found that friction (rotational) speed had the highest effect on welding strength. Beygi et al. [19] reviewed the influence of alloying elements on intermetallic formation during friction stir welding (FSW) of aluminum to steel (dissimilar metal joining). They stated that Fe-Al forms different intermetallic compounds (IMCs) such as Fe2Al5 and FeAl3 at the weld interface, whereas Si in the alloy will give solid-solution strengthening and some alloying elements like Ni and Cr in steels retard the diffusion of IMC's and acts as grain refiners. Sahin et al. [20] examined the microstructural and mechanical properties of near-nanostructured Al 5083 alloys that were subjected to rotary friction welding, which resulted in improved tensile and fatigue strength of the welded samples. However, the significant increase in hardness and mechanical properties was due to the refinement in grain size. Similarly, Pouraliakbar et al. [21] studied the impact of rotation speed (400–1600 rpm), dwell-time (3–60 s), and cooling media (air and water) on microstructural modification (crystalline size) in stir zone (SZ) while performing friction stir spot processing (FSSP) on Al-Cu-Mg alloy. They reported that the weld cooled underwater inhibited more grain growth compared to cooling taking place in the air.

Based on the author's current understanding, minimal research has been conducted on rotary friction-welded aluminum (5xxx series) alloys. The purpose of this study is to explore the alterations in the microstructure that arise during the rotary friction welding of AA5083 H116 aluminum alloys. Furthermore, the research involves examining different weld zones and analyzing several mechanical properties, including hardness, tensile strength, and fracture analysis of the weldments. Also, the H116 temper condition enhances the alloy's corrosion resistance, making it useful in a broad range of industries, including shipbuilding. The findings and insights gained from this research endeavor will contribute to the existing knowledge base and provide valuable information for future advancements in the field.

2. Experimental Materials and Methods

2.1. Parent Metal

The parent metals used for rotary friction welding were AA5083 H116 (work hardened) aluminum alloy rods with a length of 100 mm and a diameter of 20 mm. To ensure perpendicularity, these samples were face-turned. The chemical composition details of the base metal sample, AA5083 H116, are provided in Table 1.

Table 1. Chemical composition of AA5083 H116.

Element	Zn	Mn	Mg	Cu	Cr	Fe	Si	Al
Wt. %	0.11	0.53	4.9	0.03	0.09	0.27	0.04	Bal

2.2. Welding Details

A continuous drive rotary friction welding machine (ETA Technology in Bangalore, India) with a capacity of 100 KN was employed for the welding process [22]. A manually operated personal computer-controlled system displayed the welding data, such as friction pressure and rotational speed. The base metal rods (AA5083 H116) were secured in the respective chucks. As the spindle began rotating at a predetermined speed, the specimens were brought into contact with an initial force (friction pressure). The relative motion or interaction between the specimens resulted in the generation of frictional heat at the contacting region. The combination of frictional heat and a rotary mechanical force led to severe plastic deformation (flash). A sudden pressure (forging pressure) was applied to complete the rotary friction-welding process. The flash was expelled, carrying away the major impurities. The presence of flash on both sides of the weldment indicates that the heat generated during welding was sufficient to produce a sound weld [23].

Several trials were conducted to optimize the welding parameters, such as rotational speed, burn-off length, upset force, and other friction-welding parameters. The ranges of parameters for the welding trials were selected based on the literature. Due to the limited availability of raw materials, only a limited number of experiments could be carried out. Rotational speed, burn-off length, and upset force were varied one parameter at a time to achieve the best welding. The rotational speed was explored in the range of 1200–1400 rpm [24]. A rotational speed of 1400 rpm was finalized, as speeds below 1400 rpm took considerably longer time at the friction stage, and unusual cracks were observed in the flash, indicating insufficient heat generation for complete plastic deformation. Further tests were conducted while maintaining a constant speed of 1400 rpm and varying the friction burn-off in 1 mm increments. The most appropriate flash was observed at 5 mm, beyond which there was no significant advantage except for material wastage. Once these two parameters were set, the best joint was observed at an upset force of 24 kN. Methods such as visual inspection and drop tests were employed to optimize the welding parameters. The welding parameters used for the final welding are displayed in Table 2.

Table 2. Welding Parameters.

Parameters	Explored Ranges	Selected Values
Rotational speed (rev/min)	1200–1400	1400
Friction burn-off (mm)	2–5	5
Upset force (kN)	19–24	24
Soft force (kN)	(Constant)	2
Soft force time (s)	(Constant)	4
Friction force (kN)	(Constant)	12
Upset time (s)	(Constant)	3

2.3. Metallographic Examination

To section the welded specimens along the cross-section, wire-cut electrical discharge machining (EDM) was employed. Subsequently, they were cold-mounted to ensure straight alignment. The mechanical polishing of the samples was performed using SiC papers ranging from 600 to 200 grit, followed by fine cloth polishing using 3 to 0.5 µm diamond paste. For etching the samples, Keller's reagent (consisting of 1.5 mL HCl, 1 mL HF, 2.5 mL HNO_3, and 95 mL H_2O) was utilized for approximately 15–20 s. Microstructural analysis was carried out utilizing an optical microscope and a TESCAN scanning electron microscope (SEM) (Model: VEGA 3 LMU, Brno, Czech Republic). The examination was carried out at an accelerating voltage of 20 KV and a working distance of 9.97 mm. Elemental analysis across the weldment was performed using energy-dispersive X-ray spectroscopy (EDS). The X-ray diffractometer (PANalytical, Malvern, UK) was employed for XRD analysis of the weldment. The X-ray diffractor was configured with a scan current of 30 mA, a step size of 0.01 degrees, and a wavelength of 1.547 Å. A continuous scan mode was used with CuK radiation and incident split detectors. The scan range was specified to be between 5.0 and 90.0 degrees.

2.4. Mechanical Testing

To investigate the mechanical behavior of the AA5083 joints welded using rotary friction welding, several properties, including tensile strength, hardness, and fracture analysis, were examined. The samples were prepared according to ASTM standards. To measure the hardness variation across the weldment, a Vickers micro-indentation device (Shimadzu, HMV 2T E, Kyoto, Japan) was utilized. A load of 50 gm was applied for 15 s on both sides of the weld. The tensile properties of the weldment were assessed using a KAPPA 100 SS-CF universal testing machine from Austria. Tensile tests were conducted at room temperature, employing a crosshead speed of 0.5 mm/min.

3. Results and Discussion

3.1. Base Metal Microstructure

The optical micrographs of the AA5083 base metal alloy are shown in Figure 1. Figure 1a represents the transverse plane (along the cross-section), while Figure 1b represents the longitudinal plane (along the central axis). The microstructure predominantly consists of a heterogeneous mixture of unevenly scattered grains with a segregated microstructure (Figure 1a). Elongated grains can be observed along the longitudinal plane (Figure 1b) of the base metal rods. These pancake-shaped grains result from the rolling operation during manufacturing. Radetic et al. [25] reported that compression along the strain plane leads to this elongation and the formation of lamellar/pancake-shaped grains. They suggested that while these grains are stretched along the rolling direction, they contract equally in the direction perpendicular to it. Solute strengthening, dislocation strengthening, grain refinement, and precipitate hardening are some of the major strengthening mechanisms in non-heat-treatable AA5083 aluminum alloys [26,27]. The presence of point defects generated by the major alloying elements Mn and Mg in AA5083 restricts dislocation motion and enhances material strength at low strain rates. While these alloys derive strength through Mg solute atoms via solute strengthening, an increased Mg content can decrease corrosion resistance [28]. Therefore, work hardening and grain refinement remain the most efficient methods for improving the strength of AA5083 [29]. Figure 2a shows the SEM microstructure of the AA5083 base metal. Furthermore, EDS analysis (Figure 2b) of the base metal microstructure reveals the presence of Mg-rich β (Al3Mg2) precipitates/secondary phases embedded in the α (aluminum) matrix. Many reports suggest the presence of other phases such as Al3Mn2, Al-(Fe, Si, Mn, Cr), Al6Mn, and Mg2Si in this alloy [30].

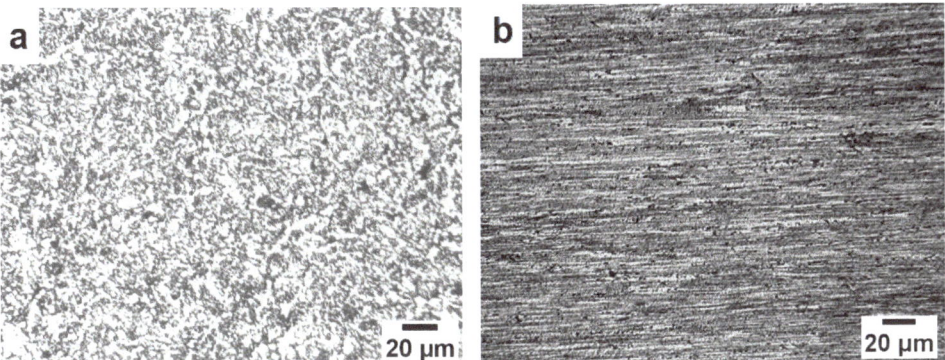

Figure 1. The optical microstructure of the AA5083 base metal in (**a**) transverse plane (along the cross-section) and (**b**) longitudinal plane (along the central axis).

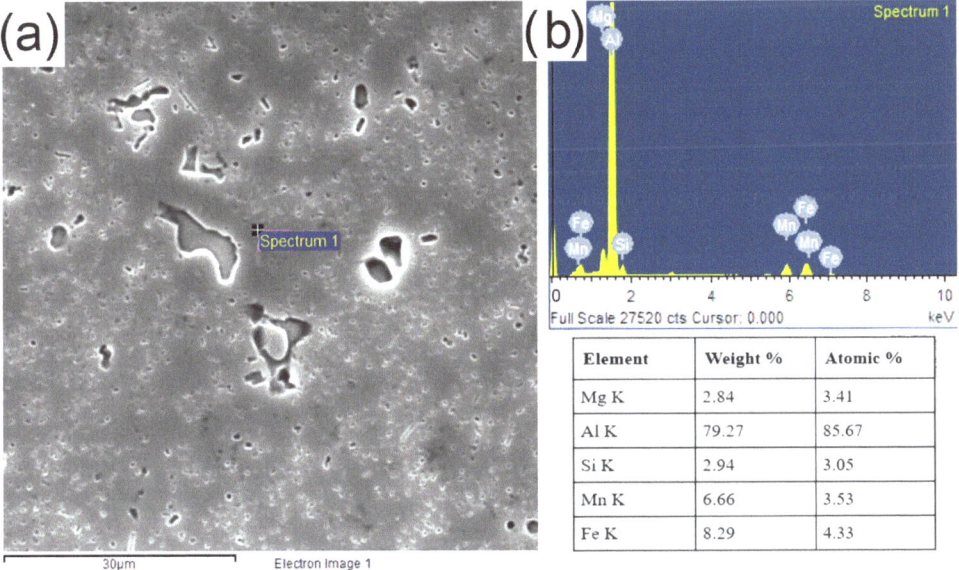

Element	Weight %	Atomic %
Mg K	2.84	3.41
Al K	79.27	85.67
Si K	2.94	3.05
Mn K	6.66	3.53
Fe K	8.29	4.33

Figure 2. SEM micrograph of the (**a**) AA5083 H116 base metal and (**b**) EDS analysis.

Figure 3 illustrates the visual view of the rotary friction-welded AA5083 H116 joint. As evident from the figure, sound welds with superior quality are produced, exhibiting an even and uniform flash on both sides of the weldment. This indicates that the heat generated during rotary friction is sufficient for the successful deformation of the base material. Figure 4 displays the macrostructure image of the rotary friction-welded AA5083 H116 joint from a cross-sectional view. It shows a balanced amount of flash on both sides, as well as grain deformation lines across the weld interface. The presence of flash assists in the removal of unwanted oxide layers and other debris, resulting in a clean and seamless weld [5]. These deformation lines are the result of severe plastic deformation that occurs during rotary friction welding [31].

Figure 3. Visual view of rotary friction-welded AA5083 H116 joint.

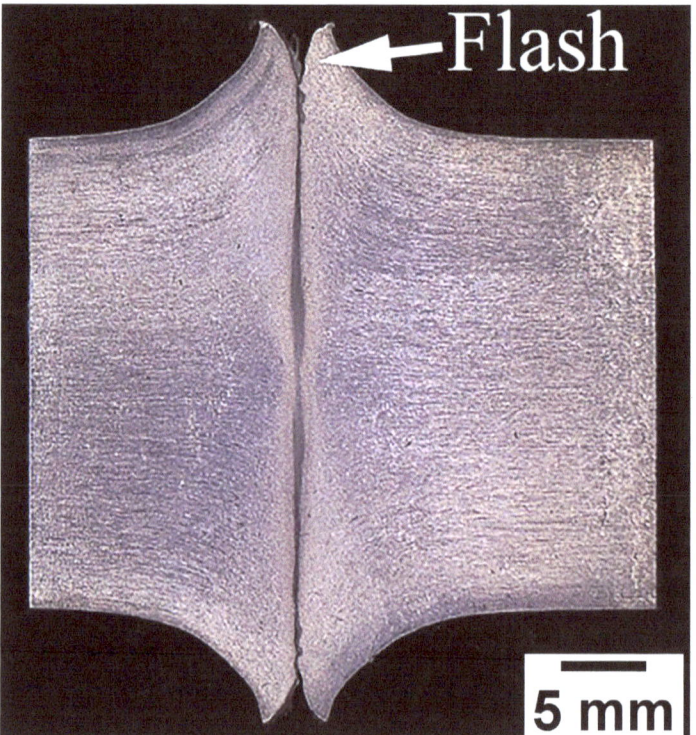

Figure 4. Macrostructure of rotary friction-welded AA5083 H116 joint.

3.2. Microstructure of Weld-Interface

During the rotary friction welding of AA5083 H116, Figure 5 illustrates the various weld zones that form along the weld interface. A detailed microstructural examination conducted across the weld interface revealed the division of the weldment into four distinct zones: (a) dynamically recrystallized zone (DRZ), (b) thermo-mechanically affected zone (TMAZ), (c) heat-affected zone (HAZ), and (d) base metal (BM) [31]. The DRZ exhibited fine (equiaxed) grains, characterized by an average grain size of 5 ± 2 μm, as revealed by microstructural analysis. The refinement of these grains can be attributed to the extreme forces and plastic deformation experienced during rotary friction welding. The continuous rotational movement of the base material promotes the formation of fine recrystallized grains, which significantly influences the final mechanical properties of the weldment [32]. It has been reported that continuous dynamic recrystallization (CDRX) plays a crucial role in the grain refinement mechanism during the friction welding of aluminum alloys [33]. CDRX involves the rearrangement of dislocations to produce subgrains, absorption of dislocations in low-angle grain boundaries, and the eventual formation of high-angle grain boundaries and new dynamically recrystallized grains [34]. It can be observed that the width of the DRZ is reduced towards the center of the weldment, indicating the differential forces experienced during the various stages of friction welding [31]. This phenomenon is observed on either side of the friction weld interface. The TMAZ consists of deformed grains arranged in lines along the direction of flash flow, which provides insight into the direction of deformation. On the other hand, the HAZ exhibits a grain arrangement similar to that of the base metal. Long, elongated grains are observed in the base metal due to deformation during the rolling process. AA5083 being a strain-hardened alloy, these grain shapes are commonly observed, highlighting the alloy's high-strength properties. However, a slight increase in grain width can be observed in the HAZ, which can be attributed to the high thermal exposure during rotary friction welding.

Figure 5. Microstructure across the various weld zones formed along the weld interface during rotary friction welding of AA5083 H116 joint.

The EBSD orientation mapping of the friction-welded AA5083 H116 zones is shown in Figure 6, where (a–d) represents the various weld zones formed during rotary friction welding of AA5083 H116, namely DRZ, TMAZ, HAZ, and BM, respectively. It is observed that the BM Al5083 alloy comprises elongated grains parallel to the rolling direction along with equiaxed grains in between the elongated grains as shown in Figure 6d. The average grain size of the equiaxed grains in the BM is 12 ± 3 μm while the larger elongated grains showed an average length of 70 ± 5 μm along the rolling direction. The inverse pole figure (IPF) showed the large portion of BM planes orientated in [101] direction. In contrast, the other zones like DRZ, TMAZ, and HAZ showed the orientation of planes in all directions.

The stir zone comprises fine equiaxed recrystallized grains (DRZ) structure (Figure 6a). The DRZ exhibited the smallest average grain size at the welding center compared to HAZ, TMAZ, and BM. The presence of dynamic recrystallization during the friction welding process accounts for this phenomenon. The DRZ exhibits an average grain size of 5 ± 2 μm. The TMAZ is situated adjacent to the DRZ on both sides, featuring an average grain size of 22 ± 4 μm. TMAZ exhibits a larger grain size when compared to DRZ. This is attributed to lower/partial recrystallization. The grains within the TMAZ exhibit an elongated morphology parallel to the welding interface, which arises from the intense plastic deformation experienced during the frictional stage of welding. Conversely, the HAZ adjacent to the TMAZ displays a coarser grain size of 32 ± 3 μm. As the HAZ is relatively far away from the weld center, it experiences less plastic deformation and heat, which results in the absence of recrystallization. Consequently, the grains formed in the HAZ are coarser compared to those in the TMAZ. Significantly, HAZ exhibits noticeable grain coarsening, which can be attributed to the annealing effect (loss of work hardening) that occurs during the welding process.

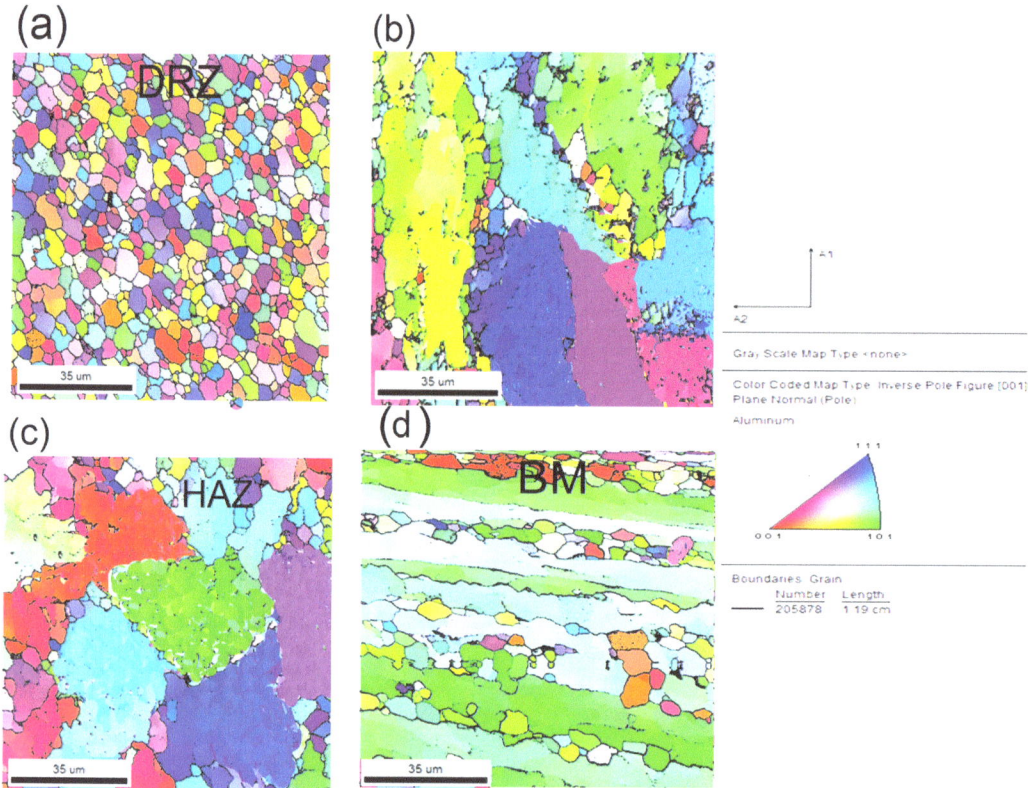

Figure 6. EBSD orientation mapping of the AA5083 H116 friction-welded samples (**a**) DRZ, (**b**) TMAZ, (**c**) HAZ, and (**d**) BM.

Figure 7a,b show the SEM micrograph of the AA5083 H116 friction-welded joint interface and EDS analysis of the weld interface, respectively. The EDS analysis reveals the presence of intermetallic compounds enriched with Fe and Mg. It also demonstrates the uniform intermixing of the main strengthening precipitates, such as β (Al3Mg2), along with various other intermetallic compounds like Al3Mn2 and Al13Fe4 in AA5083 H116. These contribute to the joint's strength, along with grain refinement during welding. Figure 8a,b show the XRD pattern of the AA5083 H116 base metal and rotary friction-welded AA5083 joint, respectively. The XRD pattern only identifies the Al phase, and the absence of Mg is observed in both conditions (base and weld). This is attributed to the development of the FCC supersaturated solid solution in the Al-Mg alloy, which leads to the diffusion of Mg atoms in the Al lattice [35]. For the welded sample, the scan area is focused on the DRZ/weld zone. The XRD spectra exhibit prominent peaks at 2θ angles of 38.36°, 44.57°, 64.79°, 77.82°, and 81.99°, corresponding to the hkl miller index planes of (111), (200), (220), (311), and (222), respectively. The (111) peak shows the highest peak intensity and texture coefficient, indicating preferred orientation growth.

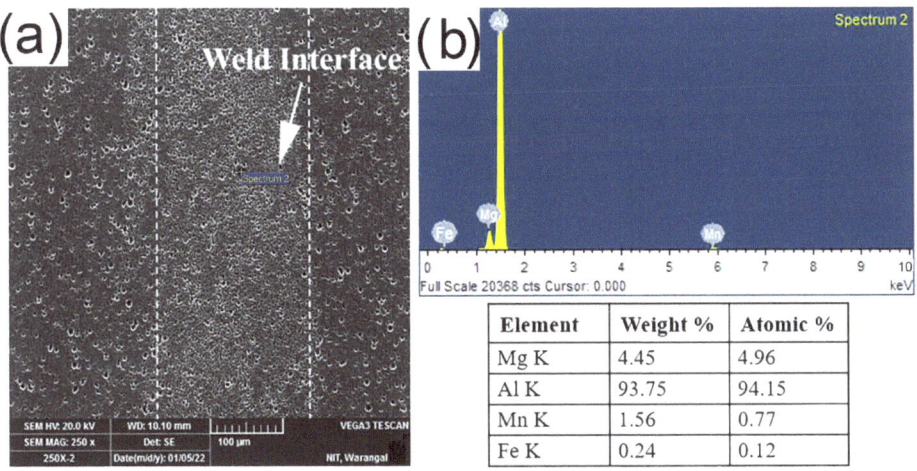

Figure 7. SEM micrograph of the AA5083 H116 friction-welded joint (**a**) weld interface and (**b**) EDS analysis of the weld interface.

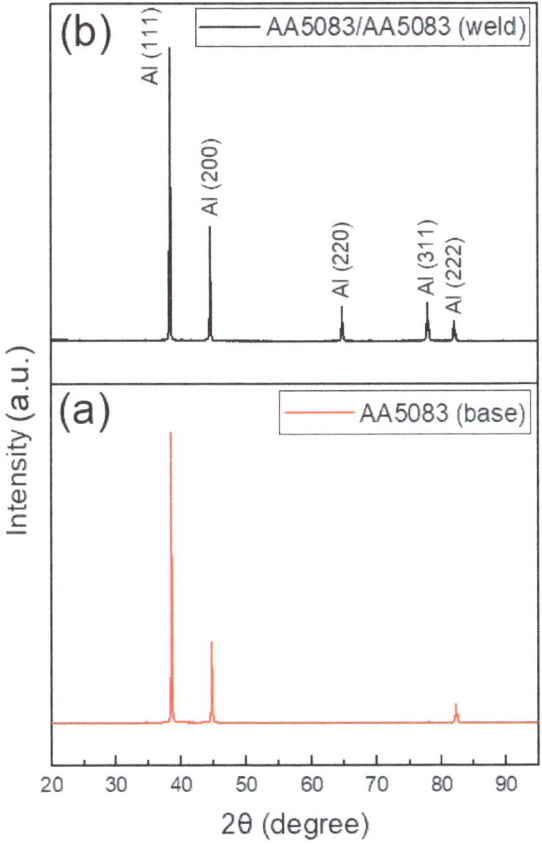

Figure 8. XRD pattern of the (**a**) AA5083 H116 base metal and (**b**) rotary friction-welded AA5083 joint.

3.3. Mechanical Properties

Figure 9 illustrates the hardness variation across the weld interface of the rotary friction-welded AA5083 H116 joint. A significant increase in hardness (107 ± 3 HV) can be observed along the DRZ of the weld interface. This can be attributed to grain refinement (recrystallization) that occurs during friction welding, resulting in fine (equiaxed) grains. This grain refinement is a consequence of the severe plastic deformation that takes place during friction welding. Additionally, a noticeable but relatively lower increase in hardness (96 ± 2 HV) can be observed in the TMAZ of the weld interface. This increase is attributed to the partial recrystallization of grains, as sufficient thermal exposure allows for partial recrystallization. The TMAZ exhibits finer grains compared to the HAZ. In the HAZ, grain coarsening and the loss of cold work result in the lowest hardness (82 ± 2 HV) along the weld interface [36]. Furthermore, a hardness of 88 ± 3 HV is observed in the BM region of the weld interface.

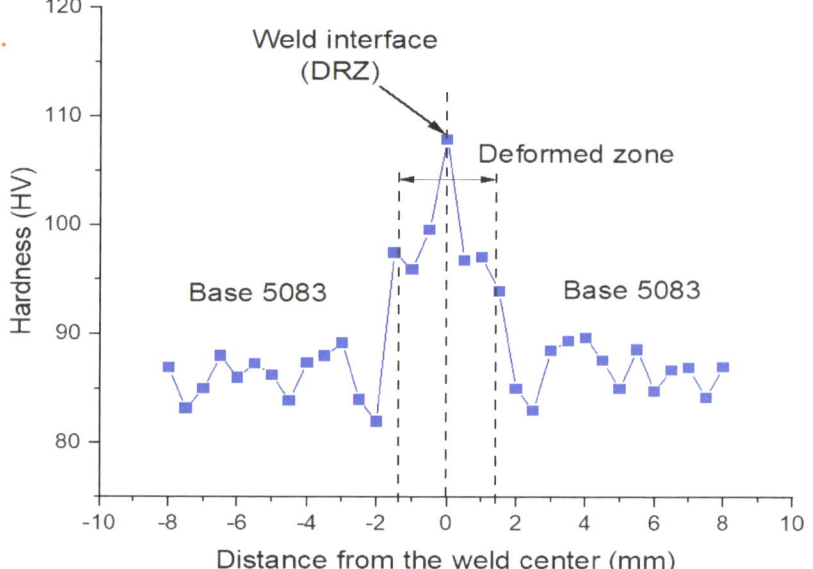

Figure 9. Microhardness distribution across the weld interface of AA5083 joint.

Figure 10 depicts a visual view of the failed tensile sample of the rotary friction-welded AA5083 H116 joint. The nature of the fracture is revealed through visual inspection of the fractured sample. Features such as limited necking and a 45° angle along the fracture axis indicate a ductile fracture.

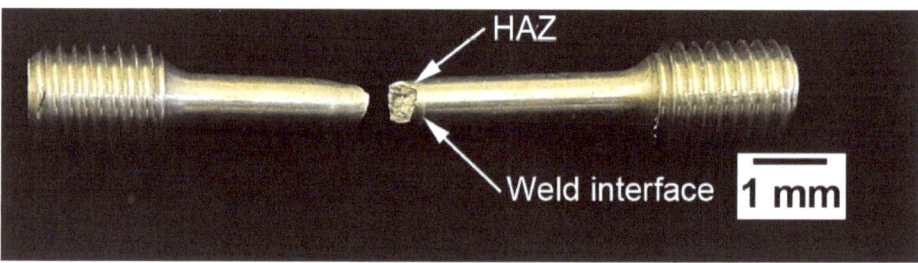

Figure 10. Failure location of tensile specimen of AA5083 H116 joint.

Furthermore, the fracture occurred in the HAZ of the weldment, suggesting that the weld joint interface formed by rotary friction welding possesses superior strength compared to the BM and HAZ. The formation of fine equiaxed grains can be attributed to dynamic recrystallization across the welding interface resulting from severe plastic deformation during friction welding [32]. Reports also indicate a significant loss of work hardening in the HAZ due to thermal exposure during welding, which leads to a reduction in strength [36]. Figure 11 displays the tensile graphs of the rotary friction-welded AA5083 H116 joint and the base metal AA5083 H116. The graph illustrates that rotary friction welding produces a sound weld with superior mechanical properties. The weldment exhibits a YS of 301 ± 6 MPa and a UTS of 425 ± 7 MPa, compared to the BM (YS: 207 ± 5 MPa; UTS: 385 ± 9 MPa). The superior strength of the weldment may be attributed to the formation of fine equiaxed grains in the DRZ. However, the weldment exhibited slightly lower % elongation (%El: 13 ± 2) compared to the BM (%El: 15 ± 3). This reduction in ductility of weldment can be attributed to the formation of various weld zones within the welded sample. The tensile graph of the base metal displays serrations across the curve, which is attributed to a common phenomenon in Al-Mg alloys known as the Portevin–Le Chatelier effect (serrated yielding). This effect is generally caused by the influence of dynamic strain aging on the macroscopic plastic flow of the material. Dislocations moving across the lattice interact with solute atoms, forming a solute atmosphere. The vacancies generated during deformation increase solute mobility in this region, leading to a decrease in dislocation velocity and a locking condition. These locking interactions result in serrations (also known as 'locking serrations') where the stress–strain curve experiences rapid spikes in flow stress followed by discontinuous fallbacks [37]. Table 3 presents the tensile properties of the BM and the rotary friction-welded AA5083 H116 joint.

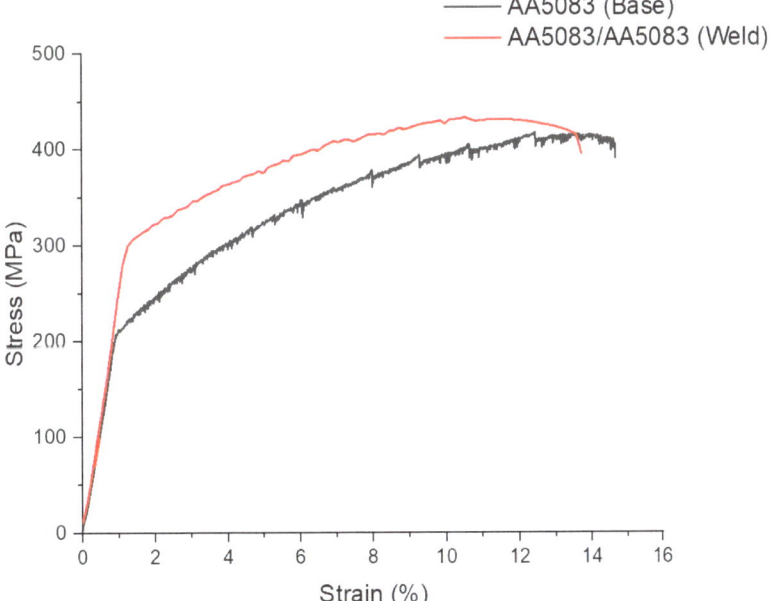

Figure 11. Typical tensile curves of AA5083 base compared with rotary friction-welded AA5083 H116 joint.

Table 3. Tensile properties of the AA5083 H116 base metal and friction welds made on AA5083 H116 alloy.

Name	Yield Stress (YS), MPa	Ultimate Tensile Strength (UTS), MPa	% Elongation (El%)
AA5083/AA5083 (weld)	301 ± 6	425 ± 7	13 ± 2
AA5083 H116 (base)	207 ± 5	385 ± 9	15 ± 3

Figure 12 depicts the fractographs of the failed rotary friction-welded AA5083 H116 joint. The fractographs clearly show a dimpled rupture as the mode of failure. The fracture surfaces exhibit features such as dimples, indicating a ductile failure mode. A detailed microscopic investigation of the fracture surfaces reveals subtle changes in overall fracture morphology and intrinsic fracture characteristics. The fractographs also suggest that the majority of the fractured surface is occupied by trans-granular bonding, characterized by the presence of dimples, with some exceptions, such as interfacial bonding resulting from rotary friction welding. The interaction of the base samples during rotary friction welding promotes interface bonding, which is evident from the pattern of large dimples on the fracture surface morphology, thereby enhancing the mechanical properties of the weld.

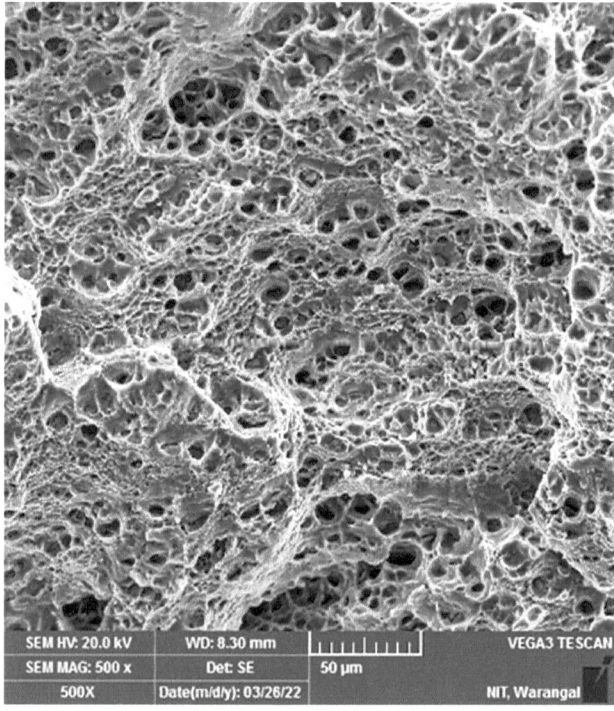

Figure 12. Fracture surface of rotary friction-welded AA5083 H116 joint.

4. Conclusions

The research on rotary friction welding of AA5083 H116 was successfully presented in this study. The investigation focused on the microstructural and mechanical properties of the weldment, and the following conclusions can be drawn:
1. Analysis using optical and electron microscopy revealed the presence of flawless welds of high quality, devoid of any defects such as cracks or porosity.
2. Close to the weld interface, two distinct microstructural zones were identified: the dynamically recrystallized zone (DRZ) and the thermomechanically affected zone

(TMAZ). The process of rotary friction welding caused significant plastic deformation in the DRZ, resulting in grain refinement. The recrystallized zone exhibited fine equiaxed grains with an average size of 5 ± 2 µm, while the TMAZ was characterized by elongated grains.

3. A notable increase in hardness was observed along the recrystallized zone (DRZ) of the weldment. This enhancement can be attributed to the formation of refined grains during the welding process.
4. The tensile test specimen failed within the heat-affected zone (HAZ) of the weldment, indicating the superior strength of the weld interface compared to the base metal and HAZ. The weldment demonstrated higher tensile values (YS: 301 ± 6 MPa; UTS: 425 ± 7 MPa) compared to the base metal (YS: 207 ± 5 MPa; UTS: 385 ± 9 MPa). The increased strength of the weldment can be attributed to the presence of fine equiaxed grains in the DRZ.
5. The tensile graph of the BM displays serrations across the curve, which is attributed to a common phenomenon in Al-Mg alloys known as the Portevin–Le Chatelier effect (serrated yielding). This effect is generally caused by the influence of dynamic strain aging on the macroscopic plastic flow of the material.
6. The weldment exhibited a slightly lower percentage of elongation (%El: 13 ± 2) compared to the base metal (%El: 15 ± 3). This decrease in ductility can be attributed to the formation of distinct weld zones within the welded sample.

Author Contributions: Conceptualization, A.M.M., K.V.K. and M.J.Q.; methodology, A.M.M., K.V.K., B.B., N.K.B., A.U.R. and M.J.Q.; formal analysis, A.M.M., B.B., N.K.B. and A.U.R.; investigation, A.M.M., K.V.K., B.B., N.K.B., A.U.R. and M.J.Q.; resources, A.M.M., K.V.K., B.B., N.K.B., A.U.R. and M.J.Q.; data curation, A.M.M., N.K.B. and A.U.R.; writing—original draft preparation, A.M.M., K.V.K., B.B., N.K.B., A.U.R. and M.J.Q.; writing—review and editing, A.M.M., N.K.B. and A.U.R.; visualization, A.M.M., N.K.B., A.U.R. and M.J.Q.; supervision, N.K.B. and A.U.R.; project administration, N.K.B. and A.U.R.; funding acquisition, A.U.R. All authors have read and agreed to the published version of the manuscript.

Funding: This research was funded by through Researchers Supporting Project number (RSPD2023R701), King Saud University, Riyadh, Saudi Arabia.

Data Availability Statement: Data are contained within this article.

Acknowledgments: The authors are thankful to King Saud University for funding this work through Researchers Supporting Project number (RSPD2023R701), King Saud University, Riyadh, Saudi Arabia.

Conflicts of Interest: The authors declare no conflict of interest.

References

1. Mathers, G. *The Welding of Aluminum and Its Alloys*; Woodhead Publishing Ltd.: Cambridge, UK, 2002.
2. Wahid, M.A.; Siddiquee, A.N.; Khan, Z.A. Aluminum Alloys in Marine Construction: Characteristics, Application, and Problems from a Fabrication Viewpoint. *Mar. Syst. Ocean Technol.* **2020**, *15*, 70–80. [CrossRef]
3. Kaufman, J.G. *Introduction to Aluminium Alloy Tempers*; ASM International: Materials Park, OH, USA, 2000.
4. Sielski, R.A. Research Needs in Aluminum Structure. *Ships Offshore Struct.* **2008**, *3*, 57–65. [CrossRef]
5. Olson, D.L.; Siewert, T.A.; Liu, S.; Edwards, G.R. *Welding, Brazing, and Soldering*; ASM International: Materials Park, OH, USA, 1993.
6. Samiuddin, M.; Li, J.; Taimoor, M.; Siddiqui, M.N.; Siddiqui, S.U.; Xiong, J. Investigation on the Process Parameters of TIG-Welded Aluminum Alloy through Mechanical and Microstructural Characterization. *Defence Technol.* **2021**, *17*, 1234–1248. [CrossRef]
7. Praveen, P.; Yarlagadda, P.K.D.V. Meeting Challenges in Welding of Aluminum Alloys through Pulse Gas Metal Arc Welding. *J. Mater. Process. Technol.* **2005**, *164–165*, 1106–1112. [CrossRef]
8. Torok, I.; Juhasz, K.; Meilinger, A.; Balogh, A. Main Characteristics of Fusion and Pressure Welding of Aluminium Alloys. *Metals* **2012**, *2*, 91.
9. Cetinel, H.; Ayvaz, M. Microstructure and Mechanical Properties of AA 5083 and AA 6061 Welds Joined with AlSi5 and AlSi12 Wires. *Mater. Test.* **2014**, *56*, 884–890. [CrossRef]
10. Babu, N.K.; Talari, M.K.; Pan, D.; Sun, Z.; Wei, J.; Sivaprasad, K. Microstructural Characterization and Grain Refinement of AA6082 Gas Tungsten Arc Welds by Scandium Modified Fillers. *Mater. Chem. Phys.* **2012**, *137*, 543–551. [CrossRef]
11. Banerjee, K.; Militzer, M.; Perez, M.; Wang, X. Nonisothermal Austenite Grain Growth Kinetics in a Microalloyed X80 Linepipe Steel. *Metall. Mater. Trans. A* **2010**, *41*, 3161–3172. [CrossRef]

12. Koilraj, M.; Sundareswaran, V.; Vijayan, S.; Koteswara Rao, S.R. Friction Stir Welding of Dissimilar Aluminum Alloys AA2219 to AA5083—Optimization of Process Parameters Using Taguchi Technique. *Mater. Des.* **2012**, *42*, 1–7. [CrossRef]
13. Thomas, M.; Nicholas, E.D.; Needham, J.C.; Murch, M.G.; Temple Smith, P.; Dawes, C. *International Patent Application No. PCT/GB92/02203 and GB Patent Application No. 9125978.8*; Weld Institute (TWI): Cambridge, UK, 1991.
14. Nicholas, E.D. Developments in the Friction Stir Welding of Metals. *Materials* **2019**, *12*, 139.
15. Pouraliakbar, H.; Pakbaz, M.; Firooz, S.; Jandaghi, M.R.; Khalaj, G. Study on the dynamic and static softening phenomena in Al–6Mg alloy during two-stage deformation through interrupted hot compression test. *Measurement* **2016**, *77*, 50. [CrossRef]
16. Lienert, T.J.; Baeslack, W.A.; Ringnalda, J.; Fraser, H.L. Inertia-Friction Welding of SiC-Reinforced 8009 Aluminium. *J. Mater. Sci.* **1996**, *31*, 2149–2157. [CrossRef]
17. Kimura, M.; Choji, M.; Kusaka, M.; Seo, K.; Fuji, A. Effect of Friction Welding Conditions on Mechanical Properties of A5052 Aluminium Alloy Friction Welded Joint. *Sci. Technol. Weld. Join.* **2006**, *11*, 209–215. [CrossRef]
18. Elumalai, B.; Omsakthivel, U.; Yuvaraj, G.; Giridharan, K.; Vijayanand, M.S. Optimization of Friction Welding Parameters on Aluminium 7068 Alloy. *Mater. Today Proc.* **2021**, *45*, 1919–1923. [CrossRef]
19. Beygi, R.; Galvao, I.; Akhavan-Safar, A.; Pouraliakbar, H.; Fallah, V.; da Silva, L.F. Effect of Alloying Elements on Intermetallic Formation during Friction Stir Welding of Dissimilar Metals: A Critical Review on Aluminum/Steel. *Metals* **2023**, *13*, 768. [CrossRef]
20. Sahin, M.; Balasubramanian, N.; Misirli, C.; Akata, H.E.; Can, Y.; Ozel, K. On Properties at Interfaces of Friction Welded Near-Nanostructured Al 5083 Alloys. *Int. J. Adv. Manuf. Technol.* **2012**, *61*, 935–943. [CrossRef]
21. Pouraliakbar, H.; Beygi, R.; Fallah, V.; Monazzah, A.H.; Jandaghi, M.R.; Khalaj, G.; da Silva, L.F.; Pavese, M. Processing of Al-Cu-Mg alloy by FSSP: Parametric analysis and the effect of cooling environment on microstructure evolution. *Mater. Lett.* **2022**, *308*, 131157. [CrossRef]
22. Masaaki, K.; Masahiro, K.; Koichi, K. Effect of Friction Welding Conditions on Joint Properties of Austenitic Stainless Steel Joints by Friction Stud Welding. *Weld. Int.* **2018**, *32*, 274–288. [CrossRef]
23. Shete, N.; Deokar, S.U. A Review Paper on Rotary Friction Welding. *IOP Conf. Ser. Mater. Sci. Eng.* **2017**, *5*, 1557.
24. Sasmito, A.; Ilman, M.N.; Iswanto, P.T.; Muslih, R. Effect of Rotational Speed on Static and Fatigue Properties of Rotary Friction Welded Dissimilar AA7075/AA5083 Aluminium Alloy Joints. *Metals* **2022**, *12*, 99. [CrossRef]
25. Radetić, T.; Popović, M.; Romhanji, E.; Milović, B.; Dodok, R. Microstructure Evolution of the Hot-Rolled Modified AA 5083 Alloys during the Two-Stage Thermal Treatment. In Proceedings of the 4th International Conference Processing and Structure of Materials, Palic, Serbia, 27–29 May 2010.
26. Huskins, E.L.; Cao, B.; Ramesh, K.T. Strengthening Mechanisms in an Al–Mg Alloy. *Mater. Sci. Eng. A* **2010**, *527*, 1292–1298. [CrossRef]
27. Engler, O.; Miller-Jupp, S. Control of Second-Phase Particles in the Al-Mg-Mn Alloy AA 5083. *J. Alloys Compd.* **2016**, *689*, 998–1010. [CrossRef]
28. Gubicza, J.; Chinh, N.Q.; Horita, Z.; Langdon, T.G. Effect of Mg Addition on Microstructure and Mechanical Properties of Aluminum. *Mater. Sci. Eng. A* **2004**, *387–389*, 55–59. [CrossRef]
29. Krishna, K.S.V.B.R.; Chandra Sekhar, K.; Tejas, R.; Naga Krishna, N.; Sivaprasad, K.; Narayanasamy, R.; Venkateswarlu, K. Effect of Cryorolling on the Mechanical Properties of AA5083 Alloy and the Portevin–Le Chatelier Phenomenon. *Mater. Des.* **2015**, *67*, 107–117. [CrossRef]
30. Yi, G.; Sun, B.; Poplawsky, J.D.; Zhu, Y.; Free, M.L. Investigation of Pre-Existing Particles in Al 5083 Alloys. *J. Alloys Compd.* **2018**, *740*, 461–469. [CrossRef]
31. Li, P.; Wang, S.; Xia, Y.; Hao, X.; Lei, Z.; Dong, H. Inhomogeneous Microstructure and Mechanical Properties of Rotary Friction Welded AA2024 Joints. *J. Mater. Res. Technol.* **2020**, *9*, 5749–5760. [CrossRef]
32. Barcellona, A.; Buffa, G.; Fratini, L. Process Parameters Analysis in Friction Stir Welding of AA6082-T6 Sheets. In Proceedings of the ESAFORM Conference, Brescia, Italy, 28 April 2004.
33. Buffa, G.; Fratini, L.; Shivpuri, R. CDRX Modelling in Friction Stir Welding of AA7075-T6 Aluminum Alloy: Analytical Approaches. *J. Mater. Process. Technol.* **2007**, *191*, 356–359. [CrossRef]
34. Galiyev, A.; Kaibyshev, R.; Gottstein, G. Correlation of Plastic Deformation and Dynamic Recrystallization in Magnesium Alloy ZK60. *Acta Mater.* **2001**, *49*, 1199–1207. [CrossRef]
35. Wagih, A. Mechanical Properties of Al–Mg/Al2O3 Nanocomposite Powder Produced by Mechanical Alloying. *Adv. Powder Technol.* **2015**, *26*, 253–258. [CrossRef]
36. Palanivel, R.; Koshy Mathews, P.; Murugan, N.; Dinaharan, I. Effect of Tool Rotational Speed and Pin Profile on Microstructure and Tensile Strength of Dissimilar Friction Stir Welded AA5083-H111 and AA6351-T6 Aluminum Alloys. *Mater. Des.* **2012**, *40*, 7–16. [CrossRef]
37. McCormick, P.G. The Portevin-Le Chatelier Effect in an Al-Mg-Si Alloy. *Acta Metall.* **1971**, *19*, 463–471. [CrossRef]

Disclaimer/Publisher's Note: The statements, opinions and data contained in all publications are solely those of the individual author(s) and contributor(s) and not of MDPI and/or the editor(s). MDPI and/or the editor(s) disclaim responsibility for any injury to people or property resulting from any ideas, methods, instructions or products referred to in the content.

Article

Study of the Microstructure and Mechanical Property Relationships of Gas Metal Arc Welded Dissimilar Protection 600T, DP450 and S275JR Steel Joints

Mustafa Elmas [1,2], Oğuz Koçar [1,*] and Nergizhan Anaç [1]

[1] Department of Mechanical Engineering, Engineering Faculty, Zonguldak Bulent Ecevit University, Zonguldak 67100, Turkey; melmas@erdemir.com.tr (M.E.); nergizhan.kavak@beun.edu.tr (N.A.)
[2] Erdemir Engineering Management & Consulting Services INC, Zonguldak 67100, Turkey
* Correspondence: oguz.kocar@yahoo.com.tr

Citation: Elmas, M.; Koçar, O.; Anaç, N. Study of the Microstructure and Mechanical Property Relationships of Gas Metal Arc Welded Dissimilar Protection 600T, DP450 and S275JR Steel Joints. *Crystals* **2024**, *14*, 477. https://doi.org/10.3390/cryst14050477

Academic Editor: Marek Sroka

Received: 28 April 2024
Revised: 13 May 2024
Accepted: 15 May 2024
Published: 19 May 2024

Copyright: © 2024 by the authors. Licensee MDPI, Basel, Switzerland. This article is an open access article distributed under the terms and conditions of the Creative Commons Attribution (CC BY) license (https://creativecommons.org/licenses/by/4.0/).

Abstract: The need for combining dissimilar materials is steadily increasing in the manufacturing industry, and the resulting products are expected to always have high performance. While there are various methods available for joining such material pairs, one of the commonly preferred techniques is fusion welding. In this study, three different steel materials (Protection 600T, DP450, and S275JR) were joined using gas metal arc welding (GMAW) in different combinations (similar/dissimilar). The microstructure and mechanical properties of the joints were evaluated. Tensile test, Vickers microhardness (HV 0.1), bending, Charpy V-notch impact testing, and microstructure examinations were conducted to analyze the weld and heat-affected zone. The tensile strengths of the base metal materials Protection 600T, DP450, and S275JR were found to be 1524.73 ± 18.7, 500.8 ± 10.4, and 508.5 ± 9.5 MPa, respectively. In welded samples of similar materials, the highest efficiency was found to be 103.05% for DP450/DP450, while in dissimilar welded joints, it was 105.5% for the DP450/S275JR pair. Hardness values for the base materials Protection 600T, DP450, and S275JR were measured as 526.5 ± 10.5, 153.8 ± 1.8, and 162.5 ± 5.2, respectively. In all welded samples, there was an increase in hardness in the weld zone (due to the welding wire) and the heat-affected zone (due to grain size refinement). While the impact energy values of similar material pairs were close to the base material impact energy values, the impact energy values of dissimilar material pairs varied according to the base materials. In addition, in joints made with similar materials, the bending force was close to the base materials, while a decrease in bending force was observed in joints formed with dissimilar materials. As a result, the welding of DP450 and S275JR materials was carried out efficiently. Protection 600T was welded with other materials, but its welding strength was limited to the strength of the material with low mechanical properties.

Keywords: dissimilar steel joints; gas metal arc; welding strength; mechanical properties; Protection 600T; DP450; S275JR steel

1. Introduction

With the development of technology around the world, the need for high-strength steel materials is steadily increasing [1]. Therefore, it is very common in the industry to enhance the quality of materials/steels by using them in conjunction with other materials. There are various structures and systems produced by using dissimilar materials together. Especially in industrial applications and academic research, it is observed that high-strength steels such as armor steels, Twinning-induced plasticity (TWIP) steels, transformation-induced plasticity (TRIP) steels, and Dual Phase (DP) steels are used together with structural steels [2,3]. Furthermore, in specialized sectors such as nuclear power plants, oil refineries, and defense industries (including military vehicles), the utilization of dissimilar materials together holds significant importance [4,5]. Moreover, combining thin yet high-strength components instead of thick ones offers the advantage of lightness [2]. Therefore, understanding the behaviors of materials in joining technologies (such as welding processes, etc.) is crucial in industries for their safe and efficient utilization [6]. In research conducted for this purpose, it has been determined that gas metal arc welding is one of the most effective techniques in joining dissimilar metal materials [7–11]. Gas metal arc welding of dissimilar metal materials typically involves joining at least two metals or alloys with different chemical compositions, melting temperatures, and thermal expansion properties [12]. Due to the change in physical properties in the weld zone during the process of joining materials with different chemical and mechanical characteristics, it is not easy to evaluate weld strength [13]. The proper and accurate execution of welding depends on selecting the appropriate welding process, process parameters, metal properties, and operating conditions [14]. Changing a parameter in the welding process significantly affects the welding properties. Due to the different thermal expansion of dissimilar metals, various undesired defects may occur during welding, such as stress corrosion cracking, residual stress formation, and disruption of stress concentration on both sides of the weld [15,16]. For this reason, some researchers focus on improving the mechanical properties of welded parts by controlling the parameters of the welding process [17]. The scientific literature on joining armor steels, which is the focus of this study, with other metals using gas metal arc welding has been carefully reviewed. Armor steels are materials with ultra-high strength and hardness, resistant to penetration by bullets and explosives. They are used in equipment resistant to ballistic impact and collision applications such as tanks, armored vehicles, and helicopter components [18].

S. Naveen Kumar et al. [19] observed that there was no study on the welding of armor steels of dissimilar qualities. Therefore, in their study, they joined Ultra-high Hard Armor (UHA) steel with Rolled Homogeneous Armor (RHA) steels using gas metal arc welding. RHA steel, which makes up 75% of armored combat vehicles, is used together with UHA steels, which have a high strength–weight advantage, to improve vehicle mobility. Austenitic Stainless Steel (ASS), Duplex Stainless Steel (DSS), and Low Hydrogen Ferritic (LHF) filler wires were used for welding. Researchers examined the effects of filler materials on the ballistic resistance of welded parts and the metallurgical characterization of both armor grade steel joints made using gas metal arc welding (GMAW).

Magudeeswaran et al. [20] evaluated the metallurgical properties of welded joints of Quenched and Tempered High Strength (Q&T) Steels used in armored vehicle manufacturing. In the welding of armor-grade Q&T steels, issues such as post-weld cold cracking caused by hydrogen in the heat-affected zone (HAZ) and softening of the HAZ due to the welding thermal cycle exist. The results revealed that these conditions adversely affect the ballistic properties of the steel.

Researchers [21] examined the effect of plate thickness on the microstructure and hardness of The Protection 500 series armor steels welded using the robotic GMAW method. The characterization of the weld using ER110S-G filler metal was completed by micro hardness tests and micro and macro structural examinations. It has been observed that as the plate thickness increases, the width of the softening zone decreases significantly, and the same amount of heat input slightly increases the microhardness. Depending on the heat changes during the welding of armor steel, their internal structure and, accordingly, their mechanical properties in different regions change.

Kaçar and Emre [22] investigated the gas metal arc welding capabilities of pairs formed from Armox 500T armor steel and AISI 304 steel. They stated that successful joining of these materials could be achieved with proper selection of welding parameters. Gas metal arc welding (GMAW) is commonly used for joining armor steels. Günen et al. compared the effects of different welding techniques (cold metal transfer arc welding (CMT) and hybrid plasma arc welding (HPAW)) on the microstructure and mechanical properties using GMAW. In all three different welding processes, both the hardness of the weld metal and the heat-affected zone were found to be higher than that of the base metal. Optimization of welding parameters has assisted in obtaining defect-free welds [23].

Qingguo Wang et al. [24] conducted multi-pass gas metal arc welding using ZGMn13Mo manganese steel and A514 low alloy steel, each having a thickness of 25 mm, with ER309L stainless steel welding wire.

In this study, Protection 600T armor steel, DP450 (Dual-phase steels), and S275JR steel (structural steel) were used. S275JR steel is a non-alloy, low carbon, mild steel grade. Dual-phase DP 450 steel, on the other hand, exhibits better cold formability and strength compared to low carbon and high-strength low-alloy steels [25]. They are preferred in the automotive industry due to their high strength and ductility [26,27]. Fusion welding processes such as gas metal arc welding (GMAW) are used to join DP steel materials [28]. These materials can be combined with each other due to their properties. As a result of these joining processes, engineering advantages such as cost and weight savings are achieved. Protection 600T material is high-resistance armor steel with ultra-high hardness against ballistic penetration. To the best knowledge of the researchers, there are no studies in the literature on the welding of Protection 600T material, which is frequently used in the manufacturing industry, with DP450 and S275JR materials. This study will contribute to the literature by presenting the joining of Protection 600T, DP450, and S275JR steels using the GMAW method.

As industrial needs change, high-strength steel types are also developing. When joining high-strength steels with other steels, it is very important to investigate the microstructure and mechanical properties in terms of welding safety. Metallurgical compatibility of strength and material pairs should be considered during the process. For this reason, the compatibility of Protection 600T, DP450, and S275JR steels was investigated in this study. Microhardness (HV 0.1), bending, impact, and tensile tests were performed for base metals and welded joints. The results were analyzed comparatively.

2. Material and Method

2.1. Properties of Steel Materials and Welding Wire

In the experiments, three different grades of steel materials were used. The first is Protection 600T steel (Miilux Oy, Manisa, Turkey), known for its ultra-high hardness and high resistance against ballistic penetration. The other material is DP450 steel (Ereğli Iron and Steel Factories, Zonguldak, Turkey) which is widely used among dual-phase steels for its high strength and good formability capability. Dual phase (DP) steels, particularly when compared to high-strength steels, have higher ultimate tensile strength (UTS) and lower yield strength. These characteristics have made DP steels indispensable in automotive applications [29]. The other material used in the study is S275JR (Yucel Pipe and Profile Industry Inc., Kocaeli, Turkey) general structural steel, which has good welding properties and strength and is used in many applications in construction and various industries, including production facilities and general buildings [30,31]. AWS A5.28:ER70S-A1 wire (Gedik Company, İstanbul, Turkey) with a diameter of 1.2 mm was used as welding wire. The yield strength of the welding wire is 460 MPa, tensile strength is 550–670 MPa and elongation is 22%. The preferred welding wire is used to join high-strength steels by gas arc welding [32–34]. The chemical compositions of steel materials and welding wire are given in Table 1.

Table 1. Chemical composition of the steels and the welding wire (wt%).

Material	C	Mn	Si	Cr	Ni	Mo	P	S	B	Fe
Protection 600T	0.45	0.80	0.70	0.50	3.0	0.60	0.015	0.004	0.003	balance
DP450	0.057	1.077	0.206	0.501	-	-	0.090	0.002	-	balance
S275JR	0.071	0.307	0.008	0.019	0.016	0.001	0.010	0.012	-	balance
ER 70 S-A1	0.1	1.1	0.6	-	-	0.5	-	-	-	balance

2.2. Preliminary Studies for the Welding Process

In this section, information about pre-welding preparations and the welding process is given. The samples to be welded were prepared in dimensions of 120 mm × 500 mm × 4 mm. The experimental design planned for welding base steels is given in Table 2, and the welding parameters are given in Table 3. The groove size created in the steel plates before welding is shown in Figure 1a. A 5 mm weld opening was opened at the joints of the plates at a V 60° groove angle, leaving a 2 mm root gap.

Table 2. Experimental design for welding.

Experimental Number	Material	Material
1	Protection 600T	Protection 600T
2	DP450	DP450
3	S275	S275
4	Protection 600T	S275
5	Protection 600T	DP450
6	DP450	S275

Table 3. Welding parameters.

GMAW Welding	Properties
Shielding gas	M21 (82% Ar + 18% CO_2)
GMAW wire	ER 70 S-A1 GEKA Ø 1.2
Welding current (A)	120–130
Welding Voltage (V)	19–20
Welding speed (mm/s)	5.7

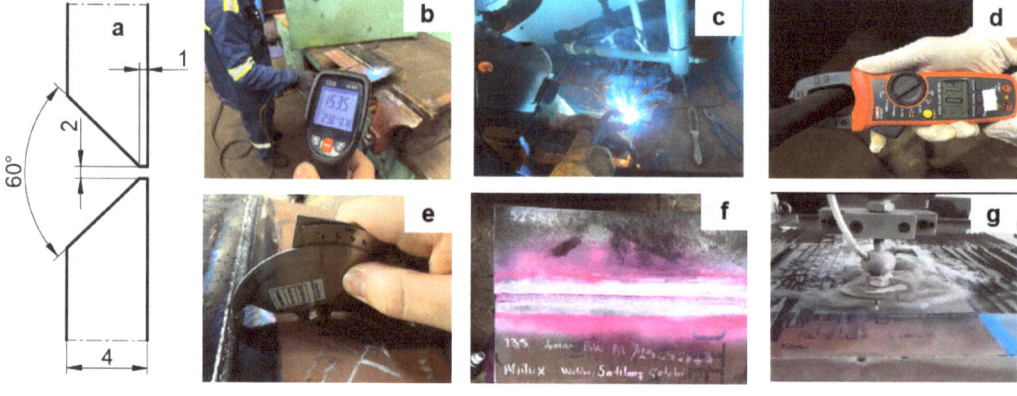

Figure 1. Preliminary preparation for welding and images taken during welding ((**a**). The groove size, (**b**). Thermal monitoring, (**c**). Welding process, (**d**). Measuring the welding voltage, (**e**). The weld cap height, (**f**). Penetrant test, (**g**). Specimen preparation process).

Due to the use of different grades of steel, carbon equivalences were examined. Since the lowest carbon equivalent (S275JR) was 0.40, the plates were annealed before welding at 150–190 °C. The pre-annealing temperature for the welding process was measured with a Cem DT-835 model (0–800 °C) thermometer (Figure 1b). A Lincoln LF-33 gas arc welding machine was used for the welding process (Figure 1c). Welding parameters are given in Table 3. The welding was conducted in four passes with 12 m^3/sec gas flow using 82% Ar + 18% CO_2 shielding gas during welding and ER 70 S-A1 brand welding wire. Verification was made by measuring the welding voltage with a clamp meter during welding (Figure 1d). Post-welding cooling was carried out in a controlled manner by wrapping the welded samples in stone wool. After welding, the weld cap height was measured using a weld cap gauge (Figure 1e). Finally, the weld seams were checked with a penetrant test (cleaner: CR60, penetrant: CR51, developer: CR70) (Figure 1f). The samples required for mechanical tests were cut using MJT4000 waterjet brand water jet with 4000 bar pressure and 300 mm/min cutting speed (Figure 1g).

2.3. Preparation of Mechanical Testing and Microstructure Samples

Figure 2 shows the dimensions of the samples used to determine the mechanical (tensile and bending, hardness, impact notch) and microstructural properties after the welding process. Hardness distribution in welded samples was conducted using the QNess 10 A+ brand micro-Vickers hardness tester by applying a 9.81 N pressure load for 15 s at 0.5 mm intervals, perpendicular to the weld line. To determine the impact energy, the impact absorption energy of the sample was measured using the JBN-300 pendulum testing machine, in accordance with the ISO 9016 standard [35], including the weld line and heat affected zone (HAZ). To determine the strength of the welded samples, a 60-ton Zwick Roell brand tensile device was used at a speed of 2 mm/min at room temperature. Finally, the material internal structure was examined with an electron microscope (Nikon Epiphot 200 Inverted Metallurgical Microscope, Artisan Technology Group, Kansas City, MO, USA). Welding efficiency is calculated using the following formula [9,36].

$$Weld\ efficiency\ (\%) = \frac{UTS\ of\ welded\ joint\ (MPa)}{UTS\ of\ base\ material\ (MPa)} \times 100 \qquad (1)$$

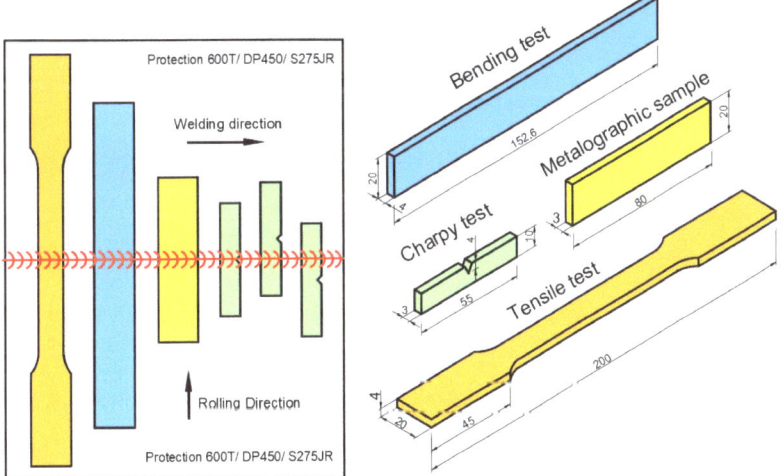

Figure 2. Dimensions of samples prepared for mechanical testing and microstructure.

3. Results

3.1. Mechanical and Microstructure Properties of the Base Materials

Tensile, microhardness, impact notch, and bending tests were performed to determine the mechanical properties of Protection 600T, DP450, and S275JR materials. As a result of the tensile test, ultimate tensile strength (UTS) was found to be 2141.98 ± 23.2 MPa for Protection 600T, 500.8 ± 10.4 MPa for DP450, and 508.15 ± 9.5 MPa for S275JR steel. Microhardness values were determined as 526.5 ± 10.5 HV, 153.8 ± 1.8 HV, and 162.5 ± 5.2 HV, respectively. Since Protection 600T has ultra-high hardness with its high carbon content, its mechanical properties were higher than the other two steel materials. Figure 3 shows the stress–strain graphs of the base materials. Accordingly, it is seen that Protection 600T has high strength but low elongation capability. When the fracture elongation of the base materials was compared, the highest elongation was obtained as 25.41 ± 1.3% in S275JR steel, while it was as 23.95 ± 1.7% in DP450 and 4.34% for Protection 600T, respectively. The tensile test, microhardness, Charpy-V, and bending test results of the base materials are given in Table 4.

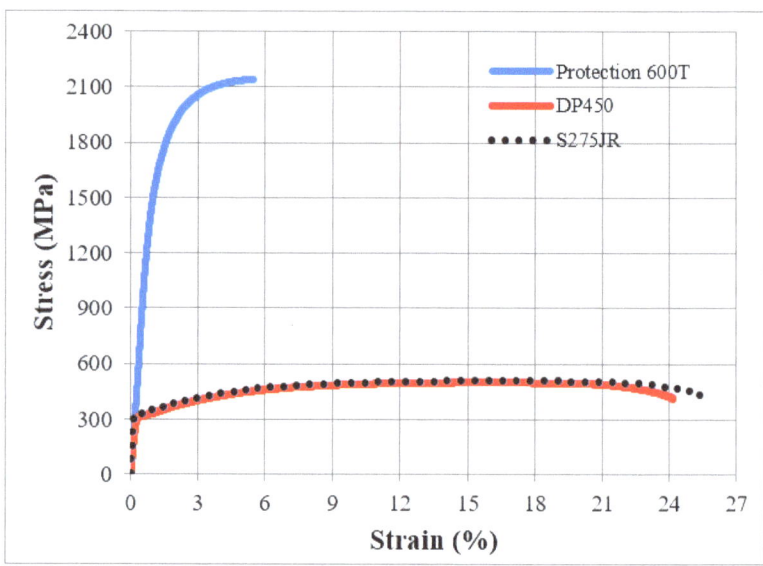

Figure 3. Stress and strain curves for base materials.

Table 4. Mechanical properties of base materials.

	Microhardness (HV)	Tensile Stress (MPa)	UTS (MPa)	Charpy-V (Joule)	Bending (kN)
Protection 600T	526.5 ± 10.5	1524.73 ± 18.7	2141.98 ± 23.2	75 ± 2.7	16.7 ± 0.4
DP450	153.8 ± 1.8	312.09 ± 11.3	500.8 ± 10.4	85 ± 3.4	11.0 ± 0.2
S275JR	162.5 ± 5.2	324.10 ± 8.2	508.5 ± 9.5	32 ± 1.2	9.3 ± 0.3

The toughness of a material is evaluated based on how much energy it absorbs. The higher the impact energy, the higher the expected toughness of the material [36]. Charpy V-notch impact energy (CVN) tests at room temperature (21 °C) revealed the impact toughness of the welded samples as follows: Protection 600T exhibited a toughness of 75 ± 2.7 J, DP450 showed a toughness of 85 ± 3.4 J, and S275JR had a toughness of 32 ± 1.2 J. The reason for the highest impact toughness observed in the DP450 material could be attributed to the dense ferrite present in its microstructure, which enhances its energy absorption capability. Additionally, the lower energy absorption capability of Protection 600T compared to DP450 can be attributed to the dense martensite structure present in Protection 600T, which contributes to its hardness. The increase in material hardness tends to reduce the energy absorption capability of materials [37–40]. S275JR general structural steel exhibits variations in ductile behavior with temperature due to its ferritic structure. While S275JR steel becomes more brittle at low temperatures, its ductility increases at high temperatures [30]. To determine the deformation of base materials and welded samples, a 90° bend test was conducted on Protection 600T, DP450, and S275JR materials and in combinations. According to the bending test results, the maximum bending forces were determined to be 16.7 ± 0.4 kN, 11.0 ± 0.2 kN, and 9.3 ± 0.3 kN, respectively.

In Figure 4, the microstructure images of Protection 600T, DP450, and S275JR steels before welding are provided. When examining the microstructure of Protection 600T, it is observed that there is martensite and retained austenite inside the prior austenite matrix (Figure 4a). Prior austenite grain boundaries (PAGB) are clearly visible and exhibit a fine-grained structure. This fine-grained martensitic structure provides high hardness and toughness to the armor steel [41]. Figure 4b–d depict the microstructure of DP450 steel. In dual-phase steels, large martensite islands are dispersed within a ferrite matrix. In dual-phase steels, mechanical properties are primarily dependent on the amount of martensite in the microstructure [42]. When examining the microstructure of DP450, it is observed that there is a dense ferrite matrix structure with a small amount of martensite. The grid method (with a grid spacing of 7.2 µm) was employed to determine the martensite phase ratio in DP450 steel (Figure 4d). The martensite phase ratio in DP450 steel was calculated as 13.3%. When examining the microstructure image of the base material of S275JR structural steel in Figure 4e,f, the black regions represent pearlite, while the lighter-colored regions represent ferrite [43]. The microstructure of S275JR steel consists of ferrite and pearlite grains, depending on the carbon content it contains [30].

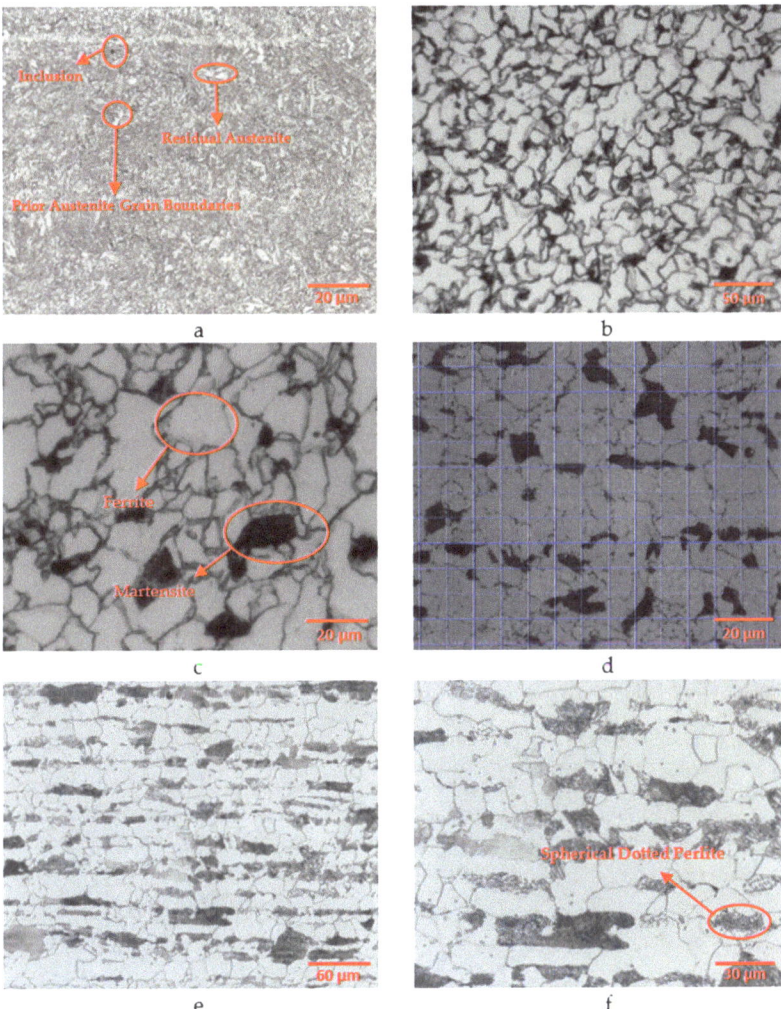

Figure 4. Base material microstructural images (**a**) Protection 600T (1000×), (**b**) DP450 (200×), (**c**) DP450 (1000×), (**d**) DP450 martensite phase ratio, (**e**) S275JR (200×), (**f**) S275JR (1000×) microstructure.

3.2. Metallographic Examination

3.2.1. Microstructure of Similar Materials after Welding

The microstructure of the HAZ and the weld zone of the Protection 600T material welded using ER70S-A1 wire is shown in Figure 5. Upon examination of the welded samples, it was determined that the transitions between the base material and the HAZ were homogeneous and smooth (Figure 5a). Although the base material has a densely tempered martensite structure, the HAZ was annealed at a medium temperature in its structure with the heat effect and tempered troostite, characterized by the gradual disappearance of the needle shape of the martensite, was observed (Figure 5b). When the microstructures shown in Figure 5c,d were examined, it was determined that the tempered martensite ratio was dominant in the weld metal, but in some parts of the martensite phase, there was a small amount of lath martensite phase aligned in parallel to form martensite beams or martensite areas. In addition, retained austenite and primary austenite grain boundaries and Weisher's tissue were detected in the weld area (Figure 5d).

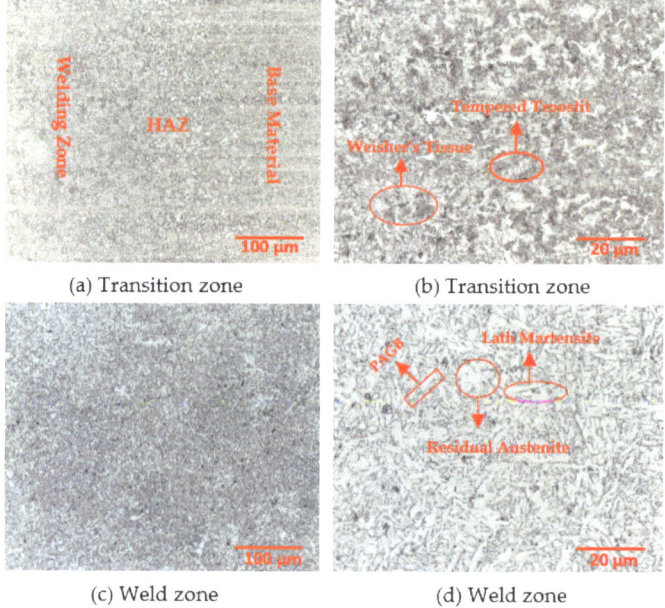

Figure 5. (**a**,**b**) Transition zone microstructures for the Protection 600T material pair (200×, 1000×) and (**c**,**d**) weld zone microstructures (200×, 1000×).

Figure 6 shows the post-welding HAZ and weld zone microstructure images of the DP450/DP450 material pair. It was determined that the martensite phase ratio was high in the HAZ (Figure 6a,b) and the ferrite ratio was high in the weld zone (Figure 6d,e). Although acicular ferrite formations were occasionally observed in the weld zone, dendritic ferrite formation was generally observed. HAZ formed a smooth transition zone with homogeneous distribution of ferrite and martensite phases. At the transition point from HAZ to the weld zone, martensite phases were arranged in columns, but upon reaching the welded structure, martensite phases were generally observed in the dendritic regions of the ferrite. The martensite phase ratio was determined to be approximately 27.2% in regions close to the melting zone (Figure 6c) and approximately 7.2% in the weld zone (Figure 6f).

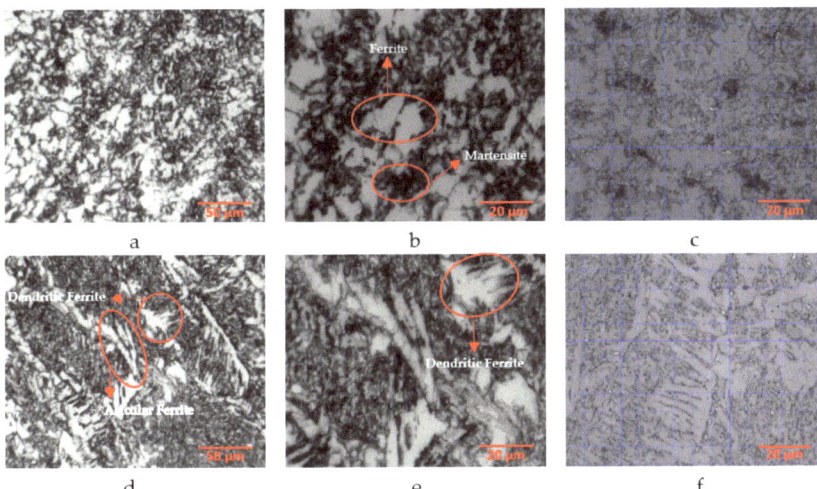

Figure 6. (**a**–**c**) HAZ region (500×, 1000×) and phase ratio, (**d**–**f**) weld zone microstructure images (500×, 1000×) and phase ratio for DP450/DP450.

In Figure 7, microstructure images of the HAZ and weld zone of the S275JR/S275JR material pair are provided. In the HAZ, a microstructure consisting of ferrite and lamellar pearlite phase (Figure 7a,b), and in the weld zone, a microstructure dominated by ferrite content (Figure 7c,d), formed. In addition, with the effect of the additional wire in the weld zone, although pearlite and Widmanstatten ferrite were observed in some areas, in general, intense acicular ferrite formation was detected. In the zones under the influence of heat, a soft transition was observed, and the ferrite and pearlite phases were distributed homogeneously, creating a soft transition zone. While pearlite and ferrite phases were arranged in columns at the transition boundary from the HAZ to the weld zone, it was observed that the ferrite phases generally turned into acicular ferrite in the weld zone.

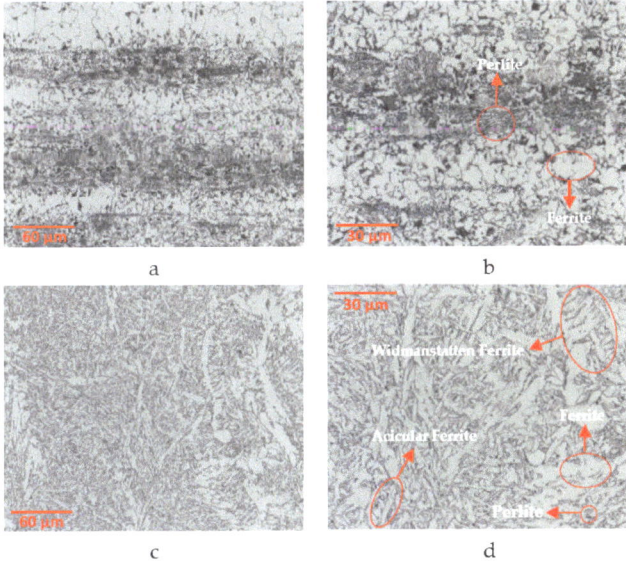

Figure 7. (**a**,**b**) HAZ region (500×, 1000×), (**c**,**d**) weld zone microstructure images (500×, 1000×) for material pair S275JR/S275JR.

3.2.2. Microstructure of Dissimilar Materials after Welding

Figure 8 shows the microstructure images of the transition zone and welding zone of the Protection 600T/S275JR material pair. It has been determined that acicular ferrite and martensite were dominant in regions in the weld zone, tempered troostite formation was seen, and primary austenite grain boundaries were observed. In the HAZ, the martensite phase was observed separately along with the ferrite and pearlite phase. In the HAZ, the martensite phase and ferrite–perlite phases were detected as not homogeneously distributed and a sharp transition was observed.

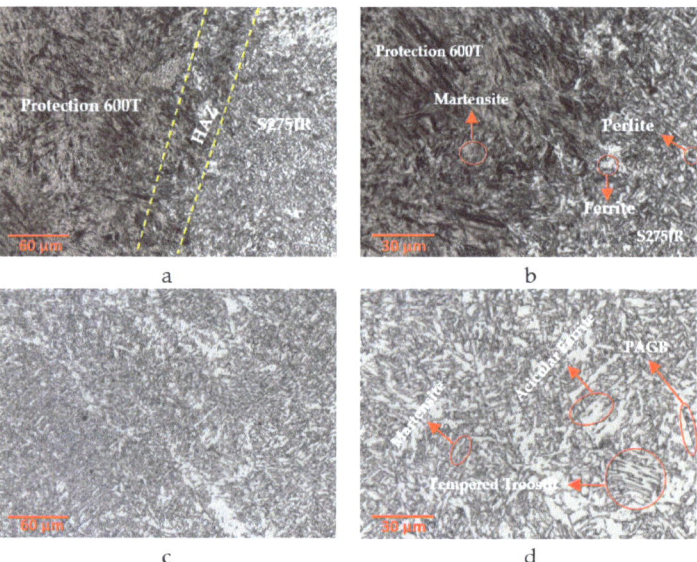

Figure 8. (**a**,**b**) HAZ (500×, 1000×), (**c**,**d**) weld zone microstructure images (500×, 1000×) for Protection 600T/S275JR material pair.

Figure 9 shows the microstructure images of the transition zone and welding zone of the DP450/Protection 600T material pair. Ferrite and martensite phases were observed in the HAZ (Figure 9a,b). In the region under the influence of heat, it has been determined that the martensite phase and ferrite phases were distributed homogeneously, and a smooth transition was observed. It was determined that ferrite and martensite were dominant in the weld zone, and residual austenite and occasionally primary austenite grain boundaries were observed (Figure 9c,d).

Figure 10 shows the microstructure images of the transition zone and welding zone of the DP450/S275JR material pair. In HAZ, a martensite phase was observed along with an acicular ferrite and pearlite phase. In the region under the influence of heat, the martensite phase and ferrite–perlite phases were detected as not homogeneously distributed and a sharp transition was observed. Therefore, a boundary was formed in the HAZ region. It has been determined that acicular ferrite and needle martenzite were dominant in the weld zone, and residual bainite and pearlite were observed. In the weld metal, intense acicular ferrite and martensite formations were observed due to the effect of the additional wire used.

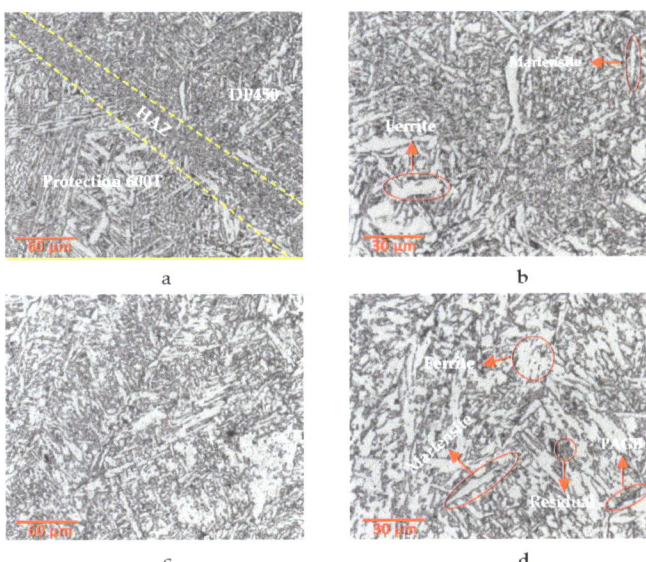

Figure 9. (**a,b**) HAZ (500×, 1000×), (**c,d**) weld zone microstructure images (500×, 1000×) for DP450/Protection 600T material pair.

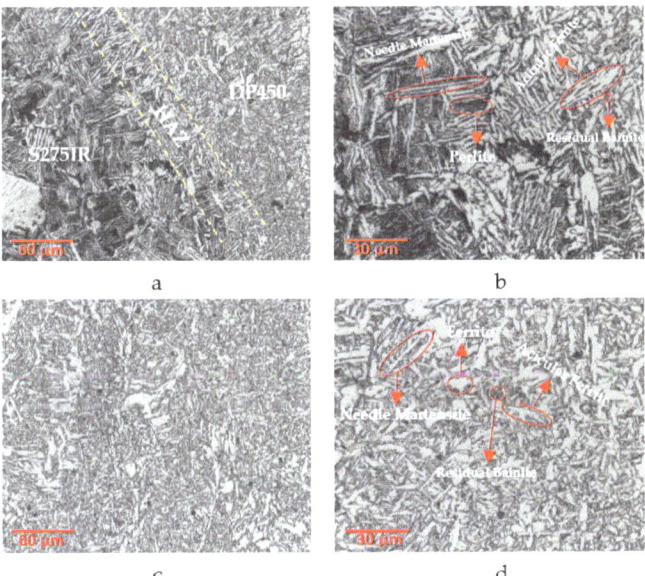

Figure 10. (**a,b**) HAZ (500×, 1000×), (**c,d**) weld zone microstructure images (500×, 1000×) for DP450-S275JR material pair.

3.3. Mechanical Test Examination

3.3.1. Microhardness

Microhardness measurements were taken after the Protection 600T, S275JR, and DP450 samples were welded using GMAW. Figure 11a–c shows the microhardness measurements of Protection 600T/Protection 600T, S275JR/S275JR, and DP450/DP450, while Figure 11d–f shows the microhardness measurements of welded samples in different combinations (Protection 600T/DP450, Protection 600T/S275JR, and DP450/S275JR). When examining

Figure 11a–c, it can be observed that the hardness increases towards the weld zone. This increase is believed to be due to the additional wire added to the weld zone during welding. Additionally, it has been determined that the microhardness value increases in the HAZ for all samples. For Protection 600T, the microhardness was determined to be 526.5 ± 10.5 HV in the base metal and 619 ± 20 HV in the weld zone. For DP450, it was 153.8 ± 1.8 HV in the base metal and 259 ± 8.1 HV in the weld zone. As for S275JR, it was 162.5 ± 5.2 HV in the base metal and 236 ± 9.3 HV in the weld zone. When examining Figure 11d,e, it is observed that microhardness values decrease from Protection 600T towards DP450/S275JR. This situation arises from the difference in mechanical properties of the materials. In Figure 11f, the highest microhardness values in the DP450/S275JR material pair were determined as follows from high to low: Heat Affected Zone (HAZ) (288 ± 26.5 HV), welding zone (232 ± 5.5 HV), and base metal (166 ± 6.3 HV).

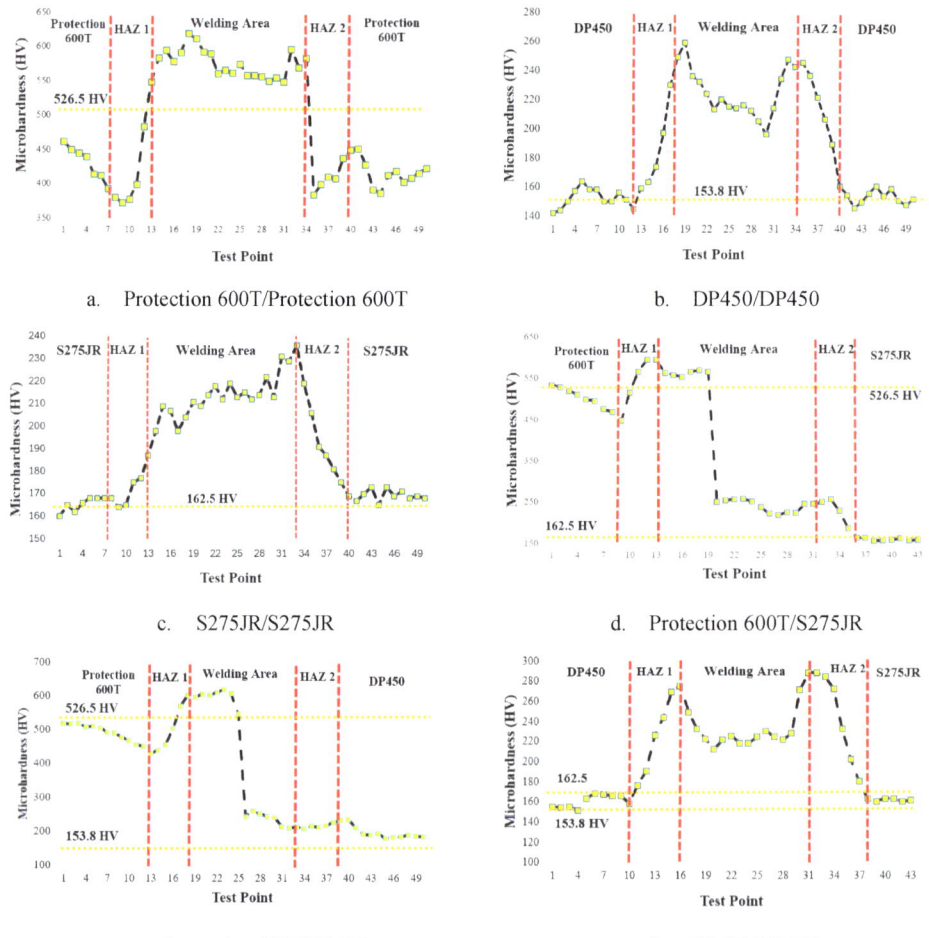

Figure 11. Post-weld microhardness measurements of Protection 600T, S275JR, and DP450 samples.

3.3.2. Tensile Test

Post-welding stress–strain graphs of Protection 600T, DP450, and S275JR materials (those close to the average value were selected) are given in Figure 11, while post-welding UTS, percentage elongation values, and welding efficiency relative to the base material are given in Table 5. Post-welding UTS values of base materials were obtained as 1083.4 ± 3.99, 516.5 ± 11 and 507.3 ± 5.7 MPa, respectively. When welding strengths were evaluated according to efficiency, they were determined as 49.4%, 103.05%, and 99.76%, respectively. It was determined that the weld strength of Protection 600T decreased compared to the base material UTS. It is thought that this is due to the additional welding wire.

Table 5. Mechanical properties after welding.

| | Material 1 | Material 2 | UTS | Elongation% | Weld Efficiency (%) | | |
					Protection 600T	DP450	S275JR
1	Protection 600T	Protection 600T	1083.4 ± 3.99	1.25 ± 0.16	49.4	-	-
2	DP450	DP450	516.5 ± 11	16.2 ± 1.46	-	103.05	-
3	S275JR	S275JR	507.3 ± 5.7	18.4 ± 0.05	-	-	99.76
4	Protection 600T	S275JR	413.6 ± 1.6	4.3 ± 0.17	19.3	-	81.34
5	Protection 600T	DP450	360.6 ± 0.4	5.3 ± 0.2	16.87	72.01	-
6	DP450	S275JR	526.3 ± 3.5	15.8 ± 1.63	-	105.5	103.5

It can be said that the welding of DP450 and S275JR materials was successful compared to the base material. In the visual inspections and penetrant tests performed for all samples, it was determined that there were no weld defects such as open weld defects or voids on the surface.

When Figure 12 is examined, it can be observed that the stress–strain diagram of welded samples (Protection 600T/Protection 600T, DP450/DP450, and S275JR/S275JR) is similar to that of the base material. The indicators given in Figure 12 correspond to the same materials as the sample numbers shown in Table 5. In the welding of dissimilar materials, the UTS values for Protection 600T/S275JR were determined to be 413.6 ± 1.6 Mpa, for Protection 600T/DP450 it was 360.6 ± 0.4 Mpa, and the percentage elongation values were determined to be 4.3% ± 0.17 and 5.3% ± 0.2, respectively. It is observed that the UTS and efficiency decreased when Protection 600T was welded with other materials.

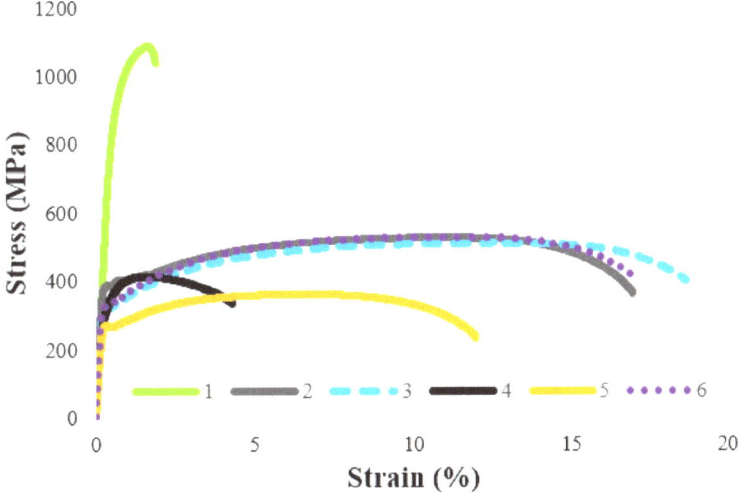

Figure 12. Post-welding stress–strain diagram.

The reason for this is likely to be the different chemical contents of the materials, the excessive coarsening of the grain structures due to the different cooling rates when passing from the base metal to the weld zone with the addition of welding wire. Similar findings are also reported in the literature. Badkoobeh et al. stated that in the joining of UNS S43000

Ferritic Stainless steel using laser welding, extremely coarse ferrite, and martensite were formed at the grain boundaries in the weld zone, and that this was responsible for the weak crystallographic texture in the zone [44].

In their study on the welding of armor steels, Çoban et al. stated that the peak temperatures and cooling rates that occur depending on the material thickness cause microstructural changes. This causes the hardness values of each zone to change. When the microstructural changes that caused this change were examined, it was stated that it caused the formation of a coarse-grained heat-affected zone in the region corresponding to the highest temperatures as well as the weld metal [21].

The welding strength of DP450/S275JR materials was determined as 526.3 ± 3.5 Mpa, the efficiency was determined as 105.5% compared to DP450 and 103.5% compared to S275JR, and the elongation was determined as 15.8 ± 1.63. It can be said that the higher mechanical properties of DP450/S275JR compared to the base material are due to the non-homogeneous distribution of martensite phase and ferrite–perlite phases in the HAZ region and the dominance of acicular ferrite and needle-like martensites in the weld zone.

The fracture surfaces of the tensile test samples are given in Figure 13. It was observed that the fracture occurred as brittle fracture in the weld pairs with S275JR material. In the DP450/DP450 and Protection 600T/DP450 material pairs, the fracture was ductile and on the DP450 side. In DP450/S275JR, the breakage occurred on the DP450 material side. The rupture in the Protection 600 T/Protection 600 T welded joints occurred in the weld area. These data showed that the welded joints were made appropriately, and the rupture occurred where it was expected according to the strength of the base material.

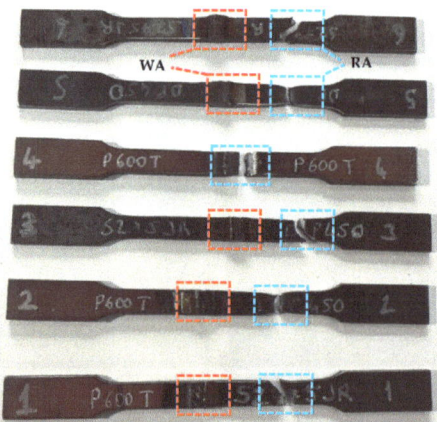

Figure 13. Tensile rupture morphologies of the welded sample (Welding Area: WA, Rapture Area: RA).

In the study, the joinability of three different materials (Protection 600T, DP450, and S275JR) was examined using ER70S-A1 welding wire. When the compatibility of the welding wire and base materials was evaluated as a result of post-welding HAZ and weld zone microstructure examinations, it was seen that the DP450/Protection 600T material pair was compatible. In the DP450/Protection 600T material pair, it was determined that while the martensite phase increased in the microstructure in the weld zone, the austenite and ferrite phases present in the structure increased the ductility relatively. However, considering the post-weld mechanical properties, the presence of phases in the material's microstructure has imparted ductility, resulting in a decrease in hardness and yield/tensile strength. In the welded samples of Protection 600T and S275JR, although tensile strength has increased, toughness has decreased, while hardness and strength have increased. When the microstructure and tensile diagrams of the DP450 and S275JR material pair are examined, it is determined that the materials and welding wire are well matched, leading to an increase in mechanical properties. In conclusion, the best results for welding the Protection 600T,

DP450, and S275JR material pairs were obtained in the following order: DP450/S275JR, Protection 600T/DP450, and Protection 600T S275JR. There has been a significant decrease in strength after welding in the Protection 600T/Protection 600T material pair. The reason for this is the lower mechanical properties of the welding wire added to the weld zone compared to Protection 600T. In the DP450/DP450 and S275JR/S275JR material pairs, there was compatibility between the welding wire and the base materials, resulting in welding strength that was the same as or higher than the base material strength.

3.3.3. Bending Tests

To determine the deformation of the weld zones and base metals, a 90° bending test was applied after welding on similar and dissimilar Protection 600T, DP450, and S275JR materials. Bending test results are given in Table 6. The bending forces of 16.7 ± 0.4, 11.0 ± 0.2, and 9.3 ± 0.3 kN for the Protection 600T, DP450, and S275JR unwelded specimens and 16.8 ± 0.1, 11.6 ± 0.9 and 10.4 ± 0.7 kN for the welded specimens, respectively, were close to each other. This indicates that the welding process was performed with high efficiency and the weld zone behaved similarly to the base material during the bend test. In dissimilar materials, however, it was determined that the bending force significantly decreased. This is likely due to the weld zone consisting of two different materials, leading to crack formation/propagation in the transition zones.

Table 6. Maximum bending test results of welded and unwelded (base material) samples.

	Material Pairs	Max. Bending Force (kN)
Base material	Protection 600T	16.7 ± 0.4
	DP450	11.0 ± 0.2
	S275JR	9.3 ± 0.3
Similar material	Protection 600 T/Protection 600T	16.8 ± 0.1
	DP450/DP450	11.6 ± 0.9
	S275JR/S275JR	10.4 ± 0.7
Dissimilar material	Protection 600 T/DP450	10.7 ± 0.1
	Protection 600 T/S275JR	9.4 ± 0.1
	DP450/S275JR	8.2 ± 0.4

3.3.4. Charpy V-Notch Tests

As a result of the CVN test, the impact energies of the base metals were found to be 75 ± 2.7 for Protection 600T, 85 ± 3.4 for DP450, and 32 ± 1.2 for S275JR. CVN test on welded samples was carried out by preparing samples from the weld zone and HAZ. Table 7 gives the impact energy values as a result of the CVN test after samples taken from HAZ 1, HAZ 2, and the welding area. While the impact energy values (87.3 ± 1.6 and 44.3 ± 1.3) for the HAZ region of Protection 600 T (1) and S275JR (3) materials were higher than the base material, the impact energy value (75.8 ± 8.4) for the HAZ region of DP450 (2) decreased. There was a decrease in the impact energy of the samples taken from the Protection 600 T, DP450, and S275JR welding area (57.0 ± 1.5, 85 ± 3.4, and 32 ± 1.2, respectively). This is due to the fact that the welding wire added to the welding zone affected the microstructure. After welding different material pairs, CVN experiments were carried out in three different regions: the weld zone and the HAZ of each material. It has been determined that in the Protection 600 T/S275JR and DP450/S275JR material pairs, the impact energy in the HAZ region of S275JR increased compared to the base material. The reason for this is that the martensite phase was formed along with the ferrite and pearlite phase in the HAZ region of S275JR under the influence of heat. A decrease in the impact energies of the samples taken from the Protection 600 T and DP450 HAZ region and the welding region of all material pairs was determined. The fracture surfaces of the selected samples after the CVN impact test are given in Figure 14. While fracture occurred in Protection 600T/Protection 600T material pair after CVN, no rupture occurred in other welded specimens. A ductile fracture was observed in all samples.

Table 7. Impact energy after HAZ 1, HAZ 2, and welding zone Charpy V-notch test.

	Material Pairs	HAZ 1	Welding Zone	HAZ 2
1	Protection 600 T/Protection 600T	87.3 ± 1.6	57.0 ± 1.5	-
2	DP450/DP450	75.8 ± 8.4	55.7 ± 0.2	-
3	S275JR/S275JR	44.3 ± 1.3	25.9 ± 2.3	
4	Protection 600 T/DP450	47.1 ± 1.0	37.9 ± 6.2	55.1 ± 7.6
5	Protection 600 T/S275JR	69.7 ± 4.6	48.79 ± 1.7	40.4 ± 1.2
6	DP450/S275JR	60.1 ± 2.0	46.47 ± 1.8	51.0 ± 0.9

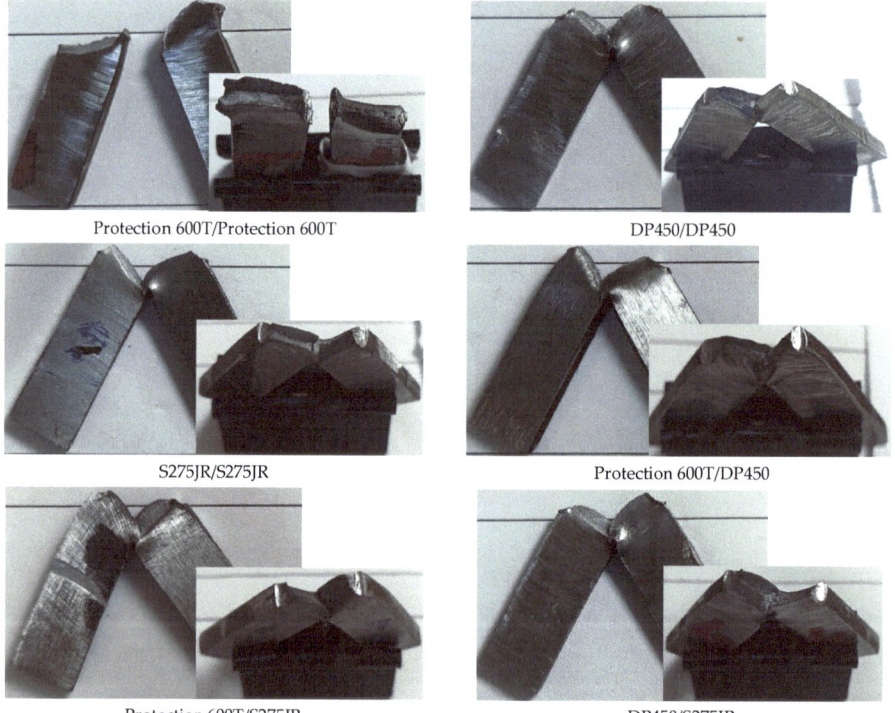

Figure 14. Fracture surfaces of selected samples after Charpy V-notch test.

4. Discussion

Joining materials with different properties is one of the significant issues in the industry since it enhances the functionality and efficiency of designs. Since dissimilar metal materials cannot be produced in the same process and do not have the same properties, they are joined using different methods. In this study, welding of three types of materials in both similar and dissimilar material combinations was aimed, and their welding capabilities were examined both mechanically and microstructurally.

The different grades of steel materials used in the experiments are Protection 600T, DP450, and S275JR general structural steel. The GMAW method was employed to join the steel materials. The thickness of the steel plates was 5 mm. Prior to welding, preparations were made by opening welding grooves (V 60° and 2 mm root gap), and welding procedures were applied using a multi-pass technique. Pre-welding preparations and post-welding quality control procedures are given in Figure 1.

In hardness measurements, the hardness values of the base materials were determined as 526.5 HV for Protection 600T, 153.8 HV for DP450, and 162.5 HV for S275JR. In similar materials (Figure 11a–c), a decrease in hardness was observed as approaching HAZ, while an increase in hardness was identified in the weld zone due to the effect of the welding wire.

The decrease in the hardness observed in the HAZ can be explained by the grain growth in the microstructure due to the influence of heat. The microhardness values in the weld zone were determined as 619 HV for Protection 600T, 259 HV for DP450, and 236 HV for S275JR. In dissimilar materials (Figure 11d,e), an increase in hardness was observed when passing from Protection 600T to the weld zone, while a decrease in hardness occurred when transiting to the other material. This decrease is believed to be due to the relatively lower hardness of the DP450 and S275JR materials. In the DP450/S275JR material pair, microhardness values were determined as weld zone (249 HV/271 HV, respectively) and base metal (168 HV/163 HV, respectively).

Zhang et al. conducted post-weld mechanical tests in their study on laser welding of Nano-Scale Precipitation-Strengthened (NPS) steels. They noted that the highest value in microhardness measurements was in the weld zone, followed by the HAZ, and the lowest value was in the base material. The reason for this is that the elements in the new phase formed in the source region do not have time to precipitate and form the second phase due to the cooling rate. As a result, the elements remaining in the phase dissolve to a large extent in the alloy, causing the solid solution to strengthen after welding [45].

Tensile test results of DP450/DP450, S275JR/S275JR, and DP450/S275JR materials show that the welding was carried out successfully. The obtained welding strength efficiency of 100% in the tensile test demonstrates the successful joining observed in both macro and microstructures. In the welded joints of DP450 and S275JR materials, fracture occurred in the DP450 material. The strength of welding area was higher than DP450 material. In this case, it can be said that the DP450 and S275JR material pair are compatible with each other and with the welding wire, resulting in improved mechanical properties. The rupture in the welded joints of the Protection 600T/Protection 600T material pair occurred in the welding area. Although Protection 600T had higher strength (2141.98 MPa), the desired strength could not be achieved in the welding area due to the mechanical properties of the welding wire (550–670 MPa).

In the case of joining Protection 600T with DP450 and Protection 600T with S275JR materials, a similar situation has been observed. In the DP450/Protection 600T material pair, an increase in the martensitic phase in the microstructure was observed in the weld area, while the existing austenite and ferrite phases in the structure had relatively increased ductility. In the samples welded with Protection 600T and S275JR, although the tensile strength increased, the toughness decreased and an increase in hardness and strength was detected. The DP450/S275JR welded joint has shown positive results that it can be used successfully in different applications (transportation, vehicle body manufacturing, etc.). Protection 600T/DP450 welded joints (with the armor feature of Protection 600T and the formability of DP450 steel) can be used for military purposes.

Bending tests provide important information about the deformation capabilities of welded joints and the ductility and toughness of the welded joints. In the bending test, the deformation resulting from the applied force is converted into data. The curvature of the deformed samples obtained as a result of the test gives an idea about the deformation ability. Bending test results were determined as 16.7 kN, 11.0 kN, and 9.3 kN for Protection 600T, DP450, and S275JR unwelded samples and 16.8 kN, 11.6 kN, and 10.4 kN for welded samples, respectively. According to these results, it was determined that the bending strength of the welded samples was better than the base material. Welding has been conducted successfully on similar materials. In dissimilar materials, the bending force remained below the base material performance. It can be said that this is because the welding area consists of two different materials and the transition zones cause crack formation/crack propagation.

Charpy impact tests were carried out at 21 °C room temperature, and impact strengths were compared with samples prepared from the weld zone and the HAZ. For similar materials (Protection 600 T and S275JR), the impact energy values in the HAZ increased compared to the base material, while in DP450, the impact energy value decreased in the HAZ. In these material pairs, there was a decrease in the impact energy of the samples taken from the welding area. In the Protection 600 T/S275JR and DP450/S275JR material pairs, it was determined that the impact energy in the HAZ of S275JR increased compared to the base material.

A decrease in the impact energies of the samples taken from the HAZ of the Protection 600 T/DP450 material pair and from the weld area of all dissimilar material pairs was determined. Impact toughness is affected by many parameters. The most important factor affecting impact toughness is the irregular distribution in the microstructure in the weld area. The impact toughness value of welded joints is directly related to ferrite, bainite content and grain size [46]. Therefore, in the study, different impact strengths were measured in impact notch samples taken from different regions.

5. Conclusions

This study examined the joinability of Protection 600T, DP450, and S275JR steels, which have different mechanical and microstructural properties, using the GMAW method. Weld joints of base materials and similar/dissimilar steels were analyzed by mechanical tests and optical examination. The obtained results are presented below.

The welding efficiency of Protection 600T, DP450, and S275JR, which are similar material pairs, was determined as 49.4%, 103.05%, and 99.76%, respectively. The reason why the efficiency is relatively low in Protection 600T is that the welding strength depends on the mechanical properties of the additional welding wire. In this study, welding of similar material pairs was successfully achieved.

In dissimilar material pairs (Protection 600T/S275JR, Protection 600T/DP450, and DP450/S275JR), the welding efficiency was determined as 19.3/81.34%, 16.87/72.01%, and 105.5/103.5%, respectively. The welding of the DP450/S275JR material pair was successfully achieved.

In the DP450/DP450, S275JR/S275JR, and DP450/S275JR pairs, it was determined that there was microstructure compatibility between the welding wire and the base materials. Therefore, the strength of the base material and welded samples was the same or higher.

The microhardness of the base materials was determined as 526.5 ± 10.5, 153.8 ± 1.8, and 162.5 ± 5.2 HV for Protection 600T, DP450, and S275JR, respectively. An increase in hardness values in the HAZ and welded zone was determined in all welded samples. After the tensile test, it was observed that the rupture in the welded joints occurred from the side with relatively low strength in the material pairs.

As a result of the CVN test, while there was an improvement in the HAZ in the welding of similar materials, there was a decrease in the absorbed energy values in the samples taken from the welding area. In dissimilar materials, there was an increase in the HAZ and welded zone compared to the S275JR material, while there was a decrease compared to the other two materials. As a result of the bending test, the bending force in welded similar material pairs was improved compared to the base material. There was a decrease in bending force in dissimilar materials.

Author Contributions: Conceptualization, O.K. and N.A.; methodology, O.K. and N.A.; investigation, M.E.; data curation, M.E., O.K. and N.A.; writing—original draft preparation, M.E., O.K. and N.A.; writing— review and editing, M.E., O.K. and N.A. All authors have read and agreed to the published version of the manuscript.

Funding: This research received no external funding.

Data Availability Statement: The original contributions presented in the study are included in the article, further inquiries can be directed to the corresponding author/s.

Conflicts of Interest: Author Mustafa Elmas was employed by the company Erdemir Engineering Management & Consulting Services INC. The remaining authors declare that the research was conducted in the absence of any commercial or financial relationships that could be construed as a potential conflict of interest.

References

1. Kocadağistan, M.E.; Çinar, O. Investigation of the weldability of AISI 430 and HARDOX 500 steels by CMT method. *Pamukkale Univ. J. Eng. Sci.* **2023**, *30*, 1–9.
2. Kılıç, S.; Öztürk, F. Comparison of Performances of Commercial TWIP900 and DP600 Advanced High Strength Steels in Automotive Industry. *J. Fac. Eng. Archit. Gazi Univ.* **2016**, *31*, 567–578.
3. Májlinger, K.; Kalácska, E.; Spena, P.R. Gas metal arc welding of dissimilar AHSS sheets. *Mater. Des.* **2016**, *109*, 615–621. [CrossRef]
4. Hanninen, H.; Aaltonen, P.; Brederholm, A.; Ehrnstén, U.; Gripenberg, H.; Toivonen, A.; Jorma, P.; Virkkunen, I. Dissimilar metal weld joints and their performance in nuclear power plant and oil refinery conditions. *VTT TIEDOTTEITA* **2006**, *2347*, 208.
5. Derya, A.; Anil, İ.; Kırık, İ. Investigation of the microstructural and mechanical properties of AISI 1020/RAMOR 500 steel couple joined by friction welding. *Dicle Univ. J. Eng.* **2022**, *13*, 51–56.
6. Rončák, J.; Adam, O.; Zobač, M. Electron beam welding of AlCoCrFeNi2. 1 high entropy alloy to EN 1.4301 austenitic steel. *J. Phys. Conf. Ser.* **2023**, *2443*, 012009. [CrossRef]
7. Abioye, T.; Olugbade, T.; Ogedengbe, T. Welding of dissimilar metals using gas metal arc and laser welding techniques: A review. *J. Emerg. Trends Eng. Appl. Sci.* **2017**, *8*, 225–228.
8. Kaçar, R.; Emre, H.E. Effects of welding methods on the mechanical properties of joining dissimilar steel couple. *J. Fac. Eng. Archit. Gazi Univ.* **2018**, *33*, 255–265.
9. Özturan, A.B.; İrsel, G.; Güzey, B.N. Study of the microstructure and mechanical property relationships of gas metal arc welded dissimilar Hardox 450 and S355J2C+ N steel joints. *Mater. Sci. Eng. A* **2022**, *856*, 143486. [CrossRef]
10. Mvola, B.; Kah, P.; Martikainen, J. Dissimilar Ferrous Metal Welding Using Advanced Gas Metal Arc Welding Processes. *Rev. Adv. Mater. Sci.* **2014**, *38*, 125–137.
11. Devaraj, J.; Ziout, A.; Qudeiri, J.E.A. Grey-based taguchi multiobjective optimization and artificial intelligence-based prediction of dissimilar gas metal arc welding process performance. *Metals* **2021**, *11*, 1858. [CrossRef]
12. Chaudhari, R.; Loharkar, P.; Ingle, A. Applications and challenges of arc welding methods in dissimilar metal joining. *IOP Conf. Ser. Mater. Sci. Eng.* **2020**, *810*, 012006. [CrossRef]
13. Erhard, A.; Munikoti, V.; Brekow, G.; Tessaro, U.; Tscharntke, D. Modeling for the improvement of dissimilar weld inspection. In Proceedings of the 16th World Conference on NDT, Montreal, QC, Canada, 30 August–3 September 2004.
14. Bayock, F.N.; Kah, P.; Mvola, B.; Layus, P. Experimental review of thermal analysis of dissimilar welds of High-Strength Steel. *Rev. Adv. Mater. Sci.* **2019**, *58*, 38–49. [CrossRef]
15. Maurya, B.K.; Pretap, B.; Kumar, A.; Rana, G. Experimental analysis of dissimilar metal welds of mild steel and stainless steel. *Int. Res. J. Eng. Technol.* **2017**, *4*, 1744–1748.
16. Bajpei, T.; Chelladurai, H.; Ansari, M.Z. Experimental investigation and numerical analyses of residual stresses and distortions in GMA welding of thin dissimilar AA5052-AA6061 plates. *J. Manuf. Process.* **2017**, *25*, 340–350. [CrossRef]
17. Singh, S.; Kumar, V.; Kumar, S.; Kumar, A. Variant of MIG welding of similar and dissimilar metals: A review. *Mater. Today Proc.* **2022**, *56*, 3550–3555. [CrossRef]
18. Saxena, A.; Kumaraswamy, A.; Reddy, G.M.; Madhu, V. Influence of welding consumables on tensile and impact properties of multi-pass SMAW Armox 500T steel joints vis-a-vis base metal. *Def. Technol.* **2018**, *14*, 188–195. [CrossRef]
19. Kumar, S.N.; Balasubramanian, V.; Malarvizhi, S.; Rahman, A.H.; Balaguru, V. Effect of failure modes on ballistic performance of Gas Metal Arc welded dissimilar armour steel joints. *CIRP J. Manuf. Sci. Technol.* **2022**, *37*, 570–583. [CrossRef]
20. Magudeeswaran, G.; Balasubramanian, V.; Reddy, G.M. Metallurgical characteristics of armour steel welded joints used for combat vehicle construction. *Def. Technol.* **2018**, *14*, 590–606. [CrossRef]
21. Çoban, O.; Gürol, U.; Erdöl, S.; Koçak, M. Effect of Plate Thickness on The Microstructure and Hardness of Robotic Fillet Welded Armour Steels. In Proceedings of the 6th International Conference on Welding Technologies and Exhibition (ICWET'21), Hatay, Turkey, 13–15 October 2021.
22. Kaçar, R.; Emre, H.E. Determination of the Mechanical Properties of Gas Metal Arc Welded Armox Steels. *AKU J. Eng. Appl. Sci.* **2018**, *1*, 15–23.
23. Günen, A.; Bayar, S.; Karakaş, M.S. Effect of different arc welding processes on the metallurgical and mechanical properties of Ramor 500 armor steel. *J. Eng. Mater. Technol.* **2020**, *142*, 021007. [CrossRef]
24. Wang, Q.; Kai, G.; Zhang, X.; Long, X.; Wei, Y. Microstructure and Properties of ZGMn13Mo/A514 Dissimilar Steel Multi-pass Welding. *Trans. Indian Inst. Met.* **2024**, *77*, 1117–1126. [CrossRef]
25. Ranji, A.R.; Gishanim, S.N.; Alirezaee, S. Fatigue assessment of welded joints and crack growth considering residual stress. *Eng. Res. Express* **2023**, *5*, 045001. [CrossRef]

26. Mukherjee, K.; Ramazani, A.; Yang, L.; Prahiet, U.; Bleck, W.; Reisgen, U.; Schleser, M.; Abdurakhmanov, A. Characterization of gas metal arc welded hot rolled DP600 steel. *Steel Res. Int.* **2011**, *82*, 1408–1416. [CrossRef]
27. Kuril, A.A.; Ram, G.J.; Bakshi, S.R. Microstructure and mechanical properties of keyhole plasma arc welded dual phase steel DP600. *J. Mater. Process. Technol.* **2019**, *270*, 28–36. [CrossRef]
28. Burns, T. *Weldability of a Dual-Phase Sheet Steel by the Gas Metal Arc Welding Process*; University of Waterloo: Waterloo, ON, Canada, 2010.
29. Ghadbeigi, H.; Pinna, C.; Celotto, S. Failure mechanisms in DP600 steel: Initiation, evolution and fracture. *Mater. Sci. Eng. A* **2013**, *588*, 420–431. [CrossRef]
30. Brnic, J.; Turkalj, G.; Niu, J.; Canadija, M.; Lanc, D. Analysis of experimental data on the behavior of steel S275JR–Reliability of modern design. *Mater. Des.* **2013**, *47*, 497–504. [CrossRef]
31. Çevik, B. Effect of welding processes on mechanical and microstructural properties of S275 structural steel joints. *Mater. Test.* **2018**, *60*, 863–868. [CrossRef]
32. Wu, W.; Zhang, T.; Chen, H.; Peng, J.; Yang, K.; Lin, S.; Wen, P.; Li, Z.; Yang, S.; Kou, S. Effect of Heat Input on Microstructure and Mechanical Properties of Deposited Metal of E120C-K4 High Strength Steel Flux-Cored Wire. *Materials* **2023**, *16*, 3239. [CrossRef]
33. Evci, C.; Işık, H.; Macar, M. Effect of welding wire and groove angle on mechanical properties of high strength steel welded joints: Einfluss von Schweißzusatzdraht und Fugenwinkel auf die mechanischen Eigenschaften von Schweißverbindungen aus hochfestem Stahl. *Mater. Und Werkst.* **2017**, *48*, 912–921. [CrossRef]
34. Mahlalela, S.S.; Pistorius, P.G.H. Influence of alloying elements and cooling rate on the presence of delta ferrite in modified 9Cr–1Mo as-welded microstructure produced by gas–metal arc welding. *Weld. World* **2023**, *67*, 1169–1180. [CrossRef]
35. EN ISO 9016:2012; Destructive Tests on Welds in Metallic Materials–Impact Tests–Test Specimen Location, Notch Orientation and Examination. ISO: Geneve, Switzerland. Available online: https://www.iso.org/standard/81122.html (accessed on 14 May 2024).
36. Yakup, K.; Atar, F. An Investigation of Joinability of Stainless Steel and The Low Carbon Steel Materials by Mig Welding Method. *Eng. Sci.* **2020**, *15*, 89–99.
37. Sirohi, S.; Gupta, A.; Pandey, C.; Vidyarthy, R.S.; Guguloth, K.; Natu, H. Investigation of the microstructure and mechanical properties of the laser welded joint of P22 and P91 steel. *Opt. Laser Technol.* **2022**, *147*, 107610. [CrossRef]
38. Metin, M.; Mehmet, Ü.; Gören, H.A. Comparison of the Hardness, Tensile and Notch Impact Properties of C95200 and C95300 Al Bronze Alloys. *Sinop Univ. J. Nat. Sci.* **2023**, *8*, 216–228.
39. Senčič, B.; Šolić, S.; Leskovšek, V. Fracture toughness–Charpy impact test–Rockwell hardness regression based model for 51CrV4 spring steel. *Mater. Sci. Technol.* **2014**, *30*, 1500–1505. [CrossRef]
40. Al-Qawabeha, U.F. Effect of heat treatment on the mechanical properties, Micro hardness, and impact energy of H13 alloy steel. *Int. J. Sci. Eng. Res.* **2017**, *8*, 100–104.
41. Gürol, U.; Karahanet, T.; Erdöl, S.; Coban, O.; Baykal, H.; Koçak, M. Characterization of Armour Steel Welds Using Austenitic and Ferritic Filler Metals. *Trans. Indian Inst. Met.* **2022**, *75*, 757–770. [CrossRef]
42. Indhu, R.; Divyaet, S.; Tak, M.; Soundarapandian, S. Microstructure development in pulsed laser welding of dual phase steel to aluminium alloy. *Procedia Manuf.* **2018**, *26*, 495–502. [CrossRef]
43. Çetinkaya, C.; Akay, A.; Arabacı, U.; Fındık, T. Effect of Primer Coating Applied To S235JR Material on Submerged Arc Weldability. *J. Polytech.* **2022**, *25*, 1335–1348.
44. Badkoobeh, F.; Mostaan, H.; Sonboli, A. On the Microstructural Transformations and Mechanical Performance of Laser Beam Welded UNS S43000 Ferritic Stainless Steel. *J. Mater. Eng. Perform.* **2024**, 1–13. [CrossRef]
45. Zhang, M.; Wang, X.; Zhu, G.; Chen, C.; Hou, J.; Zhang, S.; Jing, H. Effect of laser welding process parameters on microstructure and mechanical properties on butt joint of new hot-rolled nano-scale precipitation-strengthened steel. *Acta Metall. Sin. (Engl. Lett.)* **2014**, *27*, 521–529. [CrossRef]
46. Sirohi, S.; Pandey, S.M.; Tiwari, V.; Bhatt, D.; Fydrych, D.; Pandey, C. Impact of laser beam welding on mechanical behaviour of 2.25Cr–1Mo (P22) steel. *Int. J. Press. Vessel. Pip.* **2023**, *201*, 104867. [CrossRef]

Disclaimer/Publisher's Note: The statements, opinions and data contained in all publications are solely those of the individual author(s) and contributor(s) and not of MDPI and/or the editor(s). MDPI and/or the editor(s) disclaim responsibility for any injury to people or property resulting from any ideas, methods, instructions or products referred to in the content.

Article

Effect of Microchemistry Elements in Relation of Laser Welding Parameters on the Morphology 304 Stainless Steel Welds Using Response Surface Methodology

Kamel Touileb [1,*], Elawady Attia [2], Rachid Djoudjou [1], Abdejlil Chihaoui Hedhibi [3], Abdallah Benselama [4], Albaijan Ibrahim [1] and Mohamed M. Z. Ahmed [1,5]

[1] Department of Mechanical Engineering, College of Engineering in Al-Kharj, Prince Sattam bin Abdulaziz University, P.O. Box 655, Al-Kharj 16273, Saudi Arabia; r.djoudjou@psau.edu.sa (R.D.); i.albaijan@psau.edu.sa (A.I.); moh.ahmed@psau.edu.sa (M.M.Z.A.)

[2] Department of Industrial Engineering, College of Engineering in Al-Kharj, Prince Sattam bin Abdulaziz University, P.O. Box 655, Al-Kharj 16273, Saudi Arabia; e.attia@psau.edu.sa

[3] Laboratory of Mechanics of Sousse (LMS), National Engineering School of Sousse, University of Sousse, BP 364, Erriaydh City, Sousse 4023, Tunisia; abdejlilioueslati@yahoo.fr

[4] Department of Electrical Engineering, College of Engineering in Al-Kharj, Prince Sattam Bin Abdulaziz University, P.O. Box 655, Al-Kharj 16273, Saudi Arabia; a.benselama@psau.edu.sa

[5] Department of Metallurgical and Materials Engineering, Faculty of Petroleum and Mining Engineering, Suez University, Suez 43512, Egypt

* Correspondence: k.touileb@psau.edu.sa

Citation: Touileb, K.; Attia, E.; Djoudjou, R.; Hedhibi, A.C.; Benselama, A.; Ibrahim, A.; Ahmed, M.M.Z. Effect of Microchemistry Elements in Relation of Laser Welding Parameters on the Morphology 304 Stainless Steel Welds Using Response Surface Methodology. *Crystals* 2023, *13*, 1138. https://doi.org/10.3390/cryst13071138

Academic Editors: Reza Beygi, Mahmoud Moradi and Ali Khalfallah

Received: 28 June 2023
Revised: 13 July 2023
Accepted: 18 July 2023
Published: 21 July 2023

Copyright: © 2023 by the authors. Licensee MDPI, Basel, Switzerland. This article is an open access article distributed under the terms and conditions of the Creative Commons Attribution (CC BY) license (https://creativecommons.org/licenses/by/4.0/).

Abstract: Small differences in the contents of surface active elements can change flow direction and thus heat transfer, even for different batches of a given alloy. This study aims to determine the effects of sulfur on weld bead morphology in the laser process. The paper presents the results related to the weld bead shape of two thin AISI 304 industrial stainless steel casts. One cast contains 80 ppm (0.008%) of sulfur, considered as a high sulfur content, and the other one contains 30 ppm (0.003%) sulfur, which can be considered low sulfur. The welds were executed using a CO_2 laser. The effects of laser power (3.75, 3.67, 6 kW), welding speed (1.25, 2.40, 2.45, 3.6 m/min), focus point position (2, 7, 12 mm), and shield gas (Helium, mixed 40% helium + 60% argon and mixed 70% helium + 30% argon) with a flow rate of 10 L/min on the depth of the weld (D) and the aspect ratio (R = D/W) were investigated using RSM (response surface methodology). The experimental results show that the transfer of energy from the laser beam to the workpiece can be total in cases where the selected welding parameters prevent plasma formation. For the 304 HS cast, the focus point is the major factor in determining the depth of penetration, and its contribution is up to 52.35%. However, for 304 LS, the interaction between shield gas and focus point seems to play an important role, and the contribution of their interaction raises to 28% in relation to the laser depth of the weld. Moreover, the study shows that sulfur plays a surface active role only in the case of partial penetration beads, so that a 56% partially penetrated weld supports the hypothesis of its surface-active role in the formation of the weld pool. However, a penetration of only 36% confirms the effects of a sulfur surface-active when the bead is fully penetrated.

Keywords: laser welding; AISI 304 SS casts; surfactant elements; marangoni convection; RSM

1. Introduction

Nowadays, laser welding is widely used in many industries, such as in automobiles, aircraft, marines, shipbuilding and aviation. Laser welding is used in modern industry, and has many advantages such as less distortion, greater deep weld bead, and narrow welds compared to traditional welding technologies. As the beam can be concentrated in a small area, it provides a concentrated heat source, leading to narrow and highly penetrated welds [1,2]. Laser welding can be performed without filler materials and single-pass laser procedures have been employed in materials of up to 32 mm thick. This technique is strongly recommended for use on welding zones requiring high precision and high quality. The laser can be readily mechanized for automated, high-speed welding. Laser welding is a technique that meets the requirements of modern manufacturing in terms of the repeatability of the process and its easy automation. The lasers predominantly used in industrial material processing and welding tasks are the 1.0 µm YAG laser and the 10.6 µm CO_2 laser, with the most common active elements employed in these two types of lasers being neodymium (Nd) ion and CO_2 molecule, respectively [3]. Laser welding processes are classified as focused heat source or high-intensity and high-energy beams. Laser beam welding is one of the most promising welding techniques owing to the higher welding speeds it offers, and the lower dimensions and distortions in the welds. Moreover, its high strength-to-weld geometries and minimal heat-affected zones make it advantageous for various industrial applications. Lasers are characterized by requiring less time, which implies a lower labor cost.

The most commonly used casts of stainless alloys are austenitic steels, known as 8–18 types, which contain 16 to 18% Cr, 6–8% N and 0.03–0.1% C. Austenitic stainless steels are widely used for different applications in the chemical, petrochemical, food processing and nuclear industries, given their good mechanical and plastic properties. Austenitic stainless steels are widely used in a variety of applications owing of their corrosion resistance, good ductility, toughness, and ease of manufacture. The 300 series alloys designated by the American Iron and Steel Institute (AISI) are the most widely used of the austenitic casts. Stainless steels can generally be welded with all methods of fusion welding and solid-state welding [4,5].

The soundness of the weld is intimately related to multiple input parameters. The mastering of these inputs leads to welds with fewer defects. Finding out the optimal combinations is the main target in any research study, and finding the effective combinations is essential [6]. The optimization of welding conditions is useful in order to achieve good properties in the joints. Many researchers have applied various optimization methods to define the desired output variables, by developing mathematical models to specify the relationship between depth and welding parameters. Sampreeta et al. [7] examined bead-on-plate welding on Hastelloy C-276 plates of 1.6 mm thickness using the CO_2 laser welding process. The welding was performed based on the Taguchi L9 design, considering laser power, welding speed and the flow rate of the shielding gas as input parameters. They found that a laser power of 1600 W, welding speed of 2 m/min and shielding gas flow rate of 10 L/min gave the best results. Using analysis of variance (ANOVA), it was found that laser power is the most influential parameter on the quality of the welds.

Another study conducted by Sathiya et al. [8] was dedicated to the optimization of laser bead-on-plate welding parameters with a 3.5 kW laser using Taguchi technique. The experiments were conducted on two different shielding gases: 100% nitrogen and a gas mixture of 50% nitrogen and 50% argon. The most suitable input process parameters, such as beam power, travel speed and focus position, were selected in order to obtain the desired output, i.e., bead width and depth of penetration. They concluded that beam power (42.35%) has a greater influence on welding, followed by focus position (27.48%) and travel speed (19.07%), when 100% nitrogen shielding gas is used. However, when a mixed gas (50% N, 50% Ar) was tested, they found that the travel speed (35.06%) has more influence on the welding, followed by focus position (29.71%) and beam power (28.81%). Anawa et al. [9] successfully applied a CO_2 laser welding process and optimized the process parameters for joining dissimilar AISI 316 stainless steel and AISI 1009 low carbon steel plates. RSM was used to develop the mathematical models employed to predict the heat input and to describe the laser weld bead profile for a continuous wave 1.5 kW CO_2 laser. In their study, the laser input parameters, such as laser power, welding speed and focused position, were taken into consideration. Benyounis et al. [10] conducted a study to optimize autogenous laser welded joints to get the maximum penetration, and the minimum fusion zone width and heat-affected zone width. In order to achieve these objectives, the authors developed a mathematical model and optimized the weld bead profiles. Several studies have confirmed the preponderant role played by surface active elements such as Se, Te, O, and S in the determination of the weld bead [11,12]. These studies, most of which were carried out in TIG (tungsten inert gas), show that a sulfur-rich cast presents a narrow and deep weld appearance. Contrarily, a cast free of sulfur presents a wide and more shallow bead [13]. In spite of the large amount of literature available on the effects of surfactant elements on TIG welding, very little work has been reported so far on the effects of the above-mentioned element during laser welding. Zhao et al. [14] were interested in the effects of adding oxygen as a surfactant on convection molten metal. Stainless steel samples were subjected to laser spot welding in an environmentally controlled chamber in order to observe fluid flow behavior and examine the influence of oxygen concentration on the flow. Transitions in fluid flow direction were observed during laser spot welding, whereby an initial outward flow changed to an inward flow in the presence of an oxidizing atmosphere. This behavior can be explained in terms of changes in surface tension gradients driven by oxygen dissolution in the liquid metal. Flow reversal only occurs in the presence of sufficient environmental oxygen. CO_2 lasers are still the predominant type of laser used in the manufacturing industry, and in the higher power ranges in particular. Conduction mode welding occurs when heat is transferred from the surface into the material via thermal conduction. Conduction welding is made possible when the absorbed energy is sufficient to melt the weld zone, but insufficient for vaporization and plasma formation [15,16]. On the other hand, keyhole welding represents a type of welding wherein the laser beam energy is transferred deep into the material through a cavity filled with ionized metal vapor.

In simulations of heat conduction laser welding, the effect of the temperature-dependent surface tension coefficient (Marangoni effect) was identified as the primary driving force of the liquid melt. This coefficient is the most important driving force of the melt in conventional welding. In this study, the simulations show large differences in flow behavior caused merely by inverting the temperature dependency of $\gamma(T)$. Therefore, the Marangoni effect is very important in deep laser welding. The simulations confirm and clarify the observed variations in weld bead geometry that arise due to very different flow patterns. These differences result from the marginal fluctuations in the chemical composition within the tolerance of the standard defining the material composition of a special alloy [17]. C. R. Heiple et al. [18] confirmed that the addition of small concentrations of a surface active element (Se) to stainless steel dramatically increased the D/W ratio of a severely defocused and decoupled laser weld. A defocused laser weld strongly supports the surface tension-driven fluid flow model.

Kaul et al. [19] studied the effects of the active flux laser beam welding of austenitic stainless steel sheets on plasma plume. Laser welding using SiO_2 flux significantly modified the shape of the fusion zone (FZ), which produced narrower and deeper welds. The development of such a weld bead is caused by a reversal in the direction of the Marangoni flow caused by the oxygen-induced inversion of surface tension gradient. Ding et al. [20] reported that both the surface-activating flux and surface active element S produce significant effects on the YAG laser weld shape in terms of increased weld penetration and depth/width ratio. Su et al. [21] reported that the CO_2 LBW (laser beam welding) of AISI 304 stainless steel sheets with active flux resulted in an increased weld depth through a decrease in the electron density of the plasma plume. In contrast, active flux had little effect on the depth of Nd:YAG laser welds [22]. The direction of the convective (or Marangoni) flow in the weld pool is governed by the sign of the surface tension gradient $d\gamma/dT$.

The research focuses on weldability problems related to the cast-to-cast variation in the weld pool geometry and penetration between materials with almost the same chemical compositions. The aim of our study is to analyze the influence of the basic parameters of laser welding (i.e., laser beam power, welding speed, shield gas types and focal point position) on the weld shape. This study is conducted on two industrial casts comprising AISI 304 austenitic stainless steel sheets of 3.0 mm thickness. The RSM optimization method is used to develop a mathematical model of depth and aspect ratio in terms of welding parameters. The optimal surface design in terms of response is used to design the experiment. The model required 15 points, and another 15 points were added randomly to increase the model efficiency using Design Expert Software. After performing the multi-regression analysis using Design Expert software, equations were obtained.

The laser welding parameter plays an important role in the quality of the weld, and hence the quality of the joint produced in industry. These results reveal the welding conditions that allow the surface active elements to play their fullest role. Furthermore, the results obtained in this study offer interesting data for researchers to relate the heat applied (welding parameters) with heat the weld profile, and the ensuing changes in microstructure. The microstructural properties are strongly affected by heat transfer and metal flow in the weld pool.

2. Materials and Methods

2.1. Materials

Two casts of austenitic stainless steel, AISI 304 LS and AISI 304 HS, with thicknesses of 3 mm were investigated in this study. The chemical compositions are presented in Table 1.

Table 1. Chemical composition (wt. %) of 304 stainless steel casts.

Elements	C	Mn	Si	P	S	Cr	Ni	Fe
304HS SS	0.06	0.86	0.41	0.024	0.008	18.29	8.40	Balance
304LS SS	0.06	1.06	0.57	0.032	0.003	18.29	8.45	Balance

The sulfur content, thermal gradient and surface tension at melting temperature of the casts used are given in Table 2. Figure 1 shows the evolution of surface tension versus the temperature for both casts.

Table 2. Surface tension at melting temperature, thermal gradient and sulfur content of used casts [23].

Elements	γ (N/m) at Melting Temperature	$d\gamma/dT$ (N/m °K)	Sulfur Content (%)
304HS SS	1.62	$+8 \times 10^{-5}$	0.008
304LS SS	1.74	-10^{-5}	0.003

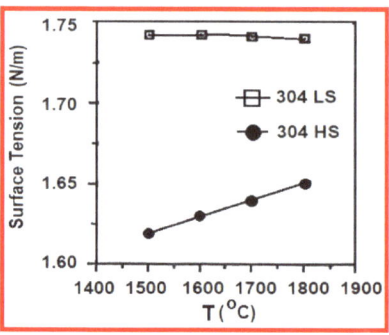

Figure 1. Surface tension against temperature of 304 SS casts [23].

The surface tension of the liquid metal is dependent upon its temperature, and therefore dependent upon the distance from the center of the weld pool, since a molten weld pool has a radial temperature gradient such that the highest temperature is directly beneath the source of heat, and the lowest is at the solid–liquid interface. The combination of the dependence of the surface tension (γ) on the temperature (T) and the presence of a temperature gradient causes a surface tension gradient ($d\gamma/dT$) even in a small region.

Figure 1 represents the variation in surface tension against temperature in the center of the weld pool. In molten metal, in the case of a 304 LS cast, a decrease in the surface tension as the temperature increases is ascribed to the absence or quasi-absence of surfactants (30 ppm of Sulfur). In the case of 304 LS, the slope is negative (the temperature coefficient of the surface tension is negative).

However, in 304 HS (with 80 ppm), by contrast, the surface tension is highest at the center. As the melting point is reached, the molten metal flows from the center to the edge. Sulfur is a strongly surface-active element that segregates in the surface layer and reduces surface tension. However, as the temperature increases, sulfur desorbs into the bulk of the liquid metal, causing an increase in surface tension. This applies to alloys with a sulfur content greater than 50 ppm [24,25]. The surface tension will be highest at the center and thus the surface flow will be radially inward, and this will induce a downward flow in the center, called inverse Marangoni convection. In the case of 304 HS, the slope is positive (positive temperature coefficient of surface tension). A shallow, wide weld bead can be obtained in low-sulfur (S) stainless steel (30 ppm); however, a deep, narrow penetration weld is obtained in the case of a sulfur (S) content of 80 ppm.

2.2. Welding Procedure

The welding tests were carried out using the laser installation of the central school of Nantes (France), with a carbon dioxide (CO_2) laser capable of producing a maximum output of 6 kW and emitting radiation in the infrared band of 10.6 µm and a pressure of 100 Pa. The laser gas was composed of $He + N_2 + CO_2$. Figure 2a depicts the CO_2 laser welding modes, and Figure 2b shows laser beam welding in progress, with a shiny dazzling light.

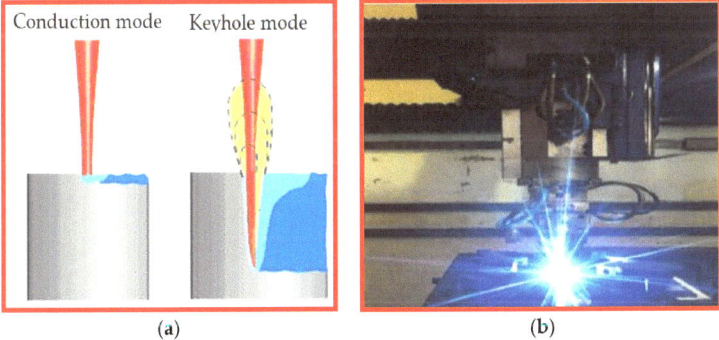

Figure 2. (**a**) Laser welding modes [26], (**b**) laser beam welding in progress [27].

Table 3 depicts the physical properties of the shield gas used. A preliminary study of the welding parameters is necessary to adjust the specific energy supplied to the sheets. The experiments involved welding a 120 mm line on a rectangular plate of 3 mm thickness, 150 mm length and 80 mm width. Before welding, the plates were cleaned with acetone.

Table 3. Physical properties of shield gas used.

Laser Welding Gas	Molecular Weight (g/mol)	Thermal Conductivity at 1 bar, 15 °C (W/m·K)	Ionization Energy (ev)	Dissociation Energy (ev)	Density Relative to Air
Helium	4	0.15363	24.6	0	0.14
Argon	40	0.01732	15.8	0	1.38

In this study, 30 tests were carried out on each cast. Four factors were considered at different levels, as presented in Table 4. Three focus lengths were tested (+2, +7, +12 mm). Four welding speeds were investigated (1.25, 2.40, 2.45, 3.6 m/min). Three levels of power were applied (3.75, 3.67, 6 kW). In addition, three levels of the shield gas were considered according to the proportion of helium (40% helium mixed with 60% of argon, 70% helium mixed with 30% argon, and 100% helium). Figure 3 depicts the focus point positions used during the CO_2 laser beam welding experiments.

Table 4. Laser welding parameters.

Runs	Focus Point Position (mm)	Welding Speed (m/min)	Power (kW)	Linear Energy (J/cm)	Shield Gas
1	2.00	1.25	3.75	180	1 (70% He + 30% Ar)
2	2.00	2.45	3.67	90	1 (70% He + 30% Ar)
3	2.00	2.40	6.00	150	1 (70% He + 30% Ar)
4	2.00	3.60	6.00	100	1 (70% He + 30% Ar)
5	7.00	1.25	3.75	180	1 (70% He + 30% Ar)
6	7.00	2.40	6.00	150	1 (70% He + 30% Ar)
7	2.00	1.25	3.75	180	2 (100% He)
8	2.00	2.45	3.67	90	2 (100% He)
9	2.00	2.40	6.00	150	2 (100% He)
10	2.00	3.60	6.00	100	2 (100% He)
11	2.00	1.25	3.75	180	3 (40% He + 60% Ar)
12	2.00	2.45	3.67	90	3 (40% He + 60% Ar)
13	2.00	3.60	6.00	100	3 (40% He + 60% Ar)
14	2.00	2.40	6.00	150	3 (40% He + 60% Ar)
15	12.00	1.25	3.75	180	2 (100% He)
16	12.00	2.45	3.67	90	2 (100% He)
17	12.00	2.40	6.00	150	2 (100% He)
18	12.00	3.60	6.00	100	2 (100% He)
19	7.00	2.40	6.00	150	3 (40% He + 60% Ar)
20	7.00	2.40	6.00	150	2 (100% He)
21	12.00	1.25	3.75	180	3 (40% He + 60% Ar)
22	12.00	2.45	3.67	90	3 (40% He + 60% Ar)
23	12.00	2.40	6.00	150	3 (40% He + 60% Ar)
24	12.00	3.60	6.00	100	3 (40% He + 60% Ar)
25	7.00	2.45	3.67	90	1 (70% He + 30% Ar)
26	7.00	2.40	6.00	150	1 (70% He + 30% Ar)
27	12.00	1.25	3.75	180	1 (70% He + 30% Ar)
28	12.00	2.45	3.67	90	1 (70% He + 30% Ar)
29	12.00	2.40	6.00	150	1 (70% He + 30% Ar)
30	12.00	3.60	6.00	100	1 (70% He + 30% Ar)

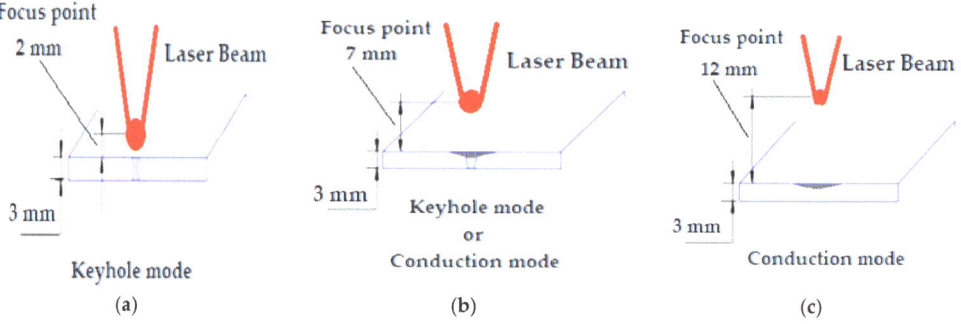

Figure 3. Scheme of the CO_2 laser beam focus positions used: (a) focus point = 2 mm, (b) focus point = 7 mm, (c) focus point = 12 mm.

The surface design yielding the optimal response has been used to design this experiment. The model required 15 points, and another 15 points were added randomly to increase the model efficiency using Design Expert software. First, a series of trials were conducted to master the welding parameters that were to be selected for our tests. In each test, three coupons were cut from the welding line to ensure the reproducibility and reliability of the results obtained. The samples were prepared by grinding, and subsequently polished, with a diamond suspension of 6 μm and 1 μm, respectively, then etched by immersion in Carpenters solution (8.5 gm $FeCl_3$, 2.4 gm $CuCl_2$, 122 mL alcohol, 122 mL HCl, 6 mL HNO_3) for macrographic analysis. The samples were observed and checked using Motic software integrated with an optical microscope. The results of the weld aspect (depth of penetration, weld width) represent an average of the three readings.

2.3. Mathematical Modeling

Using regression analysis, a mathematical model can be obtained that relates the output responses to the function of the input parameters. After the preliminary investigation, a mathematical formulation can be formed using the cubic polynomial relation. The cubic or third order polynomial function can be expressed as represented in Equation (1).

$$Y = b_o + \sum_{i=1}^{n} b_i X_i + \sum_{j>i}^{n} b_{ij} X_i X_j + \sum_{i=1}^{n} b_{ii} X_i^2 + \sum_{k>j>i}^{n} b_{ijk} X_i X_j X_k + \sum_{j>i}^{n} b_{ij} X_i X_j^2 + \sum_{i=1}^{n} b_{iii} X_i^3 + \varepsilon \quad (1)$$

where X_i represents the four input factors, such that $i \in \{f, s, p, g\}$; b represents the coefficient corresponding to the specified terms; b_o is the constant term, b_i is the coefficient of the linear terms; b_{ij} is the coefficient of the linear interaction terms and b_{ijk} is the coefficient of the interaction terms among the linear and quadratic order; b_{iii} is the coefficient of the cubic terms. We used Design Expert software, with the cubic model and auto selection of terms relying on the adjusted R^2.

3. Results and Discussion

3.1. 304 HS Stainless Steel Cast

Figure 4 represents the micrographs of the laser welding of high-sulfur content casts, carried out with a power of 6 kW, a welding speed of 2.4 m·min^{-1}, and a shield gas comprising 70% He + 30% Ar. It shows a keyhole weld bead with a focus distance of 2 mm and 7 mm. However, we see that the weld is partially penetrated when the focus point is 12 mm away from the workpiece.

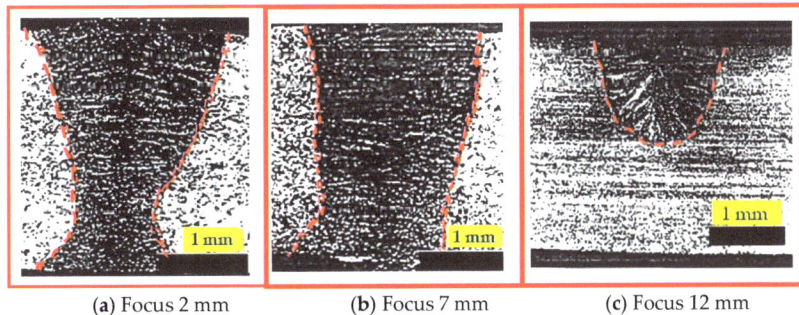

(a) Focus 2 mm   (b) Focus 7 mm   (c) Focus 12 mm

Figure 4. Effect of focus point on 304 HS laser weld carried out with a power 6 kw, welding speed 2.4 m·min^{-1}, and shield gas comprising 70% He + 30% Ar.

The effects of the laser welding parameters on the welds' morphology are gathered in Table 5, where the input (investigated) factors and responses are presented.

Table 5. Investigated factors and the responses for 304 HS.

Runs	Investigated Factors				Response Variables	
	X_f (Focus Point) (mm)	X_s (Weld Speed) (m/min)	X_p (Power) (kW)	X_g (Shield Gas Type)	Y_d Depth (mm)	Y_r R (D/W)
1	2	1.25	3.75	1 (70% He + 30% Ar)	3	1.6
2	2	2.45	3.674	1 (70% He + 30% Ar)	3	1.8
3	2	2.4	6	1 (70% He + 30% Ar)	3	1.42
4	2	3.6	6.001	1 (70% He + 30% Ar)	3	1.75
5	7	1.25	3.75	1 (70% He + 30% Ar)	3	1.17
6	7	2.4	6	1 (70% He + 30% Ar)	3	1.36
7	2	1.25	3.75	2 (100% He)	3	1.41
8	2	2.45	3.674	2 (100% He)	3	2.1
9	2	2.4	6	2 (100% He)	3	2.1
10	2	3.6	6.001	2 (100% He)	3	2.14
11	2	1.25	3.75	3 (40% He + 60% Ar)	3	1.14
12	2	2.45	3.674	3 (40% He + 60% Ar)	3	1.71
13	2	2.4	6.00	3 (40% He + 60% Ar)	3	1.76
14	2	3.6	6.001	3 (40% He + 60% Ar)	3	2.17
15	12	1.25	3.75	2 (100% He)	0.6	0.29
16	12	2.45	3.674	2 (100% He)	0.41	0.26
17	12	2.4	6.00	2 (100% He)	1.40	0.77
18	12	3.6	6.001	2 (100% He)	0.55	0.31
19	7	2.4	6.00	3 (40% He + 60% Ar)	2.01	0.82
20	7	2.4	6.00	2 (100% He)	1.92	0.71
21	12	1.25	3.75	3 (40% He + 60% Ar)	0.7	0.39
22	12	2.45	3.674	3 (40% He + 60% Ar)	0.4	0.28
23	12	2.4	6.00	3 (40% He + 60% Ar)	0.79	0.47
24	12	3.6	6.001	3 (40% He + 60% Ar)	0.41	0.23
25	7	2.45	3.674	1 (70% He + 30% Ar)	1.64	0.77
26	7	3.6	6.001	1 (70% He + 30% Ar)	1.96	0.73
27	12	1.25	3.75	1 (70% He + 30% Ar)	0.5	0.25
28	12	2.45	3.674	1 (70% He + 30% Ar)	0.48	0.31
29	12	2.4	6.00	1 (70% He + 30% Ar)	1.77	0.61
30	12	3.6	6.001	1 (70% He + 30% Ar)	1.91	1.08

Based on the results depicted in Table 5, the first 14 experiments gave full penetrated welds. All the cited runs were executed with different welding speeds and powers, but with a constant focus point of 2 mm. In this case, keyhole mode heat transfer occurred. A high-energy density laser beam vaporizes the workpiece during the welding process to form a hole, which allows the laser beam to penetrate into the metal to produce a deep, narrow melt pool, thus leading to a high aspect ratio. The range of values of the weld aspect ratio is between 1.14 and 2.17. The combination of a high ionization energy and the high thermal conductivity of helium favors higher weld aspects in comparison to mixing gas (He, Ar). Once the laser beam is defocussed at focus point 7 mm or 12 mm from the workpiece, the weld becomes partially penetrated, despite the high content of sulfur in the cast 304 HS. The effect of sulfur as a surfactant element is hidden and not obvious.

At a high energy level, helium as well as mixed gas (70% He + 30% Ar) produces larger weld beads than those produced under the shield gas mixture (40% He + 60% Ar). This is attributed to the fact that helium is characterized by a high ionization potential, which provides better protection of the weld pool by expelling the plasma and ensuring less heat loss transfer. On the other hand, at high linear energy and under a gas mixture with 60% argon, there is a risk of argon ionization, which causes the formation of plasma.

3.1.1. Regression Model for Weld Depth (Y_d)

After performing the multi-regression analysis using Design Expert software, excluding two outliers, the mathematical model shown in Equation (2) can be obtained for the responses variable Y_d.

$$Y_d = 7.03481 + 0.10019\ X_f + 0.85970\ X_s - 1.18023\ X_p - 13.46056\ X_g - 0.49642\ X_f\ X_s + 0.19418\ X_f\ X_p + 0.52264\ X_f\ X_g + 2.55656\ X_p\ X_g - 0.04246\ X_f^2 + 9.12357\ X_g^2 + 0.03329\ X_f^2\ X_s - 0.01132\ X_f^2\ X_p - 0.35694\ X_f\ X_g^2 - 1.73172\ X_p\ X_g^2 \quad (2)$$

The statistical indicators specify the significance of the proposed mathematical model. Figure 5 shows (a) the normal plot of residuals and (b) the predicted transformed data as compared to the actual transformed data. The normal plot of residuals depicts that the residuals (errors) are approximately normally distributed. This means the errors are independent and random. Also, this distribution indicates the quality of the regression model. The ANOVA results confirm the statistical significance of the proposed mathematical model, as listed in Table 6. The proposed mathematical model is significant when F-value = 73.46 is sufficiently high. The values of R^2 are as follows: obtained R^2 = 0.9875, adjusted R^2 = 0.9741, and predicted R^2 = 0.9438. The difference between R^2 and the adjusted R^2 is small (less than 0.2), which indicates the statistical significance of the equation. Moreover, the signal-to-noise ratio is sufficiently high, S/N = 21.417, which shows the adequacy of the model in representing data that can be used to design high-quality systems.

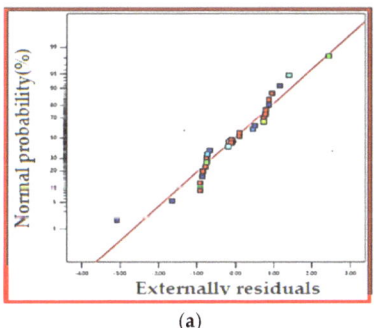
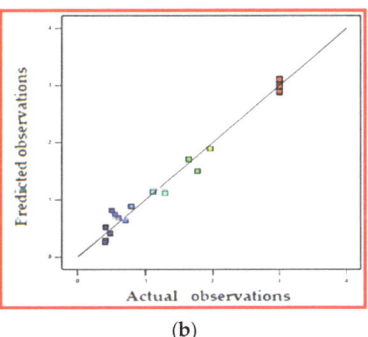

Figure 5. Statistical indicators model. (**a**) Normal plot of residuals; (**b**) predicted vs. actual observations for Y_d.

As regards the ANOVA results, the major contributor is the factor focus point (X_f), which contributes about 52.35% of the data variance with a linear effect. The second parameter is welding speed (X_s); its linear effect contributes about 9.04% of the data variance. The third factor is X_p, which shows a percentage of 5.66%. The interaction effect between the quadratic form (X_f) and X_s gives a contribution of about 5.54%. This interaction effect can be seen in Figure 6a, where the contour lines are not linear. The interaction effect between X_f and X_p is significant, with a contribution of 3.39%, as shown in Figure 6b, where the contour lines seem to be linear. Many other terms are significant; however, their contributions are small. For example, the quadratic from of X_f^2 shows 2.84%, and the interaction form ($X_f X_s$) shows 1.29%. Some terms give a contribution of about 1%, for example, ($X_p X_g^2$), ($X_f^2 X_p$), ($X_f X_g^2$), and (X_g^2). In this model, there are five terms with p-value > 0.05, which indicates the statistical insignificance of these terms, which are ($X_p X_g$), ($X_f X_g$), (X_g^2), ($X_f X_g^2$) and (X_g). However, these terms were kept in the equation to conserve the hierarchical relation of the model, with very small coefficients.

Table 6. ANOVA results of the Y_d in function of X_f, X_s, X_p and X_g.

Source	Sum of Squares	DF	Mean Square	F-Value	p-Value	Contribution (%)
Model	34.37	14	2.46	73.46	<0.0001	
X_f	8.34	1	8.34	249.44	<0.0001	52.35
X_s	1.44	1	1.44	43.06	<0.0001	9.04
X_p	0.901	1	0.901	26.96	0.0002	5.66
X_g	0.0181	1	0.0181	0.5414	0.4749	0.11
$X_f X_s$	0.2058	1	0.2058	6.16	0.0275	1.29
$X_f X_p$	0.5393	1	0.5393	16.14	0.0015	3.39
$X_f X_g$	0.0189	1	0.0189	0.5658	0.4654	0.12
$X_p X_g$	0.0329	1	0.0329	0.9857	0.3389	0.21
X_f^2	0.4529	1	0.4529	13.55	0.0028	2.84
X_g^2	0.1326	1	0.1326	3.97	0.0678	0.83
$X_f^2 X_s$	0.8819	1	0.8819	26.39	0.0002	5.54
$X_f^2 X_p$	0.1676	1	0.1676	5.02	0.0432	1.05
$X_f X_g^2$	0.1376	1	0.1376	4.12	0.0634	0.86
$X_p X_g^2$	0.1697	1	0.1697	5.08	0.0421	1.07
Residual	0.4344	13	0.0334			
Cor Total	34.8	27	15.9317			

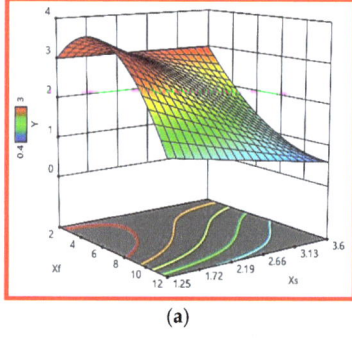 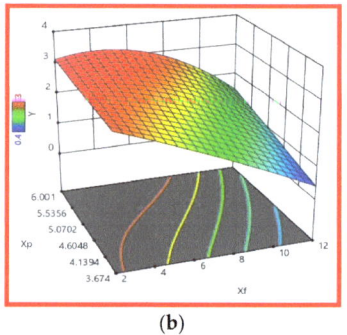

(a) (b)

Figure 6. The interaction relationships between (**a**) X_f and X_s and (**b**) X_f and X_p for Y_d.

3.1.2. Regression Model for Aspect Ratio (Y_r)

With the same methodology, the mathematical formulation of Y_r can be formulated as in Equation (3) after excluding two outlier points.

$$Y_r = 0.08185 - 0.01150\,X_f + 1.32215\,X_s - 0.06562\,X_p + 0.26042\,X_g - 0.35371\,X_f X_s + 0.14512\,X_f X_p - 0.08525\,X_s X_p - 0.00748\,X_f^2 + 0.02270\,X_f^2 X_s - 0.00942\,X_f^2 X_p \quad (3)$$

The statistical indicators show the significance of the proposed formulation. As shown in Figure 7a, the residuals are normally distributed. The residuals sit approximately around the line, which indicates that the errors are normally distributed. Moreover, Figure 7b shows a good distribution of the actual data against the predicted data. The ANOVA results for Y_r are listed in Table 7. As shown, the F-value of the model is high—F-value = 43.59—with a p-value < 0.0001, which indicates the good fit of the proposed equation for modeling the measured data. The obtained R^2 = 0.9625, the adjusted R^2 = 0.9404, and the predicted

$R^2 = 0.9145$; these good values confirm the statistical significance of the mathematical model. The signal to noise ratio $S/N = 17.725 > 4$, i.e., an adequate signal. The statistics in Figure 8a,b support the statistical significance of the proposed relation for predicting the effects of Y_r on the functions of the four parameters X_f, X_s, X_p, and X_g.

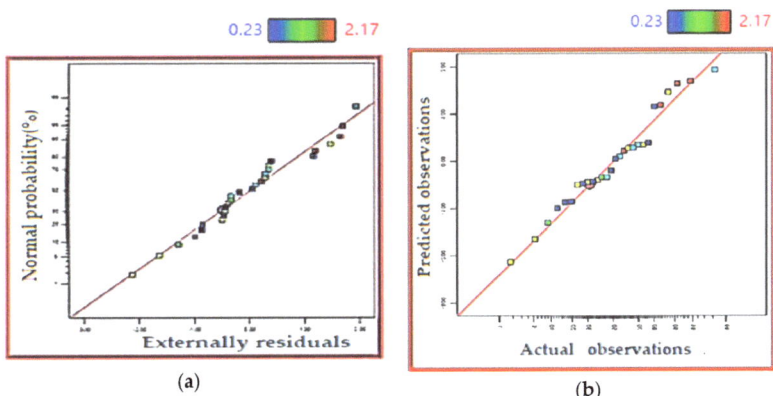

Figure 7. Model statistical indicators. (**a**) Normal plot of residuals. (**b**) The predicted vs. actual observations.

Table 7. ANOVA results of the effects of Y_r on the function of X_f, X_s, X_p and X_g.

Source	Sum of Squares	DF	Mean Square	F-Value	p-Value	Contribution (%)
Model	12.86	10	1.29	43.59	<0.0001	
X_f	11.64	1	11.64	394.76	<0.0001	79.76
X_s	0.3061	1	0.3061	10.38	0.005	2.10
X_p	0.2238	1	0.2238	7.59	0.0135	1.53
X_g	0.0977	1	0.0977	3.31	0.0865	0.67
$X_f X_s$	0.2875	1	0.2875	9.75	0.0062	1.97
$X_f X_p$	0.0738	1	0.0738	2.5	0.132	0.51
$X_s X_p$	0.0929	1	0.0929	3.15	0.0939	0.64
X_f^2	0.0085	1	0.0085	0.2886	0.5981	0.06
$X_f^2 X_s$	0.41	1	0.41	13.9	0.0017	2.81
$X_f^2 X_p$	0.134	1	0.134	4.54	0.0479	0.92
Residual	0.5014	17	0.0295			
Cor Total	13.36	27	14.5938			

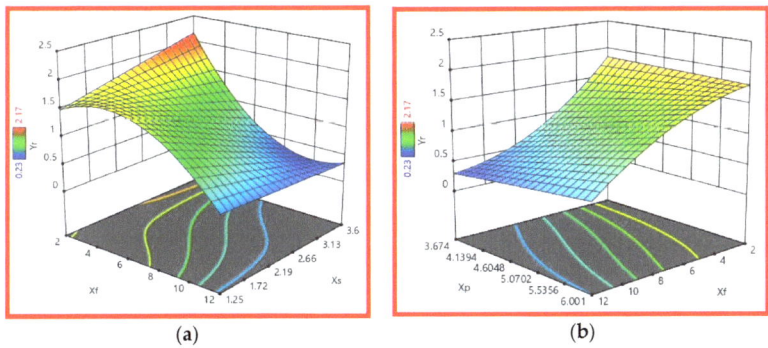

Figure 8. The interaction relationships between (**a**) X_f and X_s and (**b**) X_f and X_p for Yr.

Concerning the different terms of the equation, only ten terms are kept in the equation. The ANOVA results show the statistical significance of the linear terms. The factor X_f is responsible for the main effect, which contributes about 80% of the variance. The effects of the other variables are small compared to X_f. The second contributor is the third-order term ($X_f^2 X_s$), which contributes with only 2.81%. The interaction relationships between X_f and X_s are shown in Figure 8a. This figure shows that the contour lines are not linear. The other statistically significant terms represent X_s, $X_f X_s$, X_p, and $X_f^2 X_p$, which contribute, respectively, to 2.1%, 1.97%, 1.53%, and 0.92%. As shown in Figure 8b, the interaction relationships between X_f and X_p are weak and nonlinear. There are four insignificant terms that are kept to conserve the hierarchy of the equation, which include X_g, $X_s X_p$, $X_f X_p$, and X_f^2.

3.2. 304 LS Stainless Steel Cast

Figure 9 shows micrographs of the laser welding of a high-sulfur content cast, carried out with a power of 3.674 kw, a welding speed of 2.450 m·min^{-1}, and a shield gas comprising 70% He + 30% Ar. It produces a keyhole weld bead with focus distance of 2 mm. However, the weld is partially penetrated when the focus point is 7 mm or 12 mm from the workpiece.

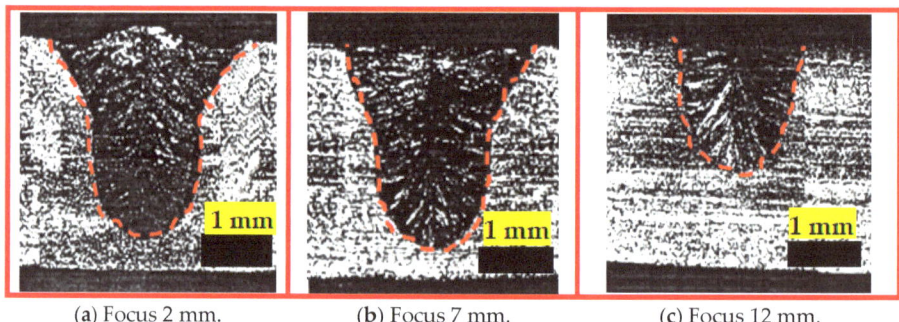

(a) Focus 2 mm. (b) Focus 7 mm. (c) Focus 12 mm.

Figure 9. Effect of focus point on 304 LS laser weld carried out with a power of 3.674 kw, a welding of speed 2.450 m·min^{-1}, and a shield gas comprising 70% He + 30% Ar.

The effects of the laser welding parameters on the welds' morphology are gathered in Table 8, which represents the investigated factors and the responses. We can specifically see that all the welds are partially penetrated. The weld is determined by the thermal conduction mode and surface tension-driven fluid flow in the weld pool. Marangoni convection forces the molten fluid to move from the center to the edges. The welds executed with a 2 mm focus point show greater penetration. Among the 30 experiments, only 7 gave aspect ratio values greater than 1, and the remaining results were all less than 1. These results confirm that the 304 LS welds were wider.

Table 8. Investigated factors and the responses for 304 LS.

Runs	Investigated Factors				Response Variables	
	X_f (Focus Point) (mm)	X_s (Weld Speed) (m/min)	X_p (Power kW)	X_g (Shield Gas Type)	Y_d Depth (mm)	Y_r R(D/W)
1	2	1.25	3.75	1 (70% He + 30% Ar)	1.80	0.84
2	2	2.45	3.674	1 (70% He + 30% Ar)	2.75	1.06
3	2	2.4	6	1 (70% He + 30% Ar)	1.27	0.66
4	2	3.6	6.001	1 (70% He + 30% Ar)	1.26	1.01
5	7	1.25	3.75	1 (70% He + 30% Ar)	1.75	0.64
6	7	2.4	6	1 (70% He + 30% Ar)	1.59	0.70

Table 8. Cont.

Runs	Investigated Factors				Response Variables	
	X_f (Focus Point) (mm)	X_s (Weld Speed) (m/min)	X_p (Power kW)	X_g (Shield Gas Type)	Y_d Depth (mm)	Y_r R(D/W)
7	2	1.25	3.75	2(100% He)	1.60	0.88
8	2	2.45	3.674	2(100% He)	0.80	0.51
9	2	2.4	6	2(100% He)	1.32	0.75
10	2	3.6	6.001	2(100% He)	1.26	1.01
11	2	1.25	3.75	3(40% He + 60% Ar)	1.07	0.61
12	2	2.45	3.674	3(40% He + 60% Ar)	1.12	0.59
13	2	2.4	6.00	3(40% He + 60% Ar)	0.82	0.65
14	2	3.6	6.001	3(40% He + 60% Ar)	1.24	1.02
15	12	1.25	3.75	2(100% He)	0.52	0.24
16	12	2.45	3.674	2(100% He)	0.37	0.22
17	12	2.4	6	2(100% He)	0.77	0.38
18	12	3.6	6.001	2(100% He)	0.39	0.21
19	7	2.4	6.00	3(40% He + 60% Ar)	1.24	0.65
20	7	2.4	6.00	2(100% He)	0.69	0.34
21	12	1.25	3.75	3(40% He + 60% Ar)	0.32	0.16
22	12	2.45	3.674	3(40% He + 60% Ar)	0.90	0.37
23	12	2.4	6	3(40% He + 60% Ar)	0.39	0.18
24	12	3.6	6.001	3(40% He + 60% Ar)	2.62	1.48
25	7	2.45	3.674	1 (70% He + 30% Ar)	2.65	1.20
26	7	3.6	6.001	1 (70% He + 30% Ar)	1.86	0.66
27	12	1.25	3.75	1 (70% He + 30% Ar)	1.53	0.58
28	12	2.45	3.674	1 (70% He + 30% Ar)	1.5	1.07
29	12	2.4	6	1 (70% He + 30% Ar)	1.76	0.66
30	12	3.6	6.001	1 (70% He + 30% Ar)	1.80	0.70

3.2.1. Regression Model for Weld Depth (Y_d)

The weld depth (Y_d) can be modeled as a function of the four parameters using the following equation, after excluding one outlier point:

$$\ln(Y_d) = +1.93616 - 0.598436\, X_f - 1.59819\, X_s + 0.123277\, X_p - 2.16420\, X_g + 0.465780\, X_f X_s - 0.146866\, X_f X_p + 1.93070\, X_f X_g + 0.145050\, X_s X_p - 0.020255\, X_f^2 + 1.74479\, X_g^2 - 0.019693\, X_f X_s X_p - 0.027802\, X_f^2 X_s + 0.015243\, X_f^2 X_p - 1.40293\, X_f X_g^2 \quad (4)$$

The obtained F-value = 21.51 and p-value < 0.0001, which indicates the statistical significant of the mathematical model. There is only a 0.01% chance that an F-value could occur due to noise. Moreover, the obtained value R^2 = 0.9556, the adjusted value R^2 = 0.9112, and the S/N = 17.388 indicate a satisfactory fitting of the proposed equation to the data. The accuracy of the mathematical model can be represented by plotting the predicted against the experimental data, as shown in Figure 10b, and by interpreting the normal plot of the residuals shown in Figure 10a. There is a small difference in some points between the experimental and predicted data.

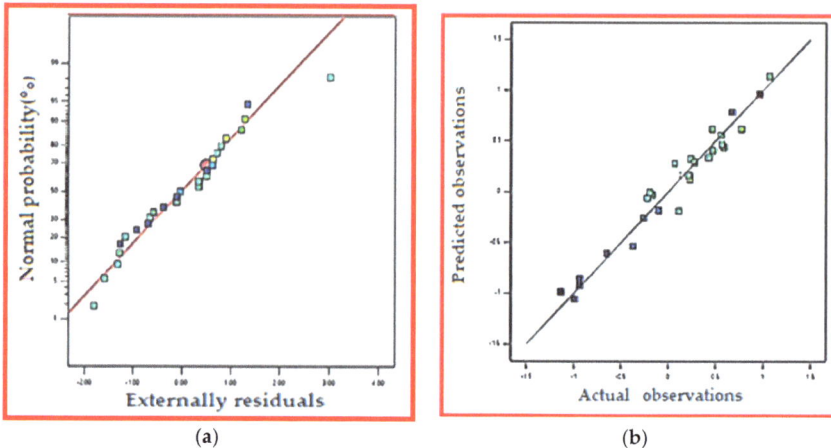

Figure 10. Model statistical indicators. (**a**) Normal plot of residuals. (**b**) The predicted observations vs. actual observations.

The ANOVA results are shown in Table 9. They indicate the statistical significance of most of the equation terms. Any term with p-value < 0.05 is considered as statistically significant. One can notice that the interactions between the variables and the higher orders are significant. In this case, x_f, $x_f x_p$, x_f^2, x_g^2, $x_f^2 x_s$, $x_f^2 x_p$ and $x_f x_g^2$ are significant model terms. Values greater than 0.1 indicate that the model terms are not significant. Also, the lack of fit is insignificant, as it confirms the statistical significance of the proposed model as regards its ability to express the effect of the four parameters on the depth.

Table 9. ANOVA results for Y_d.

Source	Sum of Squares	DF	Mean Square	F-Value	p-Value	Contribution (%)
Model	9.30	14	0.6640	21.51	<0.0001	
X_f	0.5273	1	0.5273	17.08	0.0010	6.78
X_s	0.0772	1	0.0772	2.50	0.1360	0.99
X_p	0.0752	1	0.0752	2.44	0.1409	0.97
X_g	0.0030	1	0.0030	0.0971	0.7599	0.04
$X_f X_s$	0.0798	1	0.0798	2.58	0.1302	1.03
$X_f X_p$	0.1583	1	0.1583	5.13	0.0399	2.04
$X_f X_g$	0.0419	1	0.0419	1.36	0.2633	0.54
$X_s X_p$	0.0006	1	0.0006	0.0189	0.8925	0.01
X_f^2	0.2259	1	0.2259	7.32	0.0171	2.90
X_g^2	2.86	1	2.86	92.57	<0.0001	36.77
$X_f X_s X_p$	0.1094	1	0.1094	3.54	0.0807	1.41
$X_f^2 X_s$	0.3076	1	0.3076	9.97	0.0070	3.95
$X_f^2 X_p$	0.4566	1	0.4566	14.79	0.0018	5.87
$X_f X_g^2$	2.16	1	2.16	69.84	<0.0001	27.77
Residual	0.4321	14	0.0309			
Lack of Fit	0.3837	12	0.0320	1.32	0.5094	not significant
Pure Error	0.0484	2	0.0242			
Cor Total	9.73	28				

The major contributor is factor X_g^2, which contributes about 36% of the data variance with a linear effect. The second parameter is $X_f X_g^2$; its linear effect contributes about 27% of the data variance. The third factor is X_f, which contributes a percentage of 6.78%. The interaction effect between the quadratic form X_g and X_f gives a contribution of about 2.5%. This interaction effect can be inferred from Figure 11a, where the contour lines are not linear. The interaction effect between X_f and X_p is significant, with a contribution of 3.2%, as shown by Figure 11b. The interaction effect between X_s and X_f is significant, with a contribution of 3.5%, as shown in Figure 11c.

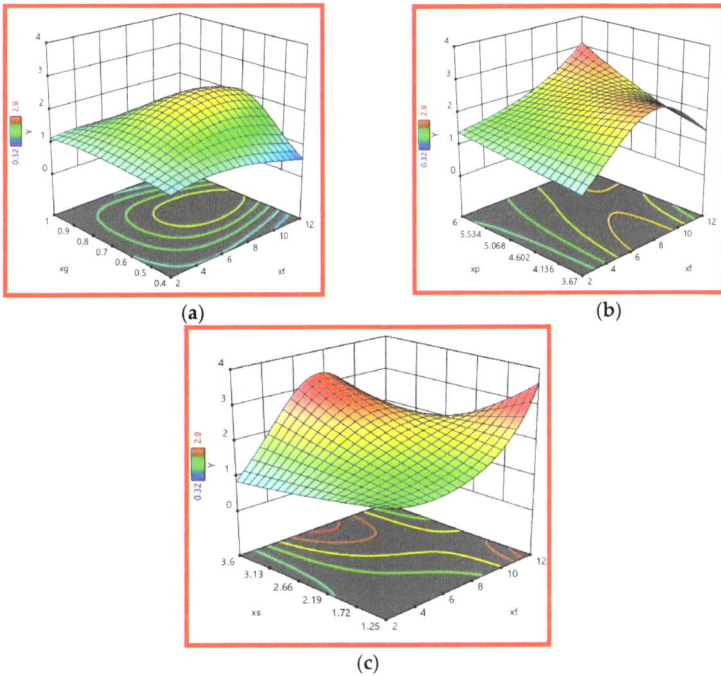

Figure 11. The interaction relationships between (**a**) X_f and X_g, (**b**) X_f and X_p and (**c**) X_f and X_s.

3.2.2. Regression Model for Aspect Ratio (Y_r)

The ratio can be modeled as a function of the four parameters, as is written in Equation (5) after excluding one outlier point:

$$Y_r = -0.611 + 0.1327\,X_f + 0.0677\,X_s + 0.0052\,X_p + 2.8902\,X_g - 0.012\,X_f^2\,X_p - 2.004 X_g^2 \quad (5)$$

The statistical indicators show the weak significance of the effects of the model's parameters on Y_r. The proposed mathematical model can be used to represent Y_r, where F-value = 6.98 and p-value = 0.0003. However, the obtained value R^2 = 0.6556, the adjusted value R^2 = 0.5617, and the signal to noise ratio S/N = 8.094 indicate a satisfactory fitting of the proposed equation to the data. The accuracy of the mathematical model can be represented by plotting the predicted against the experimental data, as shown in Figure 12b, and in the form of a normal plot of the residuals, as shown in Figure 12a. There is a difference in some points between the experimental and the predicted data. The ANOVA results are shown in Table 10, and these indicate the statistical significant of the equation terms. The main effect on Y_r is related to the variation in the focus point X_f. There is no interaction between the factors. The variation in Yr is mainly related to the linear variation of X_f and the quadratic variation of X_f and X_g.

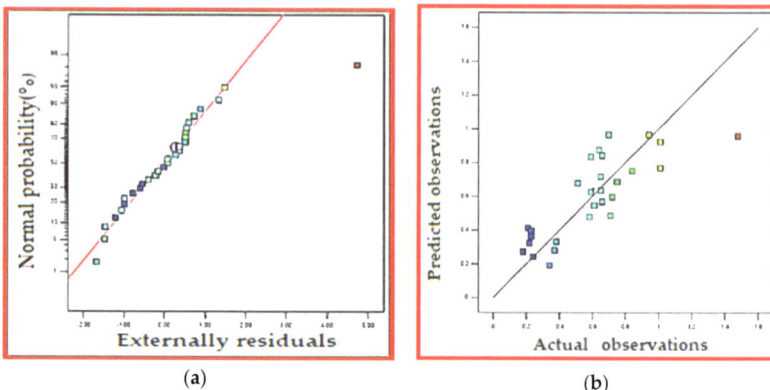

Figure 12. Model statistical indicators. (**a**) Normal plot of residuals. (**b**) The predicted observations vs. actual observations.

Table 10. ANOVA results for Y_r.

Source	Sum of Squares	df	Mean Square	F-Value	p-Value	Contribution (%)
Model	1.58	6	0.2637	6.98	0.0003	significant
X_f	0.7805	1	0.7805	20.66	0.0002	49.83
X_s	0.0469	1	0.0469	1.24	0.277	2.99
X_p	0.0006	1	0.0006	0.0157	0.9015	0.04
X_g	0.0108	1	0.0108	0.287	0.5975	0.69
X_f^2	0.235	1	0.235	6.22	0.0206	15.00
X_g^2	0.1765	1	0.1765	4.67	0.0418	11.27
Residual	0.831	22	0.0378			
Lack of Fit	0.8022	20	0.0401	2.79	0.2972	not significant
Pure Error	0.0288	2	0.0144			
Cor Total	2.41	28				

As shown in Table 10, the major contributor is the factor X_f, which contributes to almost 50% of the data variance, with a linear effect. The interaction effect between X_f and X_g can be seen in Figure 13a. It shows a contribution of 0.8%, as the contour lines are not linear. The interaction effect between X_f and X_p is not significant, with a contribution of 1%, as shown in Figure 13b.

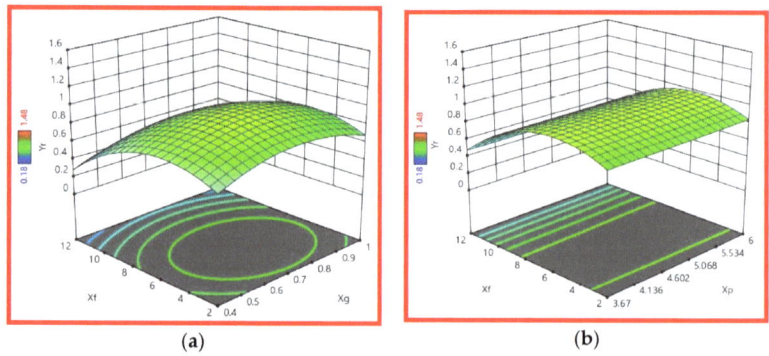

Figure 13. The interaction relationships between (**a**) X_f and X_g and (**b**) X_f and X_p.

3.3. Contributions of Surface Active Elements to AISI 304 CO_2 Laser Weld Morphology

In total, 30 tests were carried out on both casts 304 HS and 304 LS. All the tests carried out on the 304 LS led to partial-penetration welds. However, only 14 welds carried out on 304 HS were fully penetrated welds, as depicted in Table 11.

Table 11. Depth penetration and width weld bead measurements for 304 HS and LS casts.

	Cast 304 HS			Cast 304 LS			
	Inverse Marangoni Convection			Marangoni Convection			
Exp. #							Occurrences of the Mechanism Proposed by C.R Heiple et al. [18]
	Weld Depth (D) (mm)	Weld Width (W) (mm)	Full Weld Penetration (F.P)/Partial Penetration Weld (P.P)	Weld Depth (D) (mm)	Weld Width (W) (mm)	Full Weld Penetration (F.P)/Partial Penetration Weld (P.P)	
1	3	1.88	F.P	1.80	2.14	P.P	Yes
2	3	1.67	F.P	2.75	2.6	P.P	Yes
3	3	2.1	F.P	1.27	0.66	P.P	No
4	3	1.71	F.P	1.26	1.25	P.P	No
5	3	2.56	F.P	1.75	2.73	P.P	Yes
6	3	2.2	F.P	1.59	2.07	P.P	No
7	3	2.13	F.P	1.60	1.82	P.P	No
8	3	1.42	F.P	0.80	1.57	P.P	Yes
9	3	1.42	F.P	1.32	1.76	P.P	Yes
10	3	2.63	F.P	1.26	1.01	P.P	No
11	3	2.63	F.P	1.07	0.61	P.P	No
12	3	1.75	F.P	1.12	1.25	P.P	No
13	3	1.70	F.P	0.82	1.26	P.P	No
14	3	1.38	F.P	1.24	1.21	P.P	No
15	0.6	2.06	P.P	0.52	2.17	P.P	Yes
16	0.41	1.58	P.P	1.55	1.22	P.P	No
17	1.40	1.80	P.P	0.77	2.02	P.P	Yes
18	0.55	1.77	P.P	0.39	1.86	P.P	Yes
19	2.01	1.91	P.P	1.24	2.45	P.P	Yes
20	1.92	2.03	P.P	0.69	2.70	P.P	Yes
21	0.7	1.78	P.P	0.32	1.94	P.P	Yes
22	0.4	1.43	P.P	0.90	2.43	P.P	No
23	0.79	1.69	P.P	0.39	2.14	P.P	Yes
24	0.41	1.78	P.P	2.62	1.77	P.P	No
25	2.10	2.13	P.P	2.65	2.23	P.P	No
26	1.96	2.68	P.P	1.86	2.8	P.P	Yes
27	0.5	2	P.P	1.53	2.64	P.P	No
28	0.48	1.55	P.P	1.5	1.4	P.P	No
29	1.77	2.90	P.P	1.76	2.67	P.P	No
30	1.91	1.76	P.P	1.80	2.57	P.P	Yes

In total, 16 partial penetration welds were obtained for both casts under the same welding conditions (from trial 15 to 30, see Table 11). With the welding conditions chosen for experiments 1 to 14, we achieved full-penetration welds with 304 HS and partial weld penetration with 304 LS. Among the 16 cases in which both casts gave rise to partial weld penetration, the role of sulfur as a surfactant element was only obvious in 9. As such, 56% of the partial penetration welds verify the mechanism proposed by C.R Heiple et al. [18], as depicted in Table 12. We must remember that the mechanism proposed by C.R Heiple et al. [18] stipulates that casts with a high sulfur content exhibit greater weld depth than those with low sulfur content, owing to the reverse Marangoni convection leading to deeper weld beads. On the other hand, the width of the weld bead of a low-sulfur content cast is larger than that of a high-sulfur cast, due to the occurrence of Marangoni convection.

Table 12. Percentages of occurrence of the mechanism proposed by C.R Heiple et al. [18] in the laser welding process.

Types of Weld Bead	Times Getting Partial Penetration Weld (PP) for Both Casts or Full Penetration Weld (FP) for 304 HS Cast Weld Bead out of 30 Tests	Occurrences of the Mechanism Proposed by C.R Heiple et al. [18] %	Non-Occurrences of the Mechanism Proposed by C.R Heiple et al. [18] %	Percentages of Occurrences of the Mechanism Proposed by C.R Heiple et al. [18] %
Partial penetration weld (PP) for both casts.	16	9	7	56
Full penetration weld (FP) for 304 HS and partial penetration weld (PP) for 304 LS cast weld.	14	5	9	36

In the remaining 14 cases, in which the welds were fully penetrated for 304 HS and partially penetrated for 304 LS, only in 5 tests were the mechanisms proposed by C.R Heiple et al. [18] confirmed. The role of sulfur as a surfactant is disturbed or completely reduced when keyhole-mode welding occurs. In laser welding, many complicated factors are involved, especially those associated with weld pool dynamics, melt evaporation, plasma formation, keyhole instability, the aerodynamic currents of the shield gas and the interaction between the laser beam and the plasma plume. The accumulation of hot fumes at the beam focus point, under certain conditions, can turn into a plasma cloud that strongly affects the beam, and absorbs and disperses it. All the above mentioned phenomena contribute to hiding surface active role of sulfur in determining the laser weld shape. As depicted in Table 12, the rate of achievement of reverse Marangoni convection was 56% for partially penetrated welds, compared to 36% when the welds were fully penetrated (keyhole mode). When weld penetration is governed by a keyhole mechanism, the role of surface tension gradients is diminished or completely hidden.

4. Conclusions

In this work, the effects of variations in microchemistry on the laser welding of industrial austenitic stainless steel casts 304 HS and 304 LS have been studied. The effect of laser power, welding speed, focus point position, and shield gas on depth of weld (Y_d) and aspect ratio (Y_r) were investigated using RSM. An optimal response surface design was employed to design the experiment. The model required 15 points, and another 15 points were added randomly to increase the model efficiency using Design Expert software. The following conclusions can be drawn:

- For 304 HS, the focus point (X_f) is the major factor in determining the depth of penetration of the laser weld, within a range of chosen welding parameters. The last input parameter is the main input factor, such that its contribution reaches 52.35% in determining the depth of the laser weld. The second parameter is welding speed (X_s); its linear effect contributes about 9.04% of the data variance. The third factor is the X_p, which represents a percentage of 5.66%;

- For 304 LS, the depth of the laser weld is primarily determined by the focus point, with a contribution of up to 6.78% (far greater than the other input factors if we consider only the effects of singular input factors individually). However, the interaction between shield gas and focus point distance from workpiece seems to play an important role, with a contribution of up to 28%;
- The role of the shield gas in protecting the weld pool is also linked to the level of energy supplied. Thus, at a high energy level, helium as well as mixed gas (70% He + 30% Ar) produces weld beads larger than those produced with the shield gas mixture (40% He + 60% Ar). This is attributed to the fact that helium is characterized by a high ionization potential, protecting the weld pool more effectively by expelling the plasma and ensuring less heat loss transfer. On the other hand, under a gas mixture with 60% argon, at a high linear energy, there is a risk of argon ionization causing the formation of plasma;
- A statistical study shows that a partially penetrated weld rate of 56% supports the hypothesis of the role of surface active elements in the formation of the weld pool. In contrast, in the case of full penetration welds, a rate of only 36% confirms the surface active effects of sulfur. Based on all these results, we can conclude that the laser weld morphology depends on the surface active effects of sulfur only if the chosen weld parameters result in partially penetrated welds. The surface active role of sulfur is reduced, if not eliminated, owing to melt evaporation, plasma formation, keyhole instability, the aerodynamic currents of the shield gas, and the interaction between the laser beam and the plasma. The above phenomena are more pronounced when the chosen welding parameters lead to fully penetrated welds;
- The results obtained from this work constitute an interesting database for research dedicated to the numerical simulation of thermal profiles in order to validate mathematical models. The thermal profile and solidification mode can be used to predict the microstructural and mechanical properties. On the other hand, these results may be of great use in industrial applications relating to the laser welding of thin austenitic stainless steel sheets.

Author Contributions: Conceptualization, K.T.; methodology, K.T., A.I. and R.D.; software, K.T., E.A. and A.B.; validation, K.T., E.A. and A.I.; formal analysis, K.T. and A.C.H.; investigation, K.T., R.D. and E.A.; resources, A.I.; data curation, K.T. and M.M.Z.A.; writing—original draft preparation, K.T. and E.A.; writing—review and editing, K.T., R.D. and E.A.; visualization, K.T., A.B. and A.C.H.; supervision, K.T., A.I. and M.M.Z.A. All authors have read and agreed to the published version of the manuscript.

Funding: This research received no external funding.

Data Availability Statement: The data used to support the findings of this study are included within the article.

Conflicts of Interest: The authors declare no conflict of interest.

References

1. Biswas, A.R.; Banerjee, N.; Sen, A.; Maity, S.R. Applications of laser beam welding in automotive sector—A Review. In *Advances in Additive Manufacturing and Metal Joining*; Springer: Berlin/Heidelberg, Germany, 2023; pp. 43–57.
2. Narayana, R.B.; Hema, P.; Padmanabhan, G. Experimental investigation on similar and dissimilar alloys of stainless steel joints by laser beam welding. *Adv. Mater. Process. Technol.* **2022**, *8*, 13–28.
3. Hollatz, S.; Hummel, M.; Jaklen, L.; Lipnicki, W.; Olowinsky, A.; Gillner, A. Processing of keyhole depth measurement data during laser beam micro welding. *J. Mater. Des. Appl.* **2020**, *234*, 722–731. [CrossRef]
4. Olson, D.L.; Siewert, T.A.; Liu, S.; Edwards, G.R. (Eds.) *ASM Handbook, Volume 6: Welding, Brazing, Soldering*; ASM International: Materials Park, OH, USA, 2014.
5. Kurc-Lisiecka, A.; Kciuk, M. The influence of chemical composition on structure and mechanical properties of austenitic Cr-Ni steels. *J. Achiev. Mater. Manuf. Eng.* **2013**, *61*, 210–215.

6. Dong, X.; Wang, G.; Ghaderi, M. Experimental investigation of the effect of laser parameters on the weld bead shape and temperature distribution during dissimilar laser welding of stainless steel 308 and carbon steel St 37. *Infrared Phys. Technol.* **2021**, *116*, 103774. [CrossRef]
7. Sampreeta, K.R.; Vasareddy, M.; Deepan, B.K.T. Optimization of process parameters in CO2 laser welding of Hastelloy C-276. *Mater. Today Proc.* **2020**, *22*, 1572–1581. [CrossRef]
8. Sathiya, P.; Abdul Jaleel, M.Y.; Katherasan, D. Optimization of welding parameters for laser bead-on-plate welding using Taguchi method. *Prod. Eng.* **2010**, *4*, 465–476. [CrossRef]
9. Anawa, E.; Olabi, A.; Hashmi, M. Optimization of ferritic/austenitic laser welded components. In Proceedings of the AMPT2006 International Conference, Las Vegas, NV, USA, 30 July–3 August 2006.
10. Benyounis, K.Y.; Olabi, A.G.; Hashmi, M.S.J. Optimization the laser-welded butt joints of medium carbon steel using RSM. *J. Mater. Process. Technol.* **2005**, *164-165*, 986–989. [CrossRef]
11. Indacochea, J.E.; Olson, D.L. Relationship of Weld Metal Microstructure and Penetration to Weld Metal Oxygen Content. *J. Mater. Energy Syst.* **1983**, *5*, 139–148. [CrossRef]
12. Klapczynski, V.; Le Maux, D.; Courtois, M.; Bertrand, E.; Paillard, P. Surface tension measurements of liquid pure iron and 304L stainless steel under different gas mixtures. *J. Mol. Liq.* **2022**, *350*, 118558. [CrossRef]
13. Heiple, C.R.; Burgardt, P. Effect of SO_2 Shielding Gas Additions on GTA Weld Shape Welding. *Weld. J.* **1985**, *64*, 159–162.
14. Zhao, C.X.; Kwakernaak, C.; Pan, Y.; Richardson, I.M.; Saldi, Z.; Kenjeres, S.; Kleijn, R. The effect of oxygen on transitional Marangoni flow in laser spot welding. *Acta Mater.* **2010**, *5*, 6345–6357. [CrossRef]
15. Wenchao, K.; Zhi, Z.; Oliveira, J.P.; Bei, P.; Jiajia, S.; Caiwang, T.; Xiaoguo, S.; Wentao, Y. Heat transfer and melt flow of keyhole, transition and conduction modes in laser beam oscillating welding. *Int. J. Heat Mass Transf.* **2023**, *203*, 123821.
16. Coviello, D.; D'Angola, A.; Sorgente, D. Numerical Study on the Influence of the Plasma Properties on the Keyhole Geometry in Laser Beam Welding. *Front. Phys.* **2022**, *9*, 754672. [CrossRef]
17. Hou, Y.C.; Lu, W.; Wentao, Y. Influence of oxygen content on melt pool dynamics in metal additive manufacturing: High-fidelity modeling with experimental validation. *Acta Mater.* **2023**, *249*, 118824.
18. Heiple, C.R.; Roper, J.R.; Stagner, R.T.; Aden, R.J. Surface Active Element Effects on the Shape of GTA, Laser, and Electron Beam Welds. *Weld. J.* **1983**, *62*, 72–77.
19. Kaul, R.; Ganesh, P.; Singh, N.; Jagadeesh, R.; Bhagat, M.S.; Kumar, H.; Tiwari, P.; Vora, H.S.; Nath, A.K. Effect of active flux addition on laser welding of austenitic stainless steel. *Sci. Technol. Weld. Join.* **2007**, *12*, 127–137. [CrossRef]
20. Ding, F.; Ruihua, Z.; Nakata, K.; Tanaka, M.; Ushio, M. YAG laser welding with surface activating flux. *China Weld.* **2003**, *12*, 83–86.
21. Su, Y.; Aarts, R.G.K.M.; Meijer, J.; Guan, Q. Study on effect of active-fluxes in laser welding. In Proceedings of the International Congress of Applications of Lasers & Electro-Optics, Dearborn, MI, USA, 2–5 October 2000; Laser Institute of America: Orlando, FL, USA, 2000; pp. 35–41.
22. Kou, S. *Welding Metallurgy*; John Wiley & Sons: Toronto, ON, Canada, 1987; Volume 431, pp. 223–225.
23. Shahab, A. Contribution a L'etude de la Metallurgie du Soudage de L'inconel 625 et Des Aciers Inoxydables 304 et 316. Ph.D. Thesis, University of Nantes, Nantes, France, 1990.
24. Novák, V.; Řeháčková, L.; Váňová, P.; Sniego, M.; Matýsek, D.; Konecná, K.; Smetana, B.; Rosypalová, S.; Tkadlecková, M.; Drozdová, L.; et al. The Effect of Trace Oxygen Addition on the Interface Behavior of Low-Alloy Steel. *Materials* **2022**, *15*, 1592. [CrossRef] [PubMed]
25. Kamel, T.; Abdelil, C.H.; Rachid, D.; Abousoufiane, O.; Abdallah, B.; Albaijan, I.; Hany, S.A.; Mohamed, M.Z.A. Mechanical, Microstructure, and Corrosion Characterization of Dissimilar Austenitic 316L and Duplex 2205 Stainless-Steel, ATIG Welded Joints. *Materials* **2022**, *15*, 2470.
26. Besnea, D.; Dontu, O.; Gheorghe, G.I.; Ciobanu, R.; Cuta, A. Laser micro welding of mechatronics components, Romanian Review Precision Mechanics. *Opt. Mechatron.* **2014**, *46*, 7–10.
27. Narayana Reddy, B.; Hema, P.; Padmanabhan, G. Influence of CO_2 Laser Beam Welding Process Parameters on Mechanical Properties of Alloy AISI 4130 Steel Welded Joints. In *Recent Advances in Material Sciences*; Lecture Notes on Multidisciplinary Industrial Engineering; Springer: Singapore, 2019; pp. 65–76.

Disclaimer/Publisher's Note: The statements, opinions and data contained in all publications are solely those of the individual author(s) and contributor(s) and not of MDPI and/or the editor(s). MDPI and/or the editor(s) disclaim responsibility for any injury to people or property resulting from any ideas, methods, instructions or products referred to in the content.

Article

Investigation of the Microstructure and Mechanical Properties in Friction Stir Welded Dissimilar Aluminium Alloy Joints via Sampling Direction

Sipokazi Mabuwa * and Velaphi Msomi

Mechanical and Mechatronic Engineering Department, Cape Peninsula University of Technology, Bellville 7535, South Africa; msomiv@gmail.com
* Correspondence: sipokazimabuwa@gmail.com; Tel.: +27-219-538-778

Abstract: This research study investigates the influence of sampling direction on the microstructure and mechanical properties of dissimilar joints formed by friction stir welding (FSW). The specimens were cut in two directions: perpendicular (transverse) and parallel (longitudinal) to the FSW joint. The tests conducted included X-ray diffraction (XRD), macrostructure, microstructure, tensile, microhardness, and fractography analysis. Different phases were noted in the XRD patterns and explained, with the aluminum phase being the dominating one. The results further showed that the transverse dissimilar joint exhibited higher microhardness compared to the longitudinal dissimilar joint, which is consistent with the respective grain sizes. Moreover, the ultimate tensile strength of the longitudinal joint exceeded that of the transverse joints, showing a substantial 47% increase. Similarly, the elongation of the joints followed a similar trend, with the longitudinal joint displaying a significant 41% increase in elongation compared to the transverse joint. Fractographic analysis revealed ductile fracture behaviour in all joints.

Keywords: X-ray diffraction patterns; dissimilar aluminum alloys; tensile strength; microstructure

Citation: Mabuwa, S.; Msomi, V. Investigation of the Microstructure and Mechanical Properties in Friction Stir Welded Dissimilar Aluminium Alloy Joints via Sampling Direction. *Crystals* **2023**, *13*, 1108. https:// doi.org/10.3390/cryst13071108

Academic Editors: Ali Khalfallah, Mahmoud Moradi and Reza Beygi

Received: 29 June 2023
Revised: 9 July 2023
Accepted: 13 July 2023
Published: 16 July 2023

Copyright: © 2023 by the authors. Licensee MDPI, Basel, Switzerland. This article is an open access article distributed under the terms and conditions of the Creative Commons Attribution (CC BY) license (https:// creativecommons.org/licenses/by/ 4.0/).

1. Introduction

Friction stir welding (FSW) is defined as a solid-state joining technique whereby the welding process is performed below the materials melting temperature in a solid state [1]. The FSW technique was initially invented and designed for aluminum alloys by The Welding Institute (TWI) in the United Kingdom in 1991 [1,2]. Since then, the FSW technique has been normalized as a joining technique for other materials such as steel, titanium, copper, and magnesium. The application of FSW advanced into many applications, including the joining of dissimilar materials. The weld characteristics of the dissimilar FSW materials were discovered to be influenced mainly by welding parameters. Welding parameters refer to the FSW machine parameters, which include rotational speed, traverse speed, axial force, tool tilt angle, vertical force, and tool parameters such as the material of the tool, shoulder diameter, pin length, pin diameter, and pin profile [3].

There are many studies where FSW was successfully employed to join dissimilar materials, including dissimilar aluminum alloys. The AA6063 and AA7075 dissimilar aluminum alloy joint was produced using FSW to investigate the microstructure and mechanical properties of the produced joint [4]. Varying process parameters were used, and it was discovered that the sound results were obtained when the tool rotation was set at 1000 rpm and a traverse speed of 2.5 mm/s with an axial force of 8.5 KN. The results reported included the fine-grained microstructure, which was found to be responsible for the high hardness and tensile strength. A dissimilar combination of AA2024 and AA7075 was subject to FSW to evaluate the impact of welding parameters using varying parameters [5]. The varying parameters included a tool rotation range of 1000–1400 rpm, a tool traverse speed range of 20–40 mm/min, and a constant force of 4 kN. The ultimate tensile strength (UTS) was found to increase with a decrease in traverse speed, while the microstructural grain patterns were found to coarsen as the tool rotation increased due to high heat input in the deformed zone. In addition, Kailainathan et al. [6] reported a similar behavior where the UTS depreciated when the tool rotation was above 1200 rpm. This was due to the distortion in the weld region caused by the severe heat experience by the welded region during FSW.

The material positioning is considered to be crucial during the FSW of dissimilar aluminum alloys. Guo et al. [7] combined varying process parameters and material positioning to determine the impact the two factors had on the produced AA6061/AA7075 dissimilar joint characteristics. The microstructure results revealed that the material mixing was more effective when the AA6061 alloy was placed on the advancing side than when the AA7075 was positioned on the advancing side. Greater grain refinement was also found to favor the positioning of the same alloy on the advancing side. The highest UTS was obtained under the exact positioning at a high traverse speed of 5 mm/s.

To investigate the impact of material positioning and tool offset, researchers conducted a friction stir welding experiment using AA6061-T6 and AA8011-H14, as detailed in Reference [8]. The welding process involved specific parameters, including a tool rotational speed of 1070 rpm, a tool traverse speed of 50 mm/min, and a tool tilt angle of 2°. By placing the softer AA8011-H14 alloy on the advancing side of the tool, with a tool offset of 1 mm, notable improvements were achieved in the joint properties. The tensile test results revealed a maximum UTS of 77.8 MPa and an elongation of 21.96%. However, the hardness remained unaffected by the varying process parameters. The weld zone exhibited a consistent grain structure. Additionally, various studies in the literature, spanning from References [9–12], have corroborated the idea of positioning the softer material on the advancing side of the tool, irrespective of the specific combinations of dissimilar aluminum alloys.

The dissimilar AA5083/AA6083 FSW joint was sampled in two different directions, transverse and longitudinal, to assess the characteristics of the joint [13]. The aim was to analyze the variations by comparing these samples taken from different orientations with the parent materials through macro/microstructure, tensile, and micro-hardness tests. The findings indicated that the transverse samples had a hardness value of 93.90 HV0.2, while the longitudinal samples had a higher value of 119.27 HV0.2. Similarly, the transverse samples exhibited the highest tensile strength of 130.694 MPa with a strain value of 0.054, whereas the longitudinal samples had a tensile strength of 127.833 MPa and a strain value of 0.0834. However, a study conducted by Garg et al. [14] using the finite element method focused on an FSW dissimilar joint AA6061/AA7075. The results of their study showed that the longitudinally sampled specimen exhibited a higher UTS with a lower elongation compared to the transverse specimen. Despite these differences, both sampling directions maintained a ductile fracture surface morphology.

The investigation of AA6092/SiC focused on examining the microstructure and mechanical properties, considering the direction of specimen sampling [15]. The microstructure analysis revealed distinct zones within the cross-weld specimen, including the base metal, heat-affected zone, thermo-mechanically affected zone, and stir zone. On the other hand, the longitudinal specimen exhibited a fine equiaxed grain structure. In terms of tensile strength, the longitudinal specimen demonstrated both higher ductility and strength compared to the transverse specimen, which aligns with common observations in similar joints. Additionally, the hardness behavior exhibited a similar pattern to the tensile strength, attributed to the grain refinement in the longitudinal specimen.

The sampling direction in friction stir welding of dissimilar aluminum alloys holds significant importance, alongside other crucial factors, for maximizing its application. Specimens can be sampled either perpendicular or longitudinal to the weld direction. Cutting the specimens longitudinally (parallel to the direction of the joint) FSW of dissimilar aluminum alloys provides valuable information about the weld characteristics, aids in understanding the welding process, facilitates mechanical property evaluation and contributes to process optimization in dissimilar aluminum alloys. However, it is important to note that the specific novelty and significance of this approach may vary depending on the context and the research study or development being referred to.

This paper investigates the influence of sampling direction on the microstructure and mechanical properties of dissimilar aluminum alloy joints fabricated via friction stir welding.

2. Materials and Methods

This study employed two aluminum alloy grades: AA6082-T651 and AA8011-H14 plates, both with a thickness of 6 mm. The chemical composition of these alloys is provided in Table 1, while Table 2 displays their corresponding mechanical properties. The plates were cut into dimensions of 55 × 265 mm to align with the FSW fixture bed used. The FSW process was carried out using a converted LAGUN FU. 1-LA conventional universal milling machine manufactured by the LAGUN MACHINE TOOLS S.L.U. in Gipuzkoa in Spain.

In Figure 1a, a photograph of the FSW tool utilized in the study is depicted, accompanied by comprehensive tool parameters as outlined in Table 3. In addition to Table 3, it should be noted that the penetration depth is the same as the pin length, which is 5.7 mm down into the materials. To determine the FSW parameters, the Taguchi L9 method was employed [16]. The FSW setup is depicted in Figure 1b,c showcases the resulting dissimilar joint produced by FSW. Following the FSW process, test specimens were obtained from the joint, considering two sampling directions: transverse (trans) and longitudinal (long). The sampling directions are visualized in Figure 1d, with dimensions in mm.

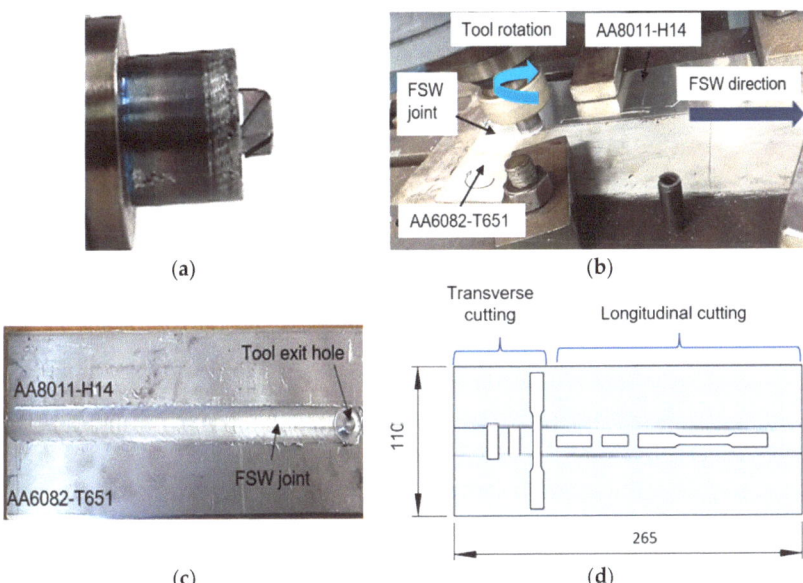

Figure 1. (**a**) FSW tool, (**b**) FSW process, (**c**) FSW produced joint, and (**d**) sampling direction.

Table 1. Chemical composition of the materials (wt %) [17].

	Mg	Zn	Ti	Cr	Si	Mn	Fe	Ni	Cu	Al
AA6082-T651	1.23	0.51	0.04	0.00	1.24	0.38	0.72	0.11	0.03	Balance
AA8011-H14	0.55	0.63	0.03	0.03	0.38	0.76	1.33	0.12	0.06	Balance

Table 2. Mechanical properties of the materials [18].

Property	Material	
	AA6082-T651	AA8011-H14
Yield strength (MPa)	270	76
Tensile strength (MPa)	307.5	94.1
Elongation (%)	26.22	40.17
Hardness (HV)	90	34

Table 3. FSW tool parameters [16].

Material	Pin Length (mm)	Pin Diameter (mm)	Tilt Angle (°)	Shoulder Diameter (mm)	Shoulder Length (mm)	Traverse Speed (mm/min)	Rotational Speed (rpm)
High-speed steel	5.7	7	2	20	15	60	1100

The conducted tests encompassed various analyses, including microstructural analysis, X-ray diffraction (XRD) analysis, tensile tests, microhardness tests, and fractographic analysis. For the dissimilar friction stir welded (FSW) joints, the chemical composition analysis was performed to determine the phases present in the XRD analysis. The specimen used for chemical and XRD analysis is depicted in Figure 2a. The Belec Compact Spectrometer HLC (Belec Spectrometry Opto-Electronics GmbH, Georgsmarienhütte, Germany) machine, utilizing high-purity Argon gas (99.99%), was employed to detect the chemical composition of the produced FSW dissimilar joints. The Belec WIN 21 software, integrated into the Belec machine, was utilized to measure the chemical composition.

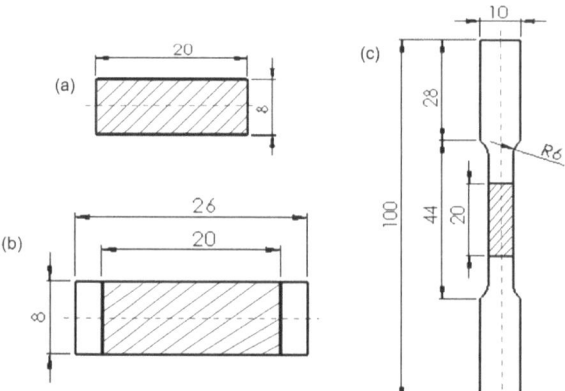

Figure 2. (a) XRD specimen—top view, (b) microstructure—top view specimen, and (c) tensile specimen (all dimensions in mm).

Subsequently, the XRD analysis was carried out on the samples following the chemical composition analysis, and the services of iThemba Laboratory Solutions company were utilized. A Bruker D8-Advance multi-purpose X-ray diffractometer manufactured by the Bruker Corporation, Massachuttes based in the United States was employed for the XRD measurements. The instrument operated in continuous θ-θ scan mode with locked coupling, employing Cu-Kα radiation. The measurement range in 2θ was determined by the user, with a typical step size of 0.034°. The diffraction data were recorded using a position-sensitive detector called Lyn-Eye, with a typical speed of 0.5 s/step, which is equivalent to an effective time of 92 s/step for a scintillation counter. The background of the data was subtracted to obtain diffraction patterns with zero background. A set of potential elements from the periodic table was selected for phase analysis, and phases were identified by comparing the calculated peaks with the measured ones until all phases were identified within the resolution limits of the results.

The microstructural analysis was conducted in accordance with the ASTME112-12 standard [19]. Cut-off specimens were mounted using hot mounting resin to prepare the specimens for microstructure analysis. Figure 2b displays the dimensions of the microstructure specimen. After mounting, the specimens underwent grinding, polishing, and subsequent etching using Weck's and Modified Keller's etchants. For Weck's etchant, a chemical mixture containing 4 g of potassium permanganate, 1 g of sodium hydroxide, and distilled water was applied to the specimens for a duration of 15 s. Modified Keller's etchant consisted of 10 mL of nitric acid, 1.5 mL of hydrochloric acid, 1.0 mL of hydrofluoric acid, and 87.5 mL of distilled water. It was applied to the joints for 20 s following Weck's etchant.

Following the etching process, microstructural images of the FSW joints were captured using the Motic AE2000 MET Trinocular 100W metallurgical light optical microscope and the Motic Images Plus 3.0 software. The microscope used in this study is manufactured by The Motic Europe S.L.U. based in Barcelona in Canada. The images were taken using a 5× objective lens for base material analysis, while a 20× objective lens was used for the stir zones. The captured images were later measured using the line intercept method in ImageJ software.

The tensile tests were conducted utilizing the Hounsfield 50 K tensile testing machine, following the guidelines of the ASTM E8M-04 standard for specimen geometry and testing [20]. The diagram of the tensile specimen can be seen in Figure 2c. For Vickers microhardness testing, the Falcon 5000 Innovatest hardness testing machine is manufactured by the INNOVATEST Europe BV Manufacturing based in Maastricht in the Netherlands. The said machine is equipped with IMPRESSIONS^{TM} software was employed. The hardness testing was carried out in accordance with the ASTM E384-11 standard [21]. The specific settings included a load of 0.3 kg, a 1 mm interval between the indents, and a total of 25 measurement indents recorded per joint. During setup, the 10× and 20× objectives were utilized for specimen focusing.

3. Results and Discussion

Figure 3 shows the XRD patterns for the FSW-Trans and FSW-Long specimens; for both XRD peaks, the presence of aluminum (α-Al), which was the dominant phase noted in both figures, is denoted by the pink dotted lines. Additional phases noted were the magnesium silicon (Mg_2Si), which is a strengthening precipitate originating from AA6082-T651 alloy [22], with the $Mg_{2.7}Fe$ and the Al_7Mn_4Fe being the iron interface base phases originating from the AA8011-H14 alloy. The iron phases, such as the Mg_2Si, are responsible for preserving the joint characteristics [23]. In light of the phases mentioned, joint failure can only result from the aluminum phase [24].

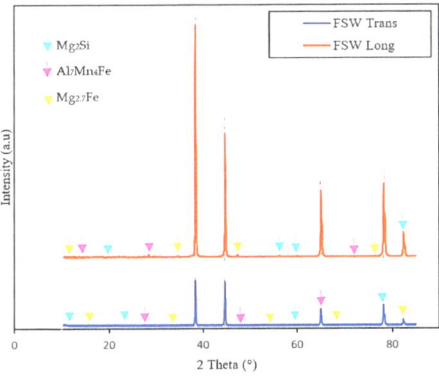

Figure 3. XRD phases for FSW-Trans and FSW-Long.

Figure 4 shows the macrographs for the FSW-Trans and FSW-Long joints. Figure 4a shows the traditional features of the FSW joint, being the single oval-shaped stir zone bands, which are commonly referred to as the onion ring structures [25,26]. The formation of the onion rings is caused by the geometric effect, which is a result of cylindrical sheets of material extruded from the retreating side of the tool during tool rotation and cutting through the sections of the two aluminum alloys [27,28]. A tunnel defect was noted as well, denoted by the red arrows. Tunnel defects are not surprising in the FSW of dissimilar material welds and are linked to insufficient heat input and signify that the tool might have traversed ahead of the welding direction before sufficient materials were deposited behind as it traversed, thereby creating a void. The void produced manifests itself in the form of a tunnel defect which plays a part in the degradation of the mechanical properties of the joint [29,30]. Comparing Figure 4b to Figure 4a, which is basically a longitudinal view of the same joint, the tunnel defect noted previously now manifests itself in the form of a line from one end to the next, denoted by the red arrow. Figure 4b appears similar to two sandwich stacked layers. These layers are identified by layer 1 (L1) and layer 2 (L2), which suggest that the microstructural arrangement in these layers may differ, thereby giving the interest to focus on microstructure examination between the two.

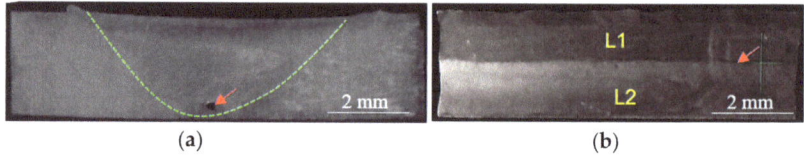

Figure 4. Macrographs, (**a**) FSW-Trans, and (**b**) FSW-Long.

Figure 5 shows the micrographs of the FSW joints produced—Figure 5a,b present the AA6082-T651 and AA8011-H14 base materials. The mean grain sizes of the two alloys were found to be 69.98 and 53.61 μm. The results obtained correlate with those reported in the literature [31]. Figure 5c shows the FSW-Trans stir zone microstructure with a mean grain size of 18.38 μm. Figure 5d depicts the FSW-Long zone, which was found to exhibit a microstructure with four layers showing the material flow of the said region. The stir zone microstructure of the FSW-Long exhibited layers in a wave-like shape due to the material mixing process during FSW. This wave-like pattern arises from the rotation and translation of the FSW tool as it traverses along the joint line. The tool exerts heat and mechanical pressure on the material, causing it to soften and mix together [32].

The stacked wave-like shape is formed by alternating regions of different material mixing concentrations. These regions represent distinct layers where the original base materials and the mixed materials are present in varying proportions. The wave-like pattern typically consists of multiple peaks and troughs, indicating areas of higher and lower mixing concentrations of the two dissimilar alloys. The presence of four different material mixing concentrations suggests that the FSW process has created four distinct layers within the stir zone of the joint. Each layer represents a combination of the original base materials and the stirred material, with varying degrees of mixing. The specific number and distribution of these layers can depend on factors such as the welding parameters, tool design, and the properties of the base materials [33,34].

However, from the four layers, two highly dominating layers were noted, those being layers L1 and L2, where the mean grain sizes were found to be 17.44 and 16.29 μm, respectively. Comparing the mean grain sizes, it was discovered that post-FSW, the grain sizes were greatly refined and equiaxed. This change was brought about by the dynamic recrystallization of the joint stir zone, where the maximum plastic deformation and thermal softening altered the grain structure completely [35,36].

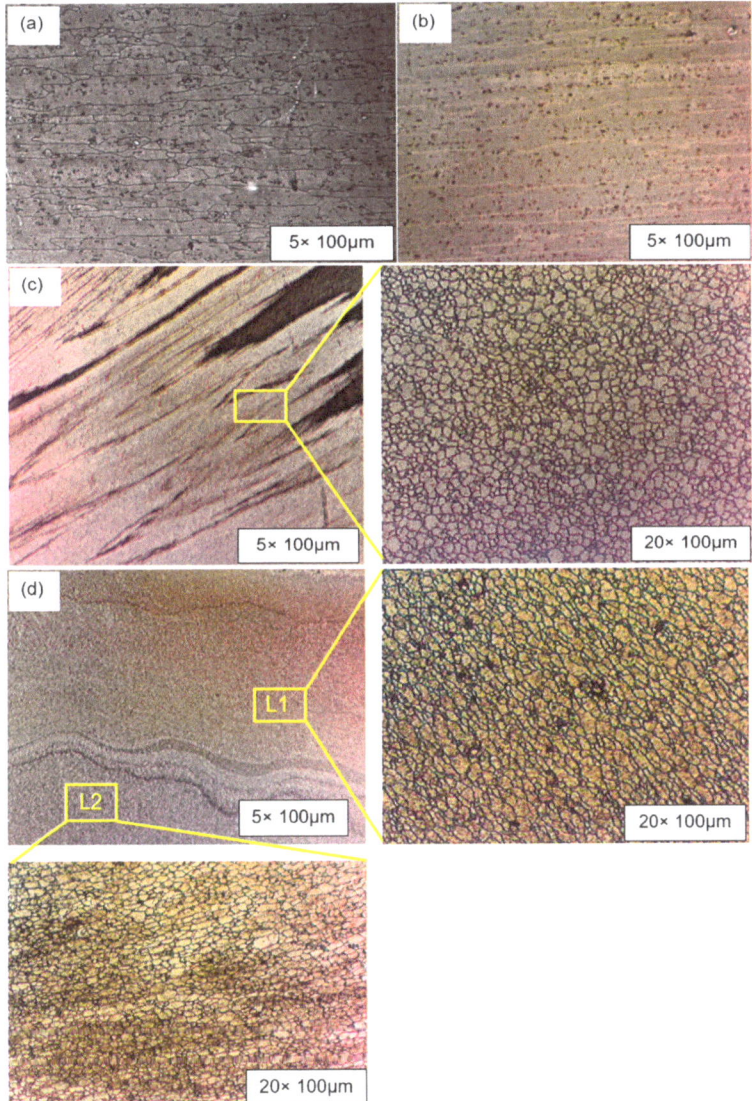

Figure 5. Base material micrographs, (**a**) AA6082-T651 BM, (**b**) AA8011-H14; stir zone micrographs, (**c**) FSW-Trans, and (**d**) FSW-Long.

Figure 6 depicts the post-tensile test specimens and produced tensile stress-strain curves with corresponding summarized tensile properties of the same in Table 4. Examining the post-tensile specimens, the stir zone (SZ) positions of fractures were noted for both specimens. The position of the fracture was influenced by the presence of the tunnel defect, as previously discussed in the macrostructural examination. Then, the tunnel defect made the stir zone the most likely region of failure [16]. Examining Figure 6c, the FSW-Long inhibited a higher ultimate tensile strength (UTS) than the FSW-Trans. Similarly, the tensile strain (elongation) followed the same behavior. This behavior is due to the FSW-Long only consisting of the stir zone material with uniform grain sizes, unlike the FSW-Trans specimen consisting of the four regions: base material, heat affected zone, thermo-mechanically affected zone, and the stir zone. An additional factor contributing to the observed increased UTS is the presence of positive residual stress in the longitudinal direction, which is greater than that in the transverse direction. Consequently, when the samples are cut along the longitudinal direction, they have the ability to release a larger amount of residual stress compared to the transverse direction. This phenomenon likely contributes to the attainment of higher strength levels in the longitudinal direction [37,38]. Additionally, in the longitudinal joint, the applied tensile load was primarily along the direction of the weld, resulting in a more direct and uniform strain distribution [15]. The uniform strain distribution promoted better load sharing among grains and contributed to higher UTS and elongation compared to the transverse joint, where the strain was distributed less uniformly.

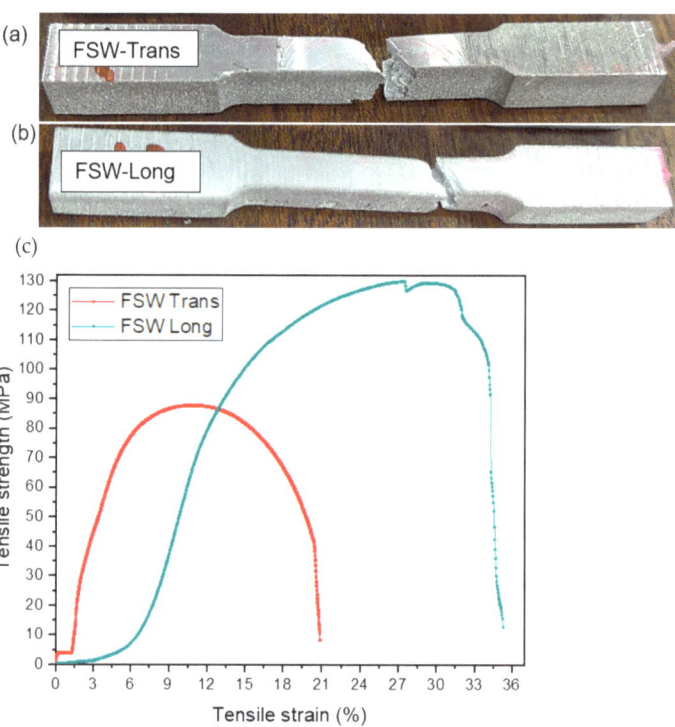

Figure 6. (**a**) FSW-Trans specimen, (**b**) FSW-Long specimen, and (**c**) tensile stress-strain curves.

Table 4. Tensile properties.

Weld Joint	Ultimate Tensile Strength (MPa)	Tensile Strain at UTS (%)	Tensile Strain at Breakpoint (%)	Position of Fracture
FSW Trans	88.53	14.89	27.49	SZ
FSW Long	129	27.38	35.23	SZ

Figure 7 shows the fractography of the FSW-Trans and FSW-Long joints. Both joints, while subjected to fractographic analysis to determine the nature of the fracture, displayed ductile fracture behaviour. This behaviour can be noted in the mentioned figure where the ductile fracture features were noted, those being equiaxed micro dimples, grain boundaries, microvoids, torn ridges, and transgranular cleavage facets [39–42]. Figure 8 shows the microhardness profiles of the FSW joints. The microhardness was analyzed similarly to the microstructure hence the similar labelling. From the figure, it was noted that the FSW-Trans microhardness showed a profile where the curve started from the AA6082-T651 alloy, heat-affected zone, thermo-mechanically affected zone, stir zone, then to the AA8011-H14 alloy. The obtained mean stir zone microhardness was found to be 44.65 HV. The FSW-Long L1 mean microhardness was 35.48 HV and 81 HV for the FSW-Long L2. It should be noted that the FSW-Long joint only consists of stir zone material. However, when studying the microhardness values obtained from layer 1 and layer 2, it was noted that layer 1 was dominated mainly by the AA8011-H14 alloy than the AA6082-T651 alloy. Layer 2, on the other side, consisted mostly of the AA6082-T651 rather than the AA8011-H14 alloy. The behaviour of the microhardness was found to correlate with the grain sizes of the mentioned regions [43,44].

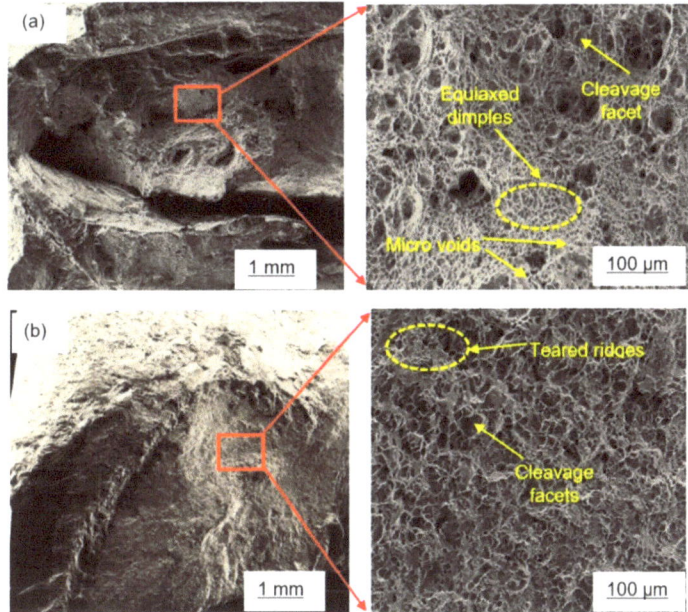

Figure 7. Fractography, (**a**) FSW-Trans, and (**b**) FSW-Long.

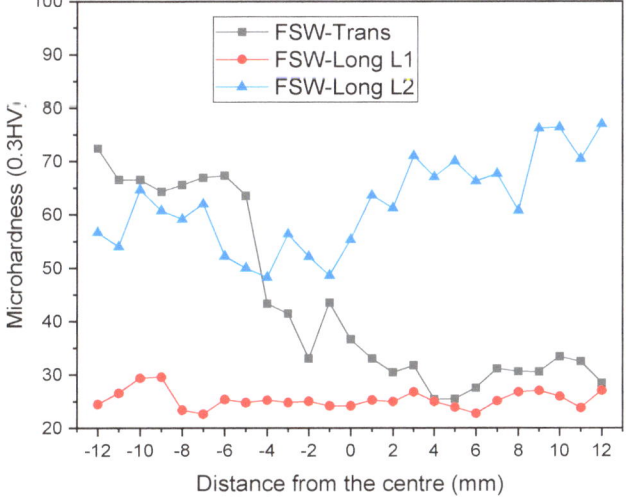

Figure 8. Microhardness profiles.

4. Conclusions

The investigation into the microstructure and mechanical properties of friction stir welded dissimilar aluminum alloy joints based on the sampling direction yielded several significant conclusions:

Firstly, the X-ray diffraction analysis indicated that aluminum was the predominant phase in the joints, accompanied by strengthening phases such as magnesium silicon and magnesium iron.

A notable observation emerged from the microstructure analysis, revealing a distinct difference between the FSW longitudinal and transverse joints. The FSW longitudinal joint exhibited a stacked sandwich-like microstructure, while the FSW transverse joint displayed a banded onion ring microstructure.

The microhardness measurements further demonstrated variations between the two joint types. The FSW transverse dissimilar joints exhibited a microhardness of 44.65 HV, whereas the FSW longitudinal joint displayed a more complex microstructural arrangement consisting of two layers. Layer 1 had a mean microhardness of 35.48 HV, while layer 2 exhibited a microhardness of 81 HV. These microhardness values aligned with the respective grain sizes observed in the joints. It is worth elaborating that FSW introduced significant plastic deformation and stirring action, leading to grain refinement in the welded zone. In the longitudinal joint, the direction of grain flow is aligned with the loading direction during hardness testing, resulting in a more direct load transmission through the grains. This alignment contributed to higher microhardness values in the longitudinal direction compared to the transverse direction, where the load may encounter a less direct path through the grains.

Furthermore, the ultimate tensile strength of the FSW longitudinal joints surpassed that of the FSW transverse joints. The FSW longitudinal joints exhibited a tensile strength of 129.9 MPa, whereas the FSW transverse joints achieved a lower tensile strength of 88.5 MPa. A similar trend was observed for elongation, with the FSW longitudinal joints demonstrating a higher tensile strain of 35.32% compared to the 20.89% tensile strain observed in the FSW transverse joints. This behavior was due to the strain and residual stresses experienced during tensile loading.

Fractographic analysis indicated that all the joints exhibited a ductile fracture behavior when subjected to mechanical testing, further supporting the presence of favorable mechanical properties.

In conclusion, the investigation successfully explored the microstructure and mechanical properties of friction stir welded dissimilar aluminum alloy joints based on the sampling direction. The findings highlighted the distinctive microstructural differences, variations in microhardness, ultimate tensile strength, and elongation between the FSW longitudinal and transverse joints. These insights contribute to a deeper understanding of the joint characteristics and performance, aiding in the development and optimization of friction stir welding processes for dissimilar aluminum alloy joints.

Author Contributions: Conceptualization, S.M. and V.M.; methodology, S.M.; software V.M.; validation, S.M. and V.M.; formal analysis, S.M.; investigation, S.M.; resources, V.M.; data curation, S.M.; writing—original draft preparation, S.M.; writing—review and editing, V.M.; visualization, S.M.; supervision, V.M.; project administration, S.M.; funding acquisition, V.M. All authors have read and agreed to the published version of the manuscript.

Funding: This research is funded by the SOUTH AFRICAN NATIONAL RESEARCH FOUNDATION—THUTHUKA GRANT 138268.

Data Availability Statement: Data is available upon request.

Acknowledgments: The authors would like to thank the Cape Peninsula the University of Technology, Department of Mechanical and Mechatronic Engineering workshop staff for their sincere support during the fabrication anda testing of the joints and Miranda Waldron from the University of Cape Town for assistance with the SEM tests. Special gratitude also goes to Remi Butcher from the iThemba Labs for assistance analyzing the XRD samples.

Conflicts of Interest: The authors declare no conflict of interest.

References

1. Mishra, R.S.; De, P.S.; Kumar, N. *Friction Stir Welding and Processing*; Springer: Berlin, Germany, 2014; pp. 259–296.
2. Mishra, R.S.; Ma, Z.Y. Friction stir welding and processing. *Mater. Sci. Eng. R Rep.* **2005**, *50*, 1–78. [CrossRef]
3. Zhang, Y.N.; Cao, X.; Larose, S.; Wanjara, P. Review of tools for friction stir welding and processing. *Canad. Metall. Q.* **2012**, *51*, 250–261. [CrossRef]
4. Varunraj, S.; Ruban, M. Investigation of the microstructure and mechanical properties of AA6063 and AA7075 dissimilar aluminium alloys by friction stir welding process. *Mater. Today Proc.* **2022**, *68*, 1654–1657. [CrossRef]
5. Gowthaman, P.S.; Saravanan, B.A. Determination of weldability study on mechanical properties of dissimilar Al-alloys using friction stir welding process. *Mater. Today Proc.* **2021**, *44*, 206–212. [CrossRef]
6. Kailainathan, S.; Sundaram, S.K.; Nijanthan, K. Influence of friction stir welding parameter on mechanical properties in dissimilar aluminium alloys. *Int. J. Innov. Res. Sci. Eng. Technol.* **2014**, *3*, 15691–15695. [CrossRef]
7. Guo, J.F.; Chen, H.C.; Sun, C.N.; Bi, G.; Sun, Z.; Wei, J. Friction stir welding of dissimilar materials between AA6061 and AA7075 Al alloys effects of process parameters. *Mater. Des.* **2014**, *56*, 185–192. [CrossRef]
8. Khanna, N.; Sharma, P.; Bharati, M.; Badheka, V. Friction stir welding of dissimilar aluminium alloys AA 6061-T6 and AA 8011-h14: A novel study. *J. Braz. Soc. Mech. Sci. Eng.* **2020**, *42*, 7. [CrossRef]
9. Karimi, N.; Nourouzi, S.; Shakeri, M.; Habibnia, M.; Dehghani, A. Effect of tool material and offset on friction stir welding of Al alloy to carbon steel. *Adv. Mater. Res.* **2012**, *445*, 747–752.
10. Palani, K.; Elanchezhian, C.; Avinash, K.; Karthik, C.; Chaitanya, K.; Sivanur, K.; Reddy, K.Y. Influence of friction stir processing parameters on tensile properties and microstructure of dissimilar AA 8011-h24 and AA 6061-t6 aluminum alloy joints in nugget zone. *IOP Conf. Ser.* **2018**, *390*, 012108. [CrossRef]
11. Cole, E.G.; Fehrenbacher, A.; Duffie, N.A.; Zinn, M.R.; Pfefferkorn, F.E.; Ferrier, N.J. Weld temperature effects during friction stir welding of dissimilar aluminum alloys 6061-T6 and 7075-T6. *Int. J. Adv. Manuf. Technol.* **2014**, *71*, 643–652. [CrossRef]
12. Rani, N.; Bishnoi, R.N.; Mahendru, A. Effect of tool pin profile on mechanical properties of single pass and double pass friction stir welded aluminium alloys AA6061 & AA8011. *Int. J. Curr. Eng. Technol.* **2015**, *5*, 3780–3783.
13. Segaetsho, M.O.M.; Msomi, V.; Moni, V. A comparative analysis between the transverse and longitudinal samples of the FSW AA5083/AA6082 Joints. *Key Eng. Mater.* **2023**, *944*, 3–12. [CrossRef]
14. Garg, A.; Raturi, M.; Bhattacharya, A. Experimental and finite element analysis of progressive failure in friction stir welded AA6061-AA7075 joints. *Procedia Struct. Integr.* **2019**, *17*, 456–463. [CrossRef]
15. Omar, S.S.; Hengan, O.U.; Xingguo, W.; Sun, W. Microstructure and mechanical properties of friction stir welded AA6092/SiC metal matrix composite. *Mater. Sci. Eng. A* **2019**, *742*, 78–88.
16. Msomi, V.; Mabuwa, S. Optimization of Normal and Submerged FSP Parameters for dissimilar aluminium joints using Taguchi technique. *MSF* **2023**, *1034*, 207–218.
17. Msomi, V.; Mabuwa, S.; Merdji, A.; Muribwathoho, O.; Motshwanedi, S.S. Microstructural and mechanical properties of submerged multi-pass friction stir processed AA6082/AA8011 TIG-welded joint. *Mater. Today Proc.* **2021**, *45*, 5702–5705. [CrossRef]
18. Mabuwa, S.; Msomi, V. The impact of submerged friction stir processing on the friction stir welded dissimilar joints. *Mater. Res. Express* **2020**, *7*, 096513. [CrossRef]
19. ASTM International. Standard Test Methods for Determining Average Grain Size (ASTM E112-12). 2012; pp. 1–27. Available online: https://www.astm.org/Standards/E112.htm (accessed on 28 June 2023).
20. ASTM International. Standard Test Methods for Tension Testing of Metallic Materials (ASTM E8M-04). 2004; pp. 1–24. Available online: https://www.astm.org/Standards/E8M.htm (accessed on 28 June 2023).
21. ASTM International. Standard Test Method for Microindentation Hardness of Materials (ASTM E384-11). 2011; pp. 1–24. Available online: https://www.astm.org/Standards/E384.htm (accessed on 28 June 2023).
22. Balakrishnan, M.; Dinaharan, I.; Palanivel, R.; Sathishkumar, R. Effect of friction stir processing on microstructure and tensile behavior of AA6061/Al3Fe cast aluminum matrix composites. *J. Alloys Comp.* **2019**, *785*, 531–541.
23. Mvola, B.; Kah, P.; Martikainen, J. Welding of dissimilar non-ferrous metals by GMAW processes. *Int. J. Mech. Mater. Eng.* **2014**, *9*, 1–11. [CrossRef]
24. Tan, L.; Katoh, Y.; Tavassoli, A.A.F.; Henry, J.; Rieth, M.; Sakasegawa, H.; Tanigawa, H.; Huang, Q. Recent status and improvement of reduced-activation ferritic martensitic steels for high-temperature service. *J. Nucl. Mater.* **2016**, *479*, 515–523. [CrossRef]
25. Chen, H.B.; Yan, K.; Lin, T.; Chen, S.B.; Jiang, C.Y.; Zhao, Y. The investigation of typical welding defects for 5456 aluminum alloy friction stir welds. *Mater. Sci. Eng. A* **2006**, *433*, 64–69. [CrossRef]

26. Palani, K.; Elanchezhian, C.; Vijaya Ramnath, B.; Bhaskar, G.B.; Naveen, E. Effect of pin profile and rotational speed on microstructure and tensile strength of dissimilar AA8011, AA01-T6 friction stir welded aluminum alloys. *Mater. Today Proc.* **2018**, *5*, 24515–24524. [CrossRef]
27. Mehdi, H.; Mishra, R.S. Effect of friction stir processing on mechanical properties and wear resistance of tungsten inert gas welded joint of dissimilar aluminum alloys. *J. Mater. Eng. Perform.* **2021**, *30*, 1926–1937. [CrossRef]
28. Mohammadzadeh, H.J.; Farahani, H.; Besharati, M.K.G.; Aghaei, M.V. Study on the effects of friction stir welding process parameters on the microstructure and mechanical properties of 5086-H34 aluminum welded joints. *Int. J. Adv. Manuf. Technol.* **2016**, *83*, 611–621. [CrossRef]
29. Agha Amini Fashami, H.; Bani Mostafa Arab, N.; Hoseinpour Gollo, M. Numerical and experimental investigation of defects formation during friction stir processing on AZ91. *SN Appl. Sci.* **2021**, *3*, 108. [CrossRef]
30. Khan, N.Z.; Siddiquee, A.N.; Khan, Z.A.; Shihab, S.K. Investigations on tunneling and kissing bond defects in FSW joints for dissimilar aluminum alloys. *J. Alloys Compd.* **2015**, *648*, 360–367. [CrossRef]
31. Mabuwa, S.; Msomi, V. The impact of material positioning towards the friction stir welded dissimilar aluminium alloy joints. *Recent Pat. Mech. Eng.* **2020**, *14*, 252–259.
32. Mabuwa, S.; Msomi, V.; Mehdi, H.; Ngonda, T. A study on the metallurgical characterization of the longitudinally sampled friction stir processed TIG welded dissimilar aluminum joints. *Proc. Inst. Mech. Eng. Part E J. Process Mech. Eng.* **2023**, 09544089231169589. [CrossRef]
33. Das, J.; Robi, P.S.; Sankar, M.R. Assessment of parameters windows and tool pin profile on mechanical property and microstructural morphology of FSWed AA2014 joints. *SN Appl. Sci.* **2020**, *2*, 1–15. [CrossRef]
34. Segaetsho, M.O.M.; Msomi, V.; Moni, V. Traverse and longitudinal analysis of AA5083/AA6082 dissimilar joint. *Eng. Res. Express* **2023**, *5*, 035004. [CrossRef]
35. He, Z.; Peng, Y.; Yin, Z.; Lei, X. Comparison of FSW and TIG welded joints in Al–Mg–Mn–Sc–Zr alloy plates. *Trans. Nonferr. Met. Soc. China* **2011**, *21*, 1685–1691. [CrossRef]
36. Peng, G.; Yan, Q.; Hu, J.; Chen, P.; Chen, Z.; Zhang, T. Effect of forced air cooling on the microstructures, tensile strength, and hardness distribution of dissimilar friction stir welded AA5A06-AA6061 joints. *Metals* **2019**, *9*, 304. [CrossRef]
37. Farhang, M.; Farahani, M.; Nazari, M.; Sam Daliri, O. Experimental correlation between microstructure, residual stresses and mechanical properties of friction stir welded 2024-T6 aluminum alloys. *Int. J. Adv. Manuf. Technol.* **2022**, *15*, 1–9.
38. Farhang, M.; Sam-Daliri, O.; Farahani, M.; Vatani, A. Effect of friction stir welding parameters on the residual stress distribution of Al-2024-T6 alloy. *J. Mech. Eng. Sci.* **2021**, *15*, 7684–7694. [CrossRef]
39. Zhao, H.J.; Wang, B.Y.; Liu, G.; Yang, L.; Xiao, W.C. Effect of vacuum annealing on microstructure and mechanical properties of TA15 titanium alloy sheets. *Trans. Nonfer. Met. Soc. China* **2015**, *25*, 1881–1888. [CrossRef]
40. Dragatogiannis, D.A.; Koumoulos, E.P.; Kartsonakis, I.; Pantelis, D.I.; Karakizis, P.N.; Charitidis, C.A. Dissimilar friction stir welding between 5083 and 6082 Al alloys reinforced with TiC nanoparticles. *Mater. Manuf. Process.* **2016**, *31*, 2101–2114. [CrossRef]
41. Msomi, V.; Moni, V. The influence of materials positioning on microstructure and mechanical properties of friction stir welded AA5083/AA6082 dissimilar joint. *Adv. Mater. Process. Technol.* **2021**, *8*, 2087–2101. [CrossRef]
42. Pouraliakbar, H.; Beygi, R.; Fallah, V.; Monazzah, A.H.; Jandaghi, M.R.; Khalaj, G.; Da Silva, L.F.M.; Pavese, M. Processing of Al-Cu-Mg alloy by FSSP: Parametric analysis and the effect of cooling environment on microstructure evolution. *Mater. Lett.* **2022**, *308*, 131157. [CrossRef]
43. Ravikumar, S.; Seshagiri-Rao, V.; Pranesh, R.V. Effect of welding parameters on macro and microstructure of friction stir welded butt joints between AA7075-T651 and AA6061-T651 alloys. *Proc. Mater. Sci.* **2014**, *5*, 1725–1735.
44. Li, K.; Liu, X.; Zhao, Y. Research status and prospect of friction stir processing technology. *Coatings* **2019**, *9*, 129. [CrossRef]

Disclaimer/Publisher's Note: The statements, opinions and data contained in all publications are solely those of the individual author(s) and contributor(s) and not of MDPI and/or the editor(s). MDPI and/or the editor(s) disclaim responsibility for any injury to people or property resulting from any ideas, methods, instructions or products referred to in the content.

MDPI AG
Grosspeteranlage 5
4052 Basel
Switzerland
Tel.: +41 61 683 77 34

Crystals Editorial Office
E-mail: crystals@mdpi.com
www.mdpi.com/journal/crystals

Disclaimer/Publisher's Note: The statements, opinions and data contained in all publications are solely those of the individual author(s) and contributor(s) and not of MDPI and/or the editor(s). MDPI and/or the editor(s) disclaim responsibility for any injury to people or property resulting from any ideas, methods, instructions or products referred to in the content.

www.ingramcontent.com/pod-product-compliance
Lightning Source LLC
LaVergne TN
LVHW072349090526
838202LV00019B/2509